W0055407

Basler Lehrbücher
A Series of Advanced Textbooks in Mathematics
Vol. 3

Edited by
Herbert Amann, Zürich
Hanspeter Kraft, Basel

Konrad Jacobs

Discrete Stochastics

Springer Basel AG

Author's address

Prof. Dr. Konrad Jacobs
Mathematisches Institut
Universität Erlangen-Nürnberg
Bismarckstr. 1 1/2
D-8520 Erlangen

Deutsche Bibliothek Cataloging-in-Publication Data

Jacobs, Konrad:
Discrete stochastics / Konrad Jacobs. – Basel ; Boston ; Berlin : Birkhäuser, 1992
(Basler Lehrbücher, a series of advanced textbooks in mathematics ; Vol. 3)
ISBN 978-3-0348-9713-6 ISBN 978-3-0348-8645-1 (eBook)
DOI 10.1007/978-3-0348-8645-1
NE: GT

This work is subject to copyright. All rights are reserved, whether the whole or part of the material is concerned, specifically those of translation, reprinting, re-use of illustrations, broadcasting, reproduction by photocopying machine or similar means, and storage in data banks. Under § 54 of the German Copyright Law, where copies are made for other than private use a fee is payable to «Verwertungsgesellschaft Wort», Munich.

© 1992 Springer Basel AG
Originally published by Birkhäuser Verlag Basel in 1992
Softcover reprint of the hardcover 1st edition 1992

Cover Design: Albert Gomm, Basel

ISBN 978-3-0348-9713-6

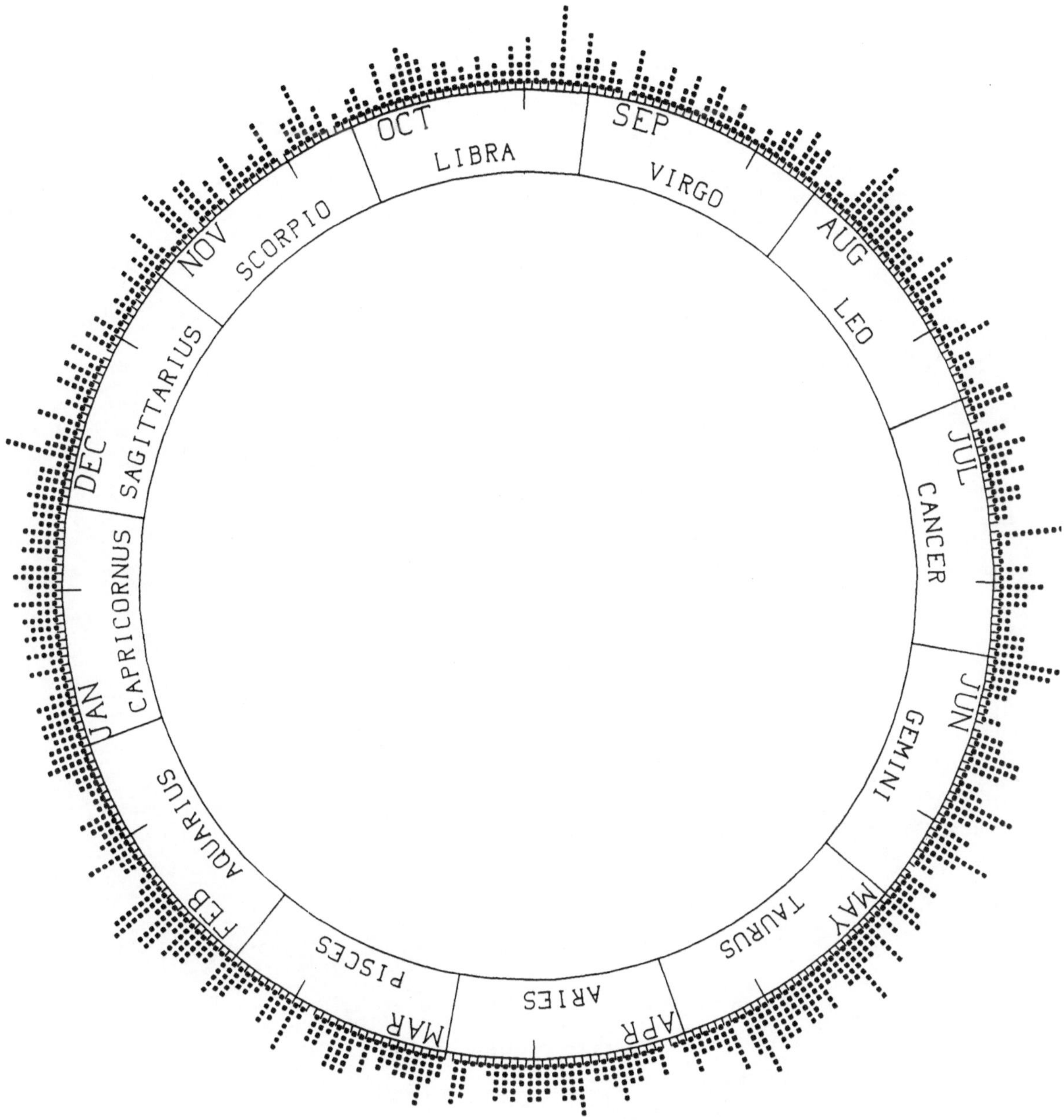

Birthdays of Mathematicians according to the membership list of the DMV of 1.1.1978

Preface

This book is based on courses in elementary stochastics which I gave several times at the Erlangen-Nürnberg university. These courses differ from full-fledged stochastics courses in that they do not use measure theory as a prerequisite. The problem thus posed to the teacher of such a course is usually settled by a marginal implicit use of measure theory. Impressive examples of such an approach are Feller [1957] and Krengel [1988]. The approach pursued here differs from those essentially only by some attempts to draw the frontier line between the measure-theory-free parts of the theory and those parts in which a bit of measure theory is inevitable, a bit more explicitly. Ch. III on discrete probability spaces provides the tools which are sufficient for the bulk of the material presented in this volume. The transgressions into measure theory are carefully motivated and explained where they are inevitable.

The editor kindly permitted me to delve in a few topics which are favorites of mine since a long time, such as fluctuation theory and casino theory ("how to gamble if you must"). Another deviation from trodden paths is ch. II on Markovian dynamics which I chose to present as a forerunner to the more classical topics, thereby relieving the Markov ch. VI from a theory which would mean a detour there. Furthermore, I have tried to counter balance the more systematic sections of the book by an intuitive presentation of striking examples of stochastic reasoning. And finally, I have tried to give an outlook on some foundational problems in ch. X.

Parts of lattice gas and percolation theory could well have been treated within the methodical framework of this book. I left these topics out in order to not overemphasize the "specialities flavor" of the book.

Although e.g. nonparametric statistics could be presented largely within the discrete probability framework used here, I finally decided to not go far into statistics, confining it to a very introductory ch. V. So the atmosphere of the book is essentially probabilistic – I would have felt stepping beyond my own limits otherwise.

I feel deeply indebted to my secretary, Helga Zech, who qualified herself as a LaTeX typesetter here for the first time. We owe much help to Klaus Mach and H.J. Schmid for quite a few diagrams and for instruction in LaTeX. In particular Klaus Mach worked out the rotund diagram displaying the birthday statistics of DMV members on page V. Ingrid Bergmann kindly provided us with the mortality density diagram on page 3. Perhaps I should also mention, that I.1 owes a lot to Davis [1986], Engel [1987] and Mosteller [1965], and VI.2 to Krengel [1988]. My cordial thanks are due to Birkhäuser for the careful editing of the final version of the book.

Erlangen, summer 1991 Konrad Jacobs

Contents

I. Introduction

1. Encountering Random

As a normal citizen, you don't know when you die. But whenever you start your car, you take into account the possibility that you die in a crash during the ride. You have to live with this uncertainty, and likewise with other uncertainties which might be less critical than life and death, but bear the same fundamental characteristics. Let us remember, in the form of questions, a few life situations involving that sort of uncertainty:

when will I die?
will I die during the voyage which I am going to undertake?
will I be robbed during my next walk in downtown Chicago?
will my airplane arrive with no more than 15 min. delay?
will I get a ticket for parking my car wrongly?
will the egg which I am just about to open, be rotten?
will the avocado seed which I am planting, germinate?
will the dice which I am throwing yield a six?
will I make at least 1.000.000 \$ next year?
am I in the files of the CIA?
did I get a malaria infection during my last trip to Africa?
...

As a rational individual, you will not be content to remain idle when experiencing uncertainty. You will try to cope with it by certain strategies, such as

transforming uncertainty into certainty by investigation

finding out essential characteristics of *bare* uncertainty

trying to quantify uncertainty resp. certainty, i.e. trying to answer questions like "how certain am I about a certain possibility"? Everyday example:

> does the fact that I had no accident for such a long period increase the probability to have one very soon?

developing a theory of rational behavior in the presence of uncertainty

Already Aristotle (384-322 BC) tried to say sometyhing about sheer chance (tyché). He said that it is inaccessible to rational investigation (parálogos) and tells a paradigm which I am roughly paraphrasing as follows:

> A went over the market (agorá) and happened to meet there his debitor B who happened to have cash money with him because his own debitor C happened to have paid back his debt just before – thus A happened to recover his money.

Today such a story might rather sound

> When driving to the airport yesterday morning, I happened to get caught in a traffic jam and therefore happened to miss my flight – luckily, as it happened to turn out, because that aircraft happened to crash. Thus I happen to be still alive, but happened to catch an infection during the trip. Fortunately I happened to have a medicament in my pocket which happened to cure me.

Up to minor contributions (see Gericke [1984][1990]), the problems of chance remained untackled until the 16th century. Girolamo Cardano (1501-1576) a universal scholar who is known to algebraists for the solution formulas for equations of degree 3 and 4 (which actually had been found prior to him by Tartaglia (= Niccoló of Brescia, 1499-1557)), devoted some competent thought to games of chance (Ore [1953], Fierz [1983]). Also Galileo Galilei (1564 –1642) became known for a clever contribution to this field. But the very birthday of modern stochastics came in 1654 when Blaise Pascal (1623-1662) wrote his famous letter to Pierre de Fermat (1601-1665) answering problems concerning games of chance again, which had been posed by a notorious gambler, the Chevalier de Mèrè (1602-1685).

Before treating these early contributions in the next section, let us continue here with a list of historical landmarks in the development of stochastics.

In 1685, Jacob Bernoulli (1654-1705) wrote down (Bernoulli [1975]) his version of what we now call the *law of large numbers* (LLN). It was published in his postumous book Ars Coniectandi (Bernoulli [1713]).

At that time, a new motive for the theory of the uncertain had come up: pension funds and insurances. The authorities of that era had quite a few public servicemen on their payroll "for lifetime" and thus faced the problem of quantifying their financial load by guessing how long those persons would live. The historically first publication of data of this kind seems to have

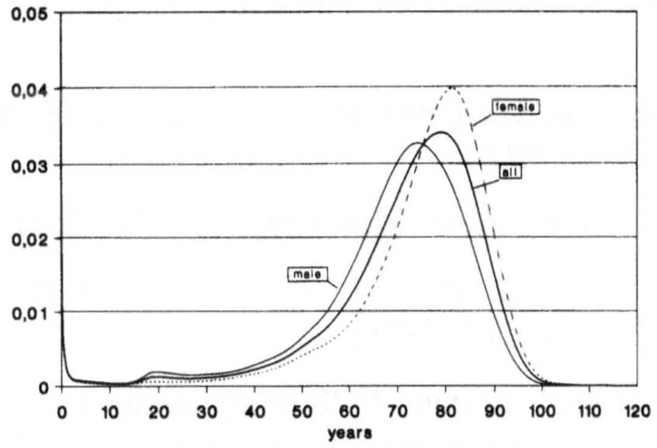

fig. I.1.1

been Graunt, I., Natural and political observations made upon the bills of mortality, London, 1662 (Graunt [1662]).

A typical life-span frequency table of our present time looks as in figure I.1.1, telling us that the average life-span in that population is around 75. Such a table doesn't answer my personal question "when shall I die?". But it gives us a very precise information on the meta-level from which we overlook the life-span phenomenon generally without regard to our individual problem. The precision of this information is illustrated by the following (not quite fictitious) conversation in an office of a big insurance company

A: Wednesday, and only five deaths so far. Eight are due every week.

B: Well, wait two more days. Number 7 and 8 usually drop in on Friday. If we get 9, its an exceptional week.

The law of large numbers if nothing but a mathematical attempt to make that precision understandable.

In 1718, Abraham de Moivre (1667-1754) published his seminal book

The doctrine of chances (Moivre [1967]),

and in 1763, notes of the late Rev. Thomas Bayes (1702-1761) were published postumously under the title

An essay towards solving a problem in the doctrine of chances (Bayes [1958])

Bayes treated problems like the following

> What is the amount of my risk if I invest 100000 £ in a slow fat ship going to East-India and how does this risk change if I learn from the captain of a fast clipper that my vessel has safely made her return as far as Cape Town?

Nowadays we speak of conditional ("if") probabilities and *Bayes' formula* here.

Further landmarks in the history of stochastics were Pierre Simon de Laplace's (1749-1827)

Théorie analytique des probabilités (Laplace [1814])

and Carl Friedrich Gauss' (1777-1855) last work on the Witwen- und Waisenkasse (see Reichel [1978]).

In 1828, Robert Brown (1773-1858) published his later famous paper " A brief account of microscopical observations made in the months of June, July and August, 1827, on the particles contained in the pollen of plants; and on the general existence of active molecules in organic and inorganic bodies" (Brown [1828]).

Irénée Bienaymé (1796-1878) and Pafnutij Tschebyshev (1821-1894) made lasting contributions to probability theory around the midst of the 19th century (see e.g. Heyde-Seneta [1977]).

The beginning of the modern development of probability theory can be characterized by a few seminal papers:

[1906] Einstein, A., Zur Theorie der Brown'schen Bewegung, Ann.Phys. IV 19 (1906) 371-381
[1919] Mises, R. von, Grundlagen der Wahrscheinlichkeitsrechnung, Math. Z. 5 (1919), 52-99
[1924] Khintchine, A., Ein Satz der Wahrscheinlichkeitsrechnung, Fund. Math. 6 (1924), 9-20
[1922] Lindeberg, I.W., Eine neue Herleitung des Exponentialgesetzes in der Wahrscheinlichkeitsrechnung, Math. Z. 15 (1922), 211-225
[1933] Neyman, J., and E.S.Pearson, On the problem of the most efficient tests of statistical hypotheses, Phil. Trans. Roy. Soc. 231 (1933), 289-

and the famous Ergebnisbericht

[1933] Kolmogorov, A.N., Grundbegriffe der Wahrscheinlichkeitsrechnung

From 1933 onward, stochastics has become a rigorous mathematical discipline of enormous width and wealth, omnipresent in many other sciences such as physics, biology and economy. One of the most fertile research themes was Brownian motion, and one of the most influential monographs was

[1953] Doob, J.L., Stochastic Processes.

The present book will inform the reader about some typical parts of this modern development. But before undertaking this in a systematic fashion, let us try to grasp the basic intuitions of stochastics via a few examples, some of which are of historical interest, too.

2. Specimens of Stochastic Reasoning

In this section, we treat 15 problems which bear the typical flavour of stochastics, some of them historically famous. We present answers in a way conformal with, and hinting at, the systematic buildup which will be developed in the subsequent chapters of this book, but will try to get along with a minimum of formalism. Professional stochasticians usually get the gist of a problem that way rather than by meticulous theory-work.

2.1. Problem I of Chevalier de Méré (1654).

What is more probable:

> to throw at least one six in a throw of 4 dice, or
> to throw at least one double-six in 24 throws with 2 dice each?

The chevaliers somewhat fuzzy idea had been

> $4 : 6 = 24 : 36$, hence both probabilities are equal,

but he had lost money relying on this equality.

Blaise Pascal (1624-1662) wrote the correct solution to Pierre de Fermat (1601-1665) in his letter of 1654 (see e.g. Pascal [1954]); in modern language, it sounds as follows:

> to *not* throw any 6 with 4 dice, amounts to throw one of the five ciphers 1,2,3,4,5 with each of them, which happens with probability $\frac{5}{6}$ for each dice, hence with probability $(\frac{5}{6})^4$ with 4 dice. Our first probability thus is
>
> $$1-(\frac{5}{6})^4 = 0.51775\ldots;$$
>
> to *not* throw a double-six with 2 dice, is an event which happens with probability $\frac{35}{36}$ for each such trial, hence with probability $(\frac{35}{36}{}^2 4$ in 24 trials. Thus our second probability is
>
> $$1-(\frac{35}{36})^2 4 = 0.4914\ldots$$
>
> which is smaller than the first one.

2.2. The three-dice-problem.

Gambling courtiers had observed that throws with 3 dice produced a sum 10 more often than 9. They found this paradoxical since both numbers can be decomposed into 3 summands in the same number of ways, namely,

$$
\begin{aligned}
9 &= 1+2+6 = 1+3+5 = 1+4+4 \\
&= 2+2+5 = 2+3+4 = 3+3+3 \\
10 &= 1+3+6 = 1+4+5 = 2+2+6 \\
&= 2+3+5 = 2+4+4 = 3+3+4
\end{aligned}
$$

The problem was solved by Girolamo Cardano (1501-1576) in his book "De ludo aleae", but since this book was published only in 1663, Galileo Galilei (1564-1642) , unaware of Cardano's solution, solved it independently sometime between 1613 and 1624 ("Sopra le scoperte dei dadi"). Both solutions are based on the remark that the above sums treat the three dice erroneously as indistinguishable. If we work with a black, a red, and a green dice, writing the outcome of the black dice as the first summand, red as the second, and green as the third, then we obtain the two lists of representations in figure I.2.2., giving 25 possibilities for sum 9, and 27 for sum 10. Actually, three different ciphers occur here in 3!=6 fashions; if two of them are equal and the third one different, 3 fashions arise; and three equal ciphers occur in only one fashion. As all triads are equally probable, sum 10 occurs with probability $\frac{27}{6^3} > \frac{25}{6^3}$ = the probability of sum 9.

2.3. Coin-in-square.

In the game "coin-in-square", players throw a coin, 2 cm in diameter, on a table covered with a design of 3cm squares. A throw wins if the coin falls entirely inside one of the squares. If it intersects any of the lines, the throw is a loss. If you bet an amount A on "win", and your opponent B on "loss" – for which choices of A and B will you break even in the long run?

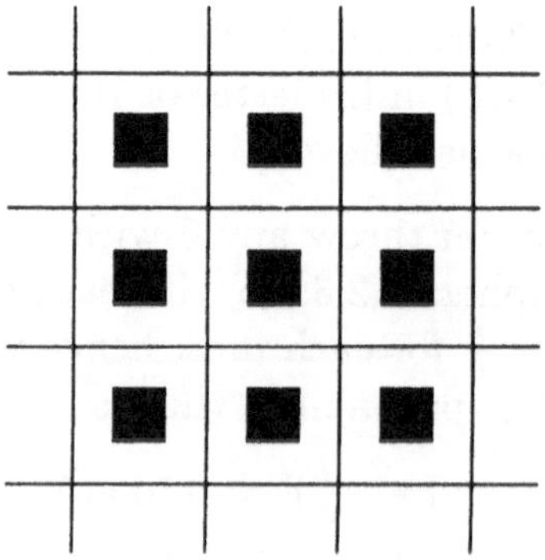

fig. I.2.1

Well, the throw wins iff the midpoint of the coin comes to rest in one of the blackened shaded areas. They cover $\frac{1}{9}$ of the total area, thus you will win in 1 out of 9 throws, and lose in 8 of 9 – roughly. That is, in $N = 9n$ throws,

9

black	red	green
1	2	6
1	6	2
2	1	6
2	6	1
6	1	2
6	2	1
1	3	5
1	5	3
3	1	5
3	5	1
5	1	3
5	3	1
1	4	4
4	1	4
4	4	1
2	2	5
2	5	2
5	2	2
2	3	4
2	4	3
3	2	4
3	4	2
4	2	3
4	3	2
3	3	3

10

black	red	green
1	3	6
1	6	3
3	1	6
3	6	1
6	1	3
6	3	1
1	4	5
1	5	4
4	1	5
4	5	1
5	1	4
5	4	1
2	2	6
2	6	2
6	2	2
2	3	5
2	5	3
3	2	5
3	5	2
5	2	3
5	3	2
2	4	4
4	2	4
4	4	2
3	3	4
3	4	3
4	3	3

fig. I.2.2

you will make nA, and your opponent $8nB$. Hence $A = 8B$ is the condition for balance – "roughly", "in the long run", "up to random fluctuations", as a gambler would put it.

2.4. Will you change your option?.

The following game appears in TV now and then: you are confronted with three closed boxes, and you are told that two of them are empty while one of them contains a 1000 $ bill. You first point to one of the boxes without opening it. Then the moderator opens one of the other two boxes, showing it empty. He permits you to change your option. Will you do it? That is, will in the long run, a "changer" be more successful than a "sticker"?

Well, in analyzing the situation, we may, without loss of generality, assume that the 1000 $ bill is in box 1. Let us draw a tree displaying all possible moves of the game in figure I.2.3.

On level 1 you choose 1 or 2 or 3 at random.
On level 2 the moderator shows you one of the other boxes to be empty: if

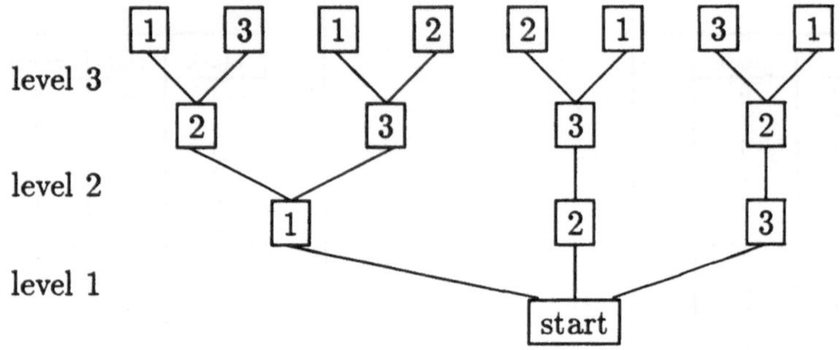

fig. I.2.3

you had pointed to lucky 1, he chooses at random between 2 and 3; if your first choice was 2, he can only show 3, and if you had selected 3, he can only show 2.

On level 3 you decide

either to stick (left branch)

or to change (right branch), of course not to the box which has just been shown to be empty.

A "sticker" wins with probability $\frac{1}{3}$ (that is, in 1 out of 3 cases, in the long run). A "changer" loses only if he had pointed to 1 in his first move: he wins with probability $\frac{2}{3}$. Surprising, isn't it?

2.5. Boy's budget in a tennis-playing family.

Dad Adam tells son Abel: "I will increase your pocket money if you win two sets in consecution, playing three sets alternatingly against Mom, eve and me." Abel's chance of winning

against Adam is a

against Eve is $e < a$ (she is very strong)

He may choose against whom to play first. What will be the better choice for him?

Look at the two diagrams in figure I.2.4.

As $e < a$, Abel decides to first play against Eve.

2.6. Unreliable information chain.

In Cheatistan, every citizen lies with probability $0 < p < \frac{1}{2}$, except for President Adams and TV speaker Zaplin who always speak the truth. Adams decides to run for president again, and Zaplin gets the information through

first against Adam	first against Eve
Abel meets Dad's condition with probability	Abel meets Dad's condition with probability
$ae + (1-a)ea = ae(2-a)$	$ae + (1-e)ae = ae(2-e)$

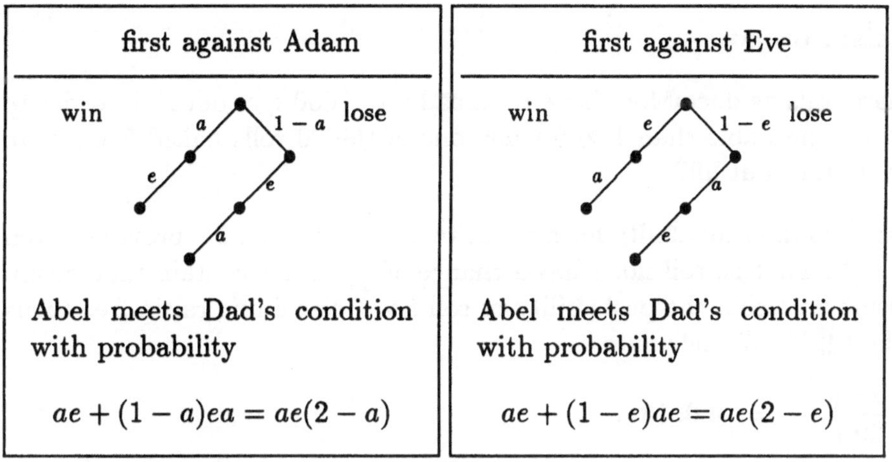

fig. I.2.4.

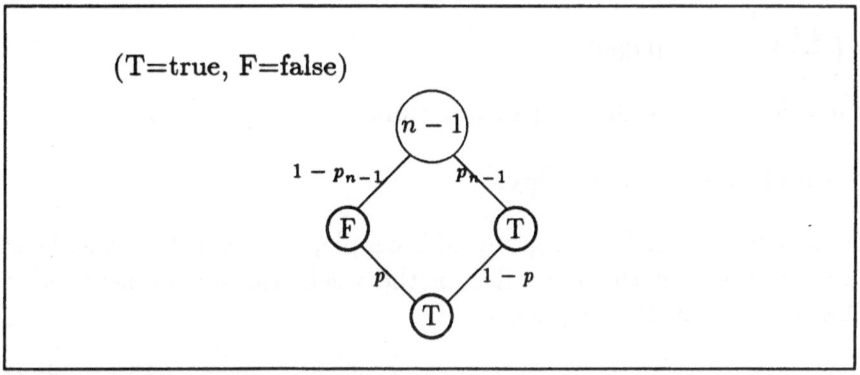

fig. I.2.5

a chain of n informants. What is the probability p_n that Zaplin announces the president's decision correctly?

Reading the graph in figure I.2.5, cleverly, we arrive at the recursion

$$
\begin{aligned}
p_n &= (1-p)p_{n-1} + p(1-p_{n-1}) \\
&= (1-2p)p_{n-1} + p.
\end{aligned}
$$

Solving this inhomogeneous difference equation as usual,

$$p_n = \frac{1 + (1-2p)^n}{2}$$

follows. Hence for large n, Zaplin might as well toss a coin.

2.7. Raisin bread.

How many raisins does Mom have to knead into 1000 g of dough in order to make it less probable than 1 % for any one of the 20 rolls baked from it, to contain no raisin at all?

Answer: $\frac{1}{20}$ is the probability for a raisin to go into roll no. 1 – provides Mom kneads well. That is, roll no. 1 has a chance of $\frac{19}{20}$ to not contain that raisin. If n raisins are used, the probability of roll no. 1 remaining raisin-free turns out to be $(\frac{19}{20})^n$. We calculate

$$\left(\frac{19}{20}\right)^{87} = 0.0115\ldots$$

$$\left(\frac{19}{20}\right)^{88} = 0.0109566\ldots$$

$$\left(\frac{19}{20}\right)^{89} = 0.0062\ldots$$

Thus $n = 89$ raisins suffice. Open questions?

2.8. Coincidence of birthdays.

Mr. Y says: whenever I see a group of 7 people, I bet 100:1 that at least two of them are born on the same day of the week. Do you understand why I have become rich in the long run?

Well, out of the 7^7 possible distributions of the 7 days of the week over 7 persons, only 7! assign a different day to every person. The probability of losing my bet is therefore

$$\frac{7!}{7^7} = 0.00612 < 0.01$$

and that does it for me.

Let me modify that problem: what is the probability p_n of meeting, in a party of n persons, at least 2 with the same birth*day*?

Answer: if we have asked already k persons and have found them all with different birthdays, the probability of finding the next person with a different birthday again, ist $\frac{365-k}{365}$. Thus, for a party of n, the probability of finding all n birthdays different is

$$1 - p_n = \frac{364}{365} \cdot \frac{363}{365} \cdots \frac{365 - n + 1}{365}$$

Here are some values

n	10	20	22	23	41	57	70	80
p_n	0.1170	0.4114	0.4757	0.5073	0.9032	0.9901	0.9992	0.9999

2.9. Investment risk.

Someone proposes you to invests in a certain venture which succeeds half the time. For each invested dollar,

you make \$ 1.60 if the venture succeeds
you lose \$ 1.00 if the venture fails

That is, your invested 1 \$ transforms itself into \$ 2.6 in the case of success, and into \$ 0 in the case of failure. In order to never get totally broke, you adopt the

Strategy: Always invest 50 % of your actual fortune.

Thus your fortune F will transform itself into

$$\tfrac{1}{2}F + 2.6 \cdot \tfrac{F}{2} = 1.8F \quad \text{in the case of success}$$

$$\tfrac{1}{2}F + 0.\tfrac{F}{2} = 0.5F \quad \text{in the case of failure}$$

Your *expected* fortune after one trial is therefore

$$\frac{1}{2}(1.8F) + \frac{1}{2}(0.5F) = 1.15F$$

This calculation might lure you into that venture. But another counselor will tell you the following: if one success if followed by one failure or vice versa, you end up with

$$0.5(1.8F) = 0.9F,$$

and if, in a sequence of $2n$ trials, you see success exactly half the time, you end up with

$$(0.9)^n F,$$

and that goes down to zero rather quickly:

n	1	2	3	4	5	6
$(0.9)^n$	0.9	0.81	0.729	0.6561	0.59049	0.348...

Sounds paradox? Well, the point is that your fortune develops by random *multiplication*. This amounts to random *addition of logarithms*. And

$$\begin{aligned}\log 1.8 + \log 0.5 &= 0.25527\ldots - 0.30103\ldots \\ &= -0.04576\ldots\end{aligned}$$

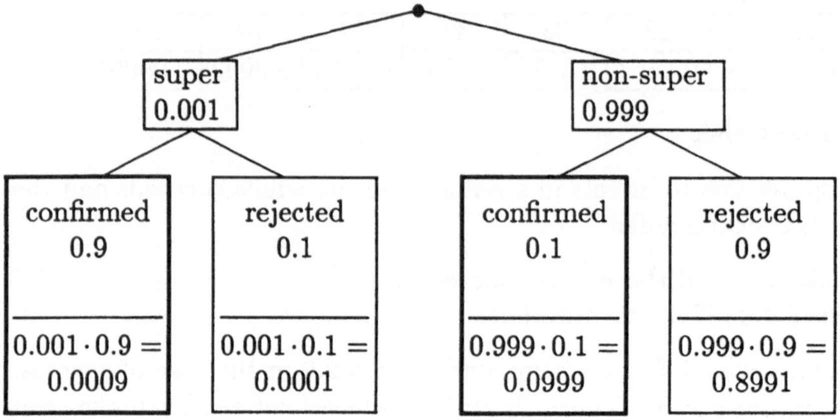

fig. I.2.6.

tells you that the expected value of these is < 0.

Sure, if the venture succeeds in more than 50 % of the trials – say in p % – you might gain in the long run. What is the minimum value of p to realize that dream? Try to calculate it!

2.10. The super growth stock.

Assume that one in 1000 stocks is a "super growth stock": it triples in a single year. You want to buy such a stock, and you first

> ask your astrologer . He tells you, stock S is a "super growth stock". But you suspect that in reality, your astrologer has chosen S at random. Thus you also
>
> ask an expert who has the reputation of being right 90 % of the time. He confirms "yes, S is a super growth stock"

What is the probability that S is really a super growth stock? Look at the diagram in figure I.2.6.

In the above situation, the two bold wall boxes represent the two possible realities: really super (left), non-super (right). The fraction of "really super" in the total probability of the two together is

$$\frac{0.0009}{0.0009 + 0.0999} = 0.0089$$

This is more than eight times higher than in the case of random selection (0.001).

2.11. The taxi problem.

Assume there are T taxi cabs in Boston, visibly numbered $1, 2, \ldots, T$. You don't know what T is, but you want to make a guess. First you ask one of the taxi drivers; he, too, doesn't know the exact value of T, but he tells you that it must be at least 3000. You want to test his claim and make a few observations of your own. Looking about at random, you observe four cabs bearing the numbers 512, 987, 355, 1200. Well, if the taxi driver was right with $T = 3000$ the probability of observing a number ≤ 1200 is $\frac{1200}{3000}$, and even smaller if $T > 3000$. To observe ≤ 1200 four times independently, bears a probability $\left(\frac{1200}{3000}\right)^4 \approx 0.0256$. That's too improbable, and you set out to make your own guess, on the basis of your four observations. After pondering for a while how to make a good guess, your intuition comes up with

Idea I: Let M be the maximum of the cab numbers observed in n trials: $n = 4, M = 1200$. Obviously, T is $\geq M$, but how much larger than M can it be? You decide to use the average of the observed gaps $355 - 0$, $512 - 355$, $987 - 512$, $1200 - 987$, that is, of 355, 157, 475, 213, that is, $\frac{1}{4}(355 + 157 + 475 + 213) = \frac{1200}{4} = 300$, as an upper estimate for $T - M$, thus you end up with $T \leq 1200 + 300 = 1500$. The probability of observing ≤ 1200 four times is now $\left(\frac{1200}{1500}\right)^4 = \left(\frac{4}{5}\right)^4 = \frac{256}{625} = 0.409\ldots$, much more as with $T \geq 3000$. – We did, by the way, some unnecessary calculations above: if you observe $x_1 < x_2 < \ldots < x_n = M$, the $\frac{M}{n}$ is the average gap length, and your estimate is $T \leq M + \frac{M}{n} = M(1 + \frac{1}{n})$.

Is this the only possible way of guessing the true value of T? You discuss the problem with a friend, and he proposes

Idea II: The probability of observing a maximal cab number M in n trials is $p_n = \left(\frac{M}{T}\right)^n$, if T is the total number of cabs. We should estimate a value of T which makes this p_n not too little – say $p_n \geq \frac{1}{20} = 0.05$. This leads to

$$\left(\frac{M}{T}\right)^n \geq \frac{1}{20} \iff T \leq M\sqrt[n]{20}$$

In our particular situation with $n = 4, M = 1200$, we end up with $\sqrt[4]{20} = 2.1147\ldots$ and hence

$$T \leq 1200 \cdot 2.1147 = 2538.$$

Had we prescribed $p_n \geq \frac{1}{3}$, the result would have been ($\sqrt[4]{3} = 1.3160\ldots$)

$$T \leq 1200 \cdot 1.3160 = 1579.28\ldots < 1580.$$

The morale of this story is: there may be a host of guessing methods, each of which with its own plausibility. You are left with the task of choosing one among them in a rational fashion, that is, to set standards for the quality of guessing methods, and to select a method which is optimal according to these standards. This is the professional task of *statisticians*, and they may have different opinions about which standards to prescribe – a non-mathematical problem in principle, although mathematics will surely enter into the dispute.

2.12. The drunkard on the cliff.

Joey the drunkard is standing one step away from the rim of a cliff. One step towards the abyss will result in death. But Joey moves at random. Will he survive?

Not for sure, but maybe with a positive probability P_1 which depends upon the probability p of Joey doing the next stumble towards the rim ($1-p$ is the probability of moving away from death).

Let Q_d denote the probability of Joey's eventual death if his initial position is d steps away from the abyss. The numbers $Q_0(=1), Q_1, Q_2, \ldots, Q_d, \ldots$ are related to each other. We have e.g.

$$\begin{aligned} Q_1 &= p + (1-p)Q_2 \\ &= pQ_0 + (1-p)Q_2 \end{aligned}$$

since Joey either dies making the first step in the fatal direction (probability p), or he moves in the safer distance 2 (probability $1-p$) and lives with a chance Q_2 of falling into the precipice later on. By the same token, the general formula

$$Q_d = pQ_{d-1} + (1-p)Q_{d+1}$$

follows, but we will concentrate on $d = 1$ here.

Now surely

$$Q_2 = Q_1^2$$

because Q_1 is also the probability of returning from 2 to 1 at some time, and then Q_1 is the chance of dying afterwards. Thus we end up with

$$Q_1 = p + (1-p)Q_1^2,$$

a quadratic equation for Q_1, which has the solutions 1 and $\frac{p}{1-p}$ (verify!). Which one should we choose? Our choice may depend upon the value of the parameter p. Now let us plot $\frac{p}{1-p}$ as a function of p, together with the constant 1 (figure I.2.7).

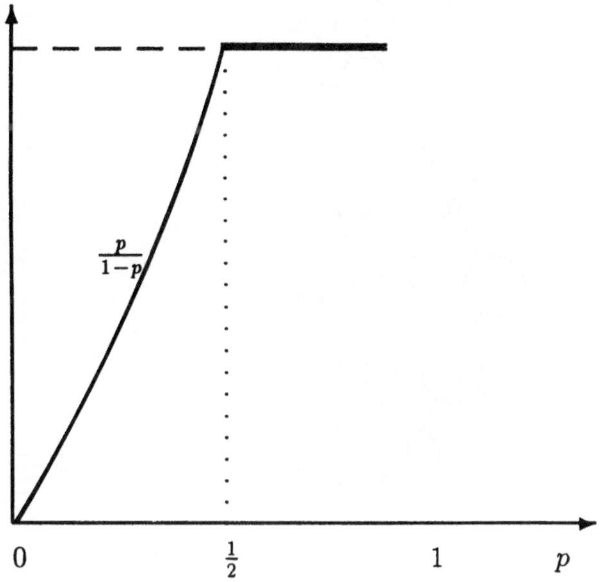

fig. I.2.7

Obviously,

for $p = 1 - p = \frac{1}{2}$, both roots coincide
for $p = 0$, only $Q_1 = 0$ is realistic
for $p = 1$, only $Q_1 = 1$ is reasonable
for $\frac{1}{2} < p < 1$, we have $\frac{p}{1-p} > 1$, hence only $Q_1 = 1$ makes sense

If we require Q_1 to be a continuous function of p, only $Q_1 = \frac{p}{1-p}$ is feasible for $0 \leq p < \frac{1}{2}$, and we end up with $Q_1 = Q_1(p) =$ the boldface line in the above picture.

2.13. Problem II of Chevalier de Méré.

In 1654 the Chevalier de Méré (1602-1685) raised the raised the following question:
two gamblers repeatedly toss a coin; gambler 1 bets on "head", 2 on "tail". The stake (16 pistols in the historical example) goes to the gambler who collects 3 wins first. Question: assume the game has to be broken off prematurely – what would be a fair splitting of the stake between the two gamblers? – Let us plot all possibilities (figure I.2.8).

In the diagram, ↗ means a win for gambler 1 (head), and ↘ a win for gambler 2 (tail). Each circle contains the probability that player 1 will be the future winner. How did I calculate these figures? Firstly, it is obvious that the circles on the middle horizontal represent situations of perfect symmetry

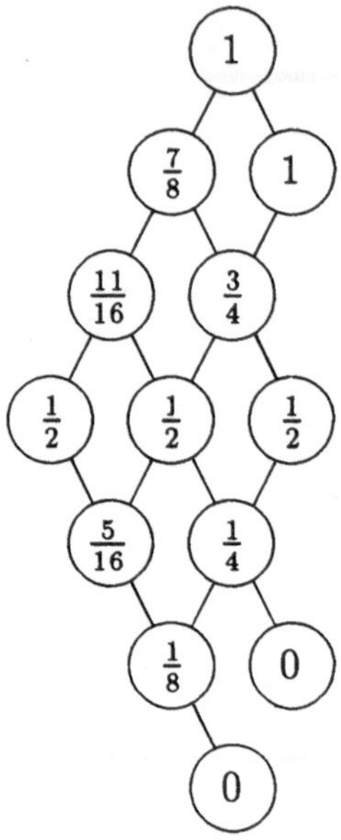

fig. I.2.8

between the two gamblers, hence have to be filled with $\frac{1}{2}$. The fillings 1 and 0 are also obvious. Now we calculate from right to left, forming arithmetical means. Why? Well, facing "head" brings gambler 1 with probability $\frac{1}{2}$ to your right upper neighbor, and hence with $\frac{1}{2}$ of the probability written there to final success. The same applies to the right lower neighbor, and thus our calculation is correct.

If we break off in a situation where 1 is the future winner with probability p, fairness demands $p \cdot 16$ pistols to be given to gambler 1, and $(1-p) \cdot 16$ to gambler 2.

2.14. Inside information.

In a district in Alaska three mines have been prospected, one for silber, one for gold, one for platinum. The federal government subdivides the district into 40 lots, none of them containing two of the mines, and opens a lottery which gives each of 20 mining companies two lots.

You own stocks of company A and, in order to assess your stocks, want to calculate the probability that A has won 2 mines. The probability is

$\frac{3}{40}$ for the first lot to contain a mine
$\frac{2}{39}$ for the second lot to also contain a mine,

thus the probability is

$$\frac{3}{40} \cdot \frac{2}{39} = \frac{6}{1560} = 0.00384\ldots$$

Similarly, the probability for exactly *one* mine is

$$\frac{3}{40}\cdot\frac{37}{30}+\frac{37}{40}\cdot\frac{3}{39}=2\cdot\frac{3}{40}\cdot\frac{37}{39}=\frac{222}{1560}=0.14230\ldots$$

(one mine in the first and none in the second trial, or none in the first and one in the second)
and for *none*:

$$\frac{37}{40}\cdot\frac{36}{39}=\frac{1332}{1560}=0.85384\ldots$$

Now some insider tells you that A got *at least one mine*. Then you calculate anew, weighing the possibility of getting two mines against the possibility of getting at least one; canceling out the common denominator 1560, you arrive at

$$\frac{6}{222+6}=\frac{6}{228}=\frac{1}{38}=0.0263$$

which is much more than the previous 0.00384. So it was really worthwhile to obtain that insider information.

What if the insider had told you that A got at least the gold mine?

2.15. Predicting rain in Tuggurt.

Rainfall in Tuggurt occurs according to the following rules (D = dry, R = rain):

today/tomorrow	D	R
D	$\frac{5}{6}$	$\frac{1}{6}$
R	$\frac{1}{3}$	$\frac{2}{3}$

or, equivalently, 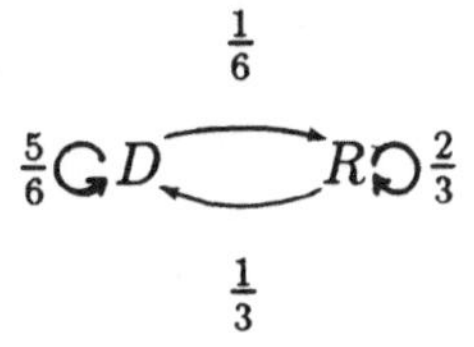

i.e. rain today entails rain tomorrow with probability $\frac{2}{3}$ etc.
Today it's a rainy Friday. What is the probability for a dry Sunday? Well, there are two possibilities

$$\begin{array}{ccccccl} \text{Fri} & & \text{Sat} & & \text{Sun} & & \\ \boxed{R} & \xrightarrow{\frac{2}{3}} & \boxed{R} & \xrightarrow{\frac{1}{3}} & \boxed{D} & & \text{probability } \frac{2}{3}\cdot\frac{1}{3}=\frac{2}{9} \\ \boxed{R} & \xrightarrow{\frac{1}{3}} & \boxed{D} & \xrightarrow{\frac{5}{6}} & \boxed{D} & & \text{probability } \frac{1}{3}\cdot\frac{5}{6}=\frac{5}{18} \end{array}$$

These probabilities add up to

$$\frac{2}{9}+\frac{5}{18}=\frac{4+5}{12}=\frac{9}{18}=\frac{1}{2}$$

i.e. you may as well toss a coin.
If, however, Friday is dry, the probability of having a dry Sunday is

$$\frac{5}{6}\cdot\frac{5}{6}+\frac{1}{6}\cdot\frac{1}{3}=\frac{25+2}{36}=\frac{27}{36}=\frac{3}{4}$$

Weather in Israel during the winter months December, January, February follows the graph

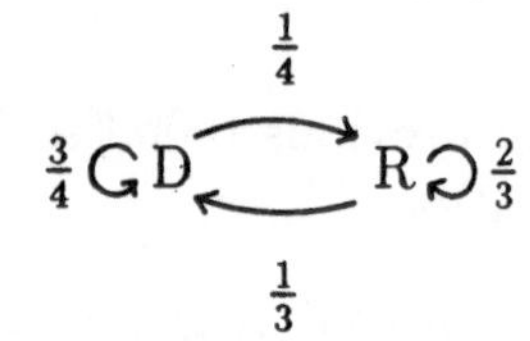

What are the above guesses in this case?

II. Markovian Dynamics

In many cases the evolution of certain phenomena can be interpreted as changes of the state of a ***dynamical system*** . A finite-state dynamical system consists of a finite "state space" D and a mapping $T : D \to D$. The interpretation is: if the system is in state $j \in D$ at time t, it will be in state $k = T(j)$ at time $t+1$. Such finite state discrete-time dynamical systems are much too primitive to describe important phenomena like the evolution of the plantetary system or the like; continuous state spaces and continuous time are a characteristic of most full-fledged dynamical systems, and calculus is the proper tool for their investigation; actually, calculus was invented for this purpose.

Primitive as it is, the above notion of a finite-state discrete-time dynamical system is sufficient to convey two fundamental ideas

> the idea of ***deterministic evolution*** : if you know where the system is at time t, you know where it will be at time $t+1$, and likewise at any future moment.
>
> the idea of ***eternal repetition***: the decreasing sequence $D \supseteq T(D) \supseteq T(T(D)) \supseteq \ldots$ of finite sets $\neq \emptyset$ becomes eventually constant $= D_0 \subseteq D$, and T acts on D_0 as a permutation which splits into cycles. Thus every ***orbit*** $D \ni j \to T(j) \to T(T(j)) \to \ldots$ becomes purely periodic after a finite number of steps.

Our derivation of the second phenomenon depends heavily upon the finiteness of D (isn't it true, by the way, that religions with a dogma of eternal return in some way or other also operate with finiteness assumptions?)

What is even more important for us here: the above notion of a finite-state discrete-time dynamical system lends itself easily to a generalization which makes precise

the idea of ***non-deterministic evolution***

in a fashion allowing to display many more non-trivial applications than the previous finite deterministic model. Moreover we will see that

the idea of ***eternal repetition***

pervades the non-deterministic model in the case of a finite state space once again (§3), and finally we will get the insight that

determinism is a special case of "Markovian" indeterminism.

In this book the "Markov theme" will be treated on two levels in two different chapters. The present chapter works on the "macroscopic" level. In §1 we introduce the notions

stochastic matrix
Markovian dynamical system,

incorporating the abovementioned ordinary dynamical systems as a special case. We will motivate the view that a study of such systems is more or less tantamount to the investigation of the asymptotic behavior of iterates of stochastic matrices. In §2 we collect the basic notions and facts which are necessary in order to deal adequately with the set W of all stochastic matrices over a finite state set D. In §3 we investigate the asymptotic behavior of iterates of stochastic matrices in some important special cases and prove the so-called ergodic theorem for stochastic matrices. In §4 we present a general method which yields the basic final result: asymptotic periodicity. This is enough for the moment; a curious reader will find further results and methods in appendices C and D of this book. In ch. VI we will take up the "Markov theme" for a second time, and on the "microscopic" level (see also ch. II §2 subsection 2.5).

1. Finite-state Markovian dynamical systems

Imagine a dynamical system that can assume states from a finite state space D and evolves nondeterministically in the following fashion: if the system is in state $j \in D$ at time t, it is in state $k \in D$ at time $t+1$, not for sure, but with a "transition probability" $P_{jk} \geq 0$ $(\sum_{k\in D} P_{jk} = 1)$. This leads to the following

Definition 1.1. Let D be a nonempty finite set. A $D \times D$-matrix $P = (P_{jk})_{j,k\in D}$ is called *stochastic* if

$$\begin{array}{ll} P_{jk} \geq 0 & (j,k \in D) \\ \sum_{k\in D} P_{jk} = 1 & (j \in D). \end{array}$$

The pair (D, P) is then called a *Markovian dynamical system* with *state space* D and *transition matrix* P. The entry P_{jk} of P is also called the *(transition) probability* for the transition $j \to k$.

Example 1.2.: determinism as a special case. . Let $D \to D$ be any mapping of the finite set $D \neq \emptyset$ into itself. Then

$$P_{jk} = \delta_{\tau(j),k} \quad (j, k \in D)$$

(δ_{jk} is the usual Kronecker symbol, i.e. $\delta_{jk} = 1$ if $j = k$, and $= 0$ else) defines a stochastic $D \times D$-matrix $P = \Delta_\tau$. The reader is invited to verify $\Delta_\sigma \Delta_\tau = \Delta_{\tau\circ\sigma}$ for any two mappings σ, τ of D into itself ($\tau\circ\sigma$ means "first σ then τ" as usual).

The above definition is easily, and in an obvious fashion, extended to the case of a denumerably infinite D. There are good motives for such an extension, as is shown by

Example 1.3.: "The drunkard on the cliff". (see ch. I §2, 2.12). Here $\mathbb{Z}_+ = \{0, 1, \ldots\}$ is the obvious choice for D. State 0 (= the rim of the cliff) is a "state of no return" (fatal, letal, absorbing ...):

$$P_{00} = 1, \quad P_{0k} = 0 \quad (k \neq 0).$$

For $j > 0$, we had, in the above-cited "specimen of stochastic reasoning",

$$P_{j,j-1} = p, \quad P_{j,j+1} = 1 - p, \; P_{j,k} = 0 \quad \text{else}$$

with some parameter $0 < p < 1$, namely, the probability of stumbling towards the rim of the cliff. We are not in the position to display the "probability of survival" investigated in ch. I §2 as a probability within our present model; see however ch. VI.

Example 1.4.: "Change of Weather in Tuggurt". (see ch. I, §2, 15). Here the transition matrix from one day to the next is

$$P = \begin{pmatrix} \frac{5}{6} & \frac{1}{6} \\ \frac{2}{3} & \frac{1}{3} \end{pmatrix}$$

where the first index is for "dry", the second for "rain".

Exercise 1.5.:"Three pistoleros". Let A, B, C be three pistoleros fighting a duel

A hits with probability α
B hits with probability β
C hits with probability γ

Shots are fired simultaneously; hits happen independently; every hit is fatal. Assume $1 > \alpha > \beta > \gamma > 0$. Let D be the set of all subsets of $\{A, B, C\}$.

	$\emptyset$	A	B	C	AB	BC	ABC
$\emptyset$	1	0	0	0	0	0	0
A	0	1	0	0	0	0	0
B	0	0	1	0	0	0	0
C	0	0	0	1	0	0	0
AB	$\alpha\beta$	$\alpha(1-\beta)$	$(1-\alpha)\beta$	0	$(1-\alpha)$ $(1-\beta)$	0	0
AC	$\alpha\gamma$	$\alpha(1-\gamma)$	0	$(1-\alpha)\gamma$	0	0	0
BC	$\beta\gamma$	0	$\beta(1-\gamma)$	$(1-\beta)\gamma$	0	$\beta\gamma$	0
ABC	0	0	0	$\alpha\beta+\alpha$ $(1-\beta)\gamma$	0	$(1-\alpha)$ $(\gamma+\beta)$ $(1-\gamma)$	$(1-\alpha)$ $(1-\beta)$ $(1-\gamma)$

$M \in D$ represents the state where precisely the pistoleros in M are still alive. Assume that every pistolero aims at his strongest opponent. Show that the transition matrix is like the table above and that AB never happens if we start with ABC.

So far we have considered transitions in one time unit only. What if we let elapse two units?

Let us imagine the elements j of D to be cells between which particles go to and fro in a diffusion process: if we put a particle into cell j ("system in state j"), we will find it, after the elapse of one time unit, in cell k, not for sure, but with probability P_{jk}. The law of large numbers (LLN, ch. IV) allows us to roughly represent probabilities by frequencies: if we put a large "unit" quantity of particles into cell j and wait for one time unit, the fraction P_{jk} of our particles will be in k. That is, our diffusion will transform the j-th unit vector "unit quantity of particles in j" into a distribution "quantity P_{jk} in k ($k \in D$)" in unit time. If we assume that particles do not interact during diffusion, we conclude that such transforms will work in a linear fashion:

if we start with a distribution of particles over the cells represented by a nonnegative vector $a = (a_j)_{j \in D}$, we will see quantity $\sum_{j \in D} a_j P_{jk}$ in cell k after unit time, that is, our diffusion sends a into aP in one time unit. One more time unit will transform aP into $aPP = aP^2$ etc. that is, we will see the ("movie") sequence $a, aP, aP^2, \ldots, aP^2, \ldots$ in discrete time. We have worked here with the tacit understanding that the law of diffusion, i.e. the matrix P, doesn't undergo any change in time; if we would drop this assumption, we might instead work with a possibly non-constant sequence $P, Q, R, \ldots$ of stochastic $D \times D$-matrices and thus have to contemplate the "movie" $a, aP, aPQ, aPQR, \ldots$; extensive study has been devoted to such non-stationary products of stochastic matrices, and even of more general operators, see e.g. Seneta [1981], Jacobs [1957a][1958], Krengel et all [1987], Martus [1989]; we will refrain from going into more details on that subject and content ourselves here with the conclusion:

> the study of Markovian dynamical systems is tantamount to the study of iterates of stochastic matrices.

Let us see what happens in some of our previous examples if we pass from P to P^2 or even to higher powers P^n.

The deterministic example 1.2. shows $(\Delta_\tau)^n = \Delta_{\tau^n}$, that is, we study $\tau^0 = \mathrm{id}_D, \tau^1 = \tau, \tau^2, \tau^3, \ldots$ via the corresponding matrices. In this sense, deterministic dynamical systems are a special case of Markovian dynamical systems.

"The drunkard on the cliff" is ruled by the iterates (write $\bar{p}$ instead of $1-p$) of

$$P = \begin{pmatrix} 1 & 0 & 0 & 0 & \ldots \\ p & 0 & \bar{p} & 0 & \ldots \\ 0 & p & 0 & \bar{p} & \ldots \\ 0 & 0 & p & 0 & \ldots \\ \cdot & \cdot & \cdot & \cdot & \ldots \end{pmatrix}$$

Exercise 1.6. Calculate P^2, P^3 for this P.

The weather in Tuggurt is governed by the iterates of

$$P = \begin{pmatrix} \frac{5}{6} & \frac{1}{6} \\ \frac{2}{3} & \frac{1}{3} \end{pmatrix}$$

Exercise 1.7. Calculate P^2, P^3, P^4 for this P.

The evolution of the "three pistoleros' duel" is steered by the iterates of the matrix given in exercise 1.5.

Exercise 1.8. Choose $\alpha = \frac{2}{3}$, $\beta = \frac{1}{2}$, $\gamma = \frac{1}{3}$ and calculate P^2 for that matrix. Show that $Q = P^n$ has $Q_{ABC,ABC} = (1-\alpha)^n(1-\beta)^n(1-\gamma)^n$.

2. The convex set of stochastic matrices

Before dealing with a stochastic $D \times D$-matrix and its iterates, we should make available a few notions and basic insights concerning the vector lattice

$\mathbf{R}^D$, the convex set V of all probability vectors in $\mathbf{R}^D$, and the convex set W of all stochastic $D \times D$-matrices.

2.1. The vector lattice $\mathbf{R}^D$.

Let D be a set of $d \geq 1$ elements and $\mathbf{R}^D$ (also written $\mathbf{R}^d$) the vector lattice of all real vectors indexed by D. Let us define resp. recall: for $x = (x_j)_{j\in D}$, $y = (y_j)_{j\in D}$,

$$\begin{aligned}
x \leq y \quad & if \quad x_j \leq y_j \quad (j \in D) \\
x < y \quad & if \quad x \leq y, x \neq y \\
x \ll y \quad & if \quad x_j < y_j \quad (j \in D) \\
x \vee y \quad & = \quad (\max[x_j, y_j])_{j\in D} \\
x \wedge y \quad & = \quad (\min[x_j, y_j])_{j\in D} \\
x_+ \quad & = \quad x \vee 0 \\
x_- \quad & = \quad (-x) \vee 0 \\
|x| \quad & = \quad x \vee (-x) \\
x \quad & = \quad x_+ - x_- \\
|x| \quad & = \quad x_+ + x_- \\
x \vee y \quad & = \quad x + (y - x)_+ \\
x \wedge y \quad & = \quad x - (x - y)_+ \\
\|x\| \quad & = \quad \sum_{j\in D} |x_j| \\
\langle x \rangle \quad & = \quad \sum_{j\in D} x_j
\end{aligned}$$

We will employ the norm $\|\cdot\|$ in order to describe the topology of $\mathbf{R}^D$. It is the same topology, whatever norm we would choose, namely, the topology of componentwise convergence; all norm bounded subsets of $\mathbf{R}^D$ are conditionally compact in this topology. We will occasionally carry over part of these notions, suitably modified, also to the space $\mathbf{R}^{D\times D}$ of all real $D \times D$-matrices.

2.2. Probability vectors and stochastic matrices over D.

Let

$$\begin{aligned}
V \quad & = \quad \{p = (p_j)_{j\in D} \mid p_j \geq 0 \quad (j \in D), \sum_{j\in D} p_j = 1\} \\
& = \quad \{p \mid p \in \mathbf{R}^D, p \geq 0, \langle p \rangle = 1\}
\end{aligned}$$

denote the set of all ***probability vectors*** over D.

As everyone knows, $D \times D$-matrices $M = (M_{jk})_{j,k\in D}$ and linear operators $M : \mathbf{R}^D \to \mathbf{R}^D$ are in one-to-one correspondence via

$$xM = (\sum_{j\in D} x_j M_{jk})_{k\in D},$$

and matrix multiplication corresponds to operator multiplication. The j-th row vector $M_{j\cdot} = (M_{jk})_{k\in D}$ of a matrix M is obtained as

$$M_{j\cdot} = e^{(j)} M$$

where $e^{(j)} = (0,\ldots,0,1,0,\ldots 0)$ (unique 1 at place j) denotes the j-th unit vector.

The set W of all ***stochastic*** (or ***Markov***) ***$D \times D$ matrices*** may be defined resp. described as

$$\begin{aligned} W &= \{P \mid P = (P_{jk})_{j,k\in D},\ P_{jk} \geq 0 \ \ (j,k \in D), \\ &\qquad \sum_{i\in D} P_{ji} = 1 \ \ (j \in D)\} \\ &= \{P \mid P = (P_{jk})_{j,k\in D},\ P_{j\cdot} \in V \ \ (j \in D)\} \\ &= \{P \mid P : \mathbf{R}^D \to \mathbf{R}^D \text{ linear},\ VP \subseteq V\} \end{aligned}$$

The unit matrix $I = (\delta_{jk})_{j,k\in D}$ (we employ Kronecker's symbol $\delta_{jk}(= 0$ if $j \neq k, = 1$ if $j = k)$ as usual) is a stochastic matrix, and so is every matrix Δ_τ associated to an arbitrary mapping $\tau : D \to D$ via

$$\Delta_\tau = (\delta_{\tau(j),k})_{j,k\in D}$$

If τ is bijective, i.e. a permutation of D, we call Δ_τ the corresponding ***permutation matrix***.

Let us investigate the basic facts about convexity and extremal points of V and W.

A subset K of a real linear space H is called *convex* if

(1) $\quad x, y \in K,\ 0 \leq \alpha \leq 1 \Longrightarrow \alpha x + (1-x)y \in K$

or equivalently,

$$n \geq 1,\ x^{(1)}, \ldots, x^{(n)} \in K, \alpha_1, \ldots, \alpha_n \geq 0,\ \alpha_1 + \ldots + \alpha_n = 1$$

$$\Longrightarrow \alpha_1 x^{(1)} + \ldots + \alpha_n x^{(n)} \in K$$

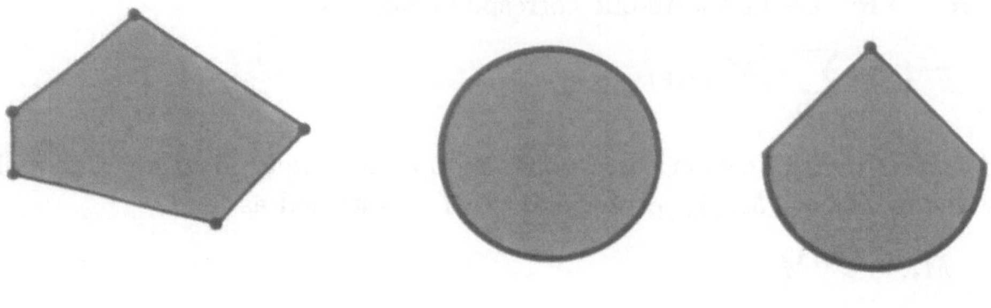

fig. II.2.1

Such linear combinations with nonnegative real coefficients summing up to 1 are called *convex combinations* of $x^{(1)}, \ldots, x^{(n)}$. If we interpret $\{\alpha x+(1-\alpha)y \mid 0 \leq \alpha \leq 1\}$ as the (line) segment xy joining x and y, we may restate (1) as

(2) $x, y \in K \Longrightarrow xy \subseteq K$

and say: K is convex if any two points in K "see" each other within K (i.e. even if $H \backslash K$ is opaque).

A point e in a convex set $K \subseteq H$ is said to be *extremal* in K or an *extremal point* of K if

$$x, y \in K,\ 0 < \alpha < 1,\ e = \alpha x + (1-\alpha)y \Longrightarrow x = y = e,$$

visually: $e \in K$ is extremal in K iff one may not push a segment of positive length through e without making at least one of its ends stick out of K. In figure 2.2.1 are a few plane convex sets with extremal points marked bold.

Proposition 2.1.

1. V is a convex subset of $H = \mathbf{R}^D$, and the unit vectors $e^{(j)}$ ($j \in D$) form the set of all of its extremal points.
2. W is a convex subset of $H = \mathbf{R}^{D \times D}$, and $\{\Delta_\tau \mid \tau : D \to D\}$ is the set of its extremal points.

PROOF. The proof rests on the fact that 0 and 1 are the extremal points of the unit interval $[0,1]$: you can't convex-combine 0 from numbers in $[0,1]$ unless the numbers you really employ are all 0, and for 1 likewise. The details are easy and left to the reader. □

It is easy to convex-combine arbitrary elements of V resp. W from extremal points. For V we have

$$V \ni p = (p_j)_{j\in D} \Longrightarrow p = \sum_{j\in D} p_j e^{(j)}$$

and this ("barycentric") representation is unique. In the case of a $P \in W$, all row sums are $= 1$, hence for every row j, there is at least one $\tau_1(j) \in D$ such that $P_{j,\tau_1(j)} > 0$. With $\alpha_1 = \min_{j\in D} P_{j,\tau_1(j)}$ we have $P \geq \alpha_1 \Delta_\tau$ componentwise, and $P - \alpha\Delta_\tau$ has all row sums $= 1 - \alpha_1$, and at least one entry 0 more than P. If $\alpha_1 = 1$, we have $P = \Delta_{\tau_1}$ and are through. If $\alpha_1 < 1$, we may apply the same procedure to $\frac{1}{a-\alpha_1}(p - \alpha_1\Delta_{\tau_1})$, ending up with $P = \alpha_1\Delta_{\tau_1} + \ldots + \alpha_r\Delta_{\tau_r}, \alpha_1, \ldots, \alpha_r > 0$, $\alpha_1 + \ldots + \alpha_r = 1$ after a finite number of iterations.

Although we will not need it elsewhere in this book, we prove the following interesting result here:

Proposition 2.2.(Birkhoff [1946]). Call a $D \times D$-matrix P *doubly stochastic* if both P and its transpose P^T are stochastic, i.e. iff $P = (P_{jk})_{j,k\in D}$, $P_{jk} \geq 0$ and $\sum_{i\in D} P_{ji} = 1 = \sum_{i\in D} P_{ik}$ $(j, k \in D)$, and denote by W_2 the set of all such matrices. Then the *permutation* matrices Δ_τ form the set of all extremal points of W_2.

PROOF. Every permutation matrix Δ_τ is doubly stochastic: it contains one 1 in every row as well as in every column, and only 0 else. As these Δ_τ are extremal even in $W \supseteq W_2$, they are the more extremal in W_2. If we can show that every $P \in W_2$ is a convex combination of permutation matrices, the proof that there are no extremal points other than permutation matrices in W_2 is an easy consequence: if $P \in W_2$ is representable in the form $P = \alpha_1\Delta_{\tau_1} + \ldots + \alpha_r\Delta_{\tau_r}$ with r pairwise different permutations $\tau_1, \ldots, \tau_r$ of D and $\alpha_1, \ldots, \alpha_r > 0$, $\alpha_1 + \ldots + \alpha_r = 1$, then $r \geq 2$ iff P is not itself a permutation matrix. But if $r \geq 2$, then $\alpha_1, \ldots, \alpha_r < 1$ and we may choose $\epsilon > 0$ such that $0 \leq \alpha_i - \epsilon < \alpha_i + \epsilon \leq 1$ for $i = 1, \ldots, r$. The matrices

$$\begin{aligned} P_+ &= (\alpha_1 + \epsilon)\Delta_{\tau_1} + (\alpha_2 - \epsilon)\Delta_{\tau_2} + \alpha_3\Delta_{\tau_3} + \ldots + \alpha_r\Delta_{\tau_r} \\ P_- &= (\alpha_1 - \epsilon)\Delta_{\tau_1} + (\alpha_2 + \epsilon)\Delta_{\tau_2} + \alpha_3\Delta_{\tau_3} + \ldots + \alpha_r\Delta_{\tau_r} \end{aligned}$$

are still in W_2 but differ since otherwise $2\epsilon\Delta_{\tau_1} = 2\epsilon\Delta_{\tau_2}$ i.e. $\Delta_{\tau_1} = \Delta_{\tau_2}$ would follow by subtraction. As $P = \frac{1}{2}P_+ + \frac{1}{2}P_-$, P cannot be an extremal point in W_2. Our proof will certainly be complete if we can prove that every doubly-stochastic P has a componentwise minorant $\alpha\Delta_\tau$ with $\alpha > 0$ and τ a permutation; in fact we then can apply the obvious exhaustion method which we applied to stochastic matrices previously, in order to obtain the desired representation. Now, in order to obtain a minorant $\alpha\Delta_\tau$ of P, we proceed as follows: for every $j \in D$ define $F(j) = \{k \mid P_{jk} > 0\}$ – the set of all "friends" of j. This "friendship system" fulfils the hypothesis of the *marriage theorem* (see appendix A): however we select a subset J of D, we have $|\bigcup_{j \in J} F(j)| \geq |J|$. In fact, if this would fail for some $J \subseteq D$, we would conclude that all non-zero entries of P in the rows $j \in J$ could be found within the columns k from a set $K \subseteq D$ with $|K| < |J|$. The non-zero entries of P clearly sum up to $d = |D|$; but we collect them all if we sum over all columns $k \in K$ (yielding a contribution $|K|$ as P is doubly-stochastic) and over all rows $j \in D \backslash J$ (yielding a contribution $d - |J|$); the sum of the two contributions is $|K| + d - |J| < |J| + d - |J| = d$, a contradiction. Thus the marriage theorem applies and yields a permutation $\tau : D \to D$ such that $\min_{j \in D} P_{j,\tau(j)} = \alpha > 0$. Now clearly $\alpha \Delta_\tau \leq P$. □

The reader should recall the following basic

Theorem 2.3. (Minkowski [1911], Carathéodory). Every convex compact subset K of a finite-dimensional real vector space is the convex hull of the set ∂K of its extremal points: every $x \in K$ can be represented as a finite convex combination of some (and even at most $d+1$) extremal points of K.

For a proof see appendix B.

Proposition 2.4. The set of W of all stochastic $D \times D$-matrices is a semigroup, (that is: closed under matrix multiplication), commutative iff $d = 1$.

PROOF. As we have seen, W can be characterized via $V = \{p \mid \mathbf{R}^D \ni p \geq 0, \langle p \rangle = 1\}$, the set of all probability vectors over D:

$$W = \{P \mid VP \subseteq V\}$$

Thus $P, Q \in W \Longrightarrow$

$$V(PQ) = (VP)Q \subseteq VQ \subseteq V$$

$\Longrightarrow PQ \in W$.— If $d > 1$, then even the Δ_τ don't all commute. □

Exercise 2.5. Prove that W_2 is a sub-semigroup of W, commutative iff $d \leq 2$.

3. The asymptotic behavior of P^n: some special cases

The fundamental aim of this chapter is the investigation of the asymptotic behavior of P^n for an arbitrary single stochastic $D \times D$-matrix P and $n \to \infty$. It seems therefore appropriate to first display some easy results for special types of stochastic matrices.

Special case I:. all rows of P equal.

If all rows $P_{j.}$ of P are the same probability vector $p \in V$, then P sends all $q \in V$ into that same p, and p into itself:

$$VP = \{p\} = VP^2 = VP^3 = \dots,$$

or, equivalently,

$$P = P^2 = P^3 = \dots$$

– the simplest case of asymptotic behavior.

Exercise 3.1. Let $D = D_1 \cup \ldots \cup D_r$ be a disjoint decomposition of our finite index set D and $p^{(\rho)}$ a probability vector living on $D_{(\rho)}$, i.e. $p_j^{(\rho)} = 0$ ($j \notin D_\rho, \rho = 1, \dots, r$). Let $P = (P_{jk})_{j,k\in D}$ be a stochastic matrix such that

$$P_{j.} = p^{(\rho)} \quad (j \in D_\rho, \rho = 1, \dots, r)$$

Prove that $P = P^2 = P^3 = \dots$

Special case II:. all $P_{jk} > 0$.

If $P = (P_{jk})_{j,k\in D}$ is a stochastic matrix with all entries > 0, say

(1) $P_{jk} \geq \epsilon > 0 \quad (j, k \in D)$,

then for every $p \in V$, the probability vector pP has all entries $\geq \epsilon > 0$ as well:

$$(2) \quad (pP)_k = \sum_{j\in D} p_j P_{jk} \geq \epsilon \sum_{j\in D} p_j = \epsilon.$$

VP lies thus entirely in the interior of V, with a positive distance from the boundary of V, and we may imagine that this phenomenon reappears at every passage $VP^n \to VP^{n+1} \subseteq VP^n$, leading to a contraction of $V \supseteq VP \supseteq VP^2 \supseteq \dots$ to a single point $\bar{p} \in V$. We may in fact argue as follows: For any two reals $\alpha, \beta \geq 0$ we have (prove it!)

$$|\alpha - \beta| = \alpha + \beta - 2\min\{\alpha, \beta\}.$$

Applying this to the entries of two probability vectors p', q' we find

$$\begin{aligned} \|p' - q'\| &= \|p'\| + \|q'\| - 2\sum_{j\in J} \min\{p'_j, q'_j\} \\ &= 2\left(1 - \sum_j \min\{p'_j, q'_j\}\right) \end{aligned}$$

Applying this in turn to $p' = e^{(i)}P$, $q' = e^{(k)}P$ for two different unit vectors (=extremal points) $e^{(i)}$, $e^{(k)} \in V$, we infer from (2) the estimate

$$\|e^{(i)}P - e^{(k)}P\| \leq 2(1 - |D|\epsilon)$$

Let $p'', q'' \in V$ be such that $p'' \wedge q'' = 0$, that is, for every $j \in D$, at least one of p''_j, q''_j is $= 0$. This entails a decomposition $D = D_+ \cup D_-$, $D_+ \cap D_- = 0$ such that $j \in D_- \Rightarrow p''_j = 0$ and $k \in D_+ \Rightarrow q''_k = 0$. In particular

$$\begin{aligned}
\|p'' - q''\| &= \sum_j |p''_j - q''_j| = \sum_{j \in D_+} |p''_j - 0| + \sum_{k \in D_-} |0 - q''_k| \\
&= \sum_{j \in D_+} p''_j + \sum_{k \in D_-} q''_k = \sum_{j \in D} p''_j + \sum_{k \in D} q''_k \\
&= 1 + 1 = 2
\end{aligned}$$

Now, assuming (1), we estimate

$$\begin{aligned}
\|p''P - q''P\| &= \|\sum_{j \in D} p''_j \, e^{(j)}P - \sum_{j \in D} q''_j \, e^{(j)}P\| \\
&= \|\sum_{j \in D_+} p''_j \, e^{(j)}P - \sum_{k \in D_-} q''_k \, e^{(k)}P\| \\
&= \| \sum_{j \in D_+, \, k \in D_-} \left[p''_j \, q''_k \, e^{(j)}P - p''_j \, q''_k \, e^{(k)}P \right] \| \\
&\leq \sum_{j \in D_+, \, k \in D_-} p''_j \, q''_k \|e^{(j)}P - e^{(k)}P\| \\
&\leq \sum_{j \in D_+, \, k \in D_-} p''_j \, q''_k 2(1 - |D|\epsilon) \\
&= 2(1 - |D|\epsilon) \\
&= (1 - |D|\epsilon)\|p'' - q''\|
\end{aligned}$$

This estimate remains true if we pass to $\alpha p'' - \alpha q''$ with an arbitrary $\alpha \geq 0$. But every vector x with $\langle x \rangle = \sum_{j \in J} x_j = 0$ can be represented this way: if $x \neq 0$, then $x = x_+ - x_-$ with $\langle x_+ \rangle = \langle x_- \rangle > 0$, and we have only to choose $\alpha = \langle x_+ \rangle$, $p = \frac{1}{\alpha} x_+$, $q = \frac{1}{\alpha} x_-$. Thus

$$\|xP\| \leq (1 - |D|\epsilon)\|x\| \qquad (\langle x \rangle = 0)$$

Finally, whenever $p, q \in V$, $\langle p - q \rangle = 0$, and thus, putting $x = p - q$, we arrive at

$$\|pP - qP\| \leq (1 - |D|\epsilon)\|p - q\|.$$

Thus P acts strictly contracting on the metric defined by $\| \cdot \|$ in V. By the Banach fixed point theorem, $V \supseteq VP \supseteq VP^2 \supseteq \ldots$ contracts to the only

fixed point $\bar{p}$ of P in V. For those who are not familiar with Banach's fixed point theorem, we carry out the proof completely:
Let $p \in V$ be arbitrary and $\theta = (1 - |D|\epsilon)$, hence $\theta \leq \theta < 1$. For the sequence $p, pP, pP^2, \ldots$ we have

$$\begin{aligned} \|pP^{n+1} - pP^n\| &= \|(pP^n - pP^{n-1})P\| \\ &\leq \theta \|pP^n - pP^{n-1}\| \\ &\leq \theta^n \|pP - p\| \quad (n = 1, 2, \ldots). \end{aligned}$$

and thus, for $m < n$,

$$\begin{aligned} \|pP^n - pP^m\| &\leq \|pP^n - pP^{n-1}\| + \ldots + \|pP^{m+1} - pP^m\| \\ &\leq (\theta^n + \theta^{n-1} + \ldots + \theta^{m+1})\|pP - p\| \\ &\leq \frac{\theta^m}{1-\theta}\|pP - p\| \to 0 \end{aligned}$$

for $m \to \infty$, independently of n. As $\mathbf{R}^D$ is complete metric with $\|\cdot\|$ (a direct consequence of the completeness of $\mathbf{R}$), the existence of a unique $\bar{p} \in V$ with $\lim_{n\to\infty} \|pP^n - \bar{p}\| = 0$ follows. As $P : V \to V$ is obviously norm continuous, the norm convergence $pP^n \to \bar{p}$ entails $pP^{n+1} \to \bar{p}P$. But as $pP^{n+1} \to \bar{p}$, we obtain $\bar{p}P = \bar{p}$, i.e. $\bar{p}$ is a fixed point of P. If $\bar{p}' \in V$ is another fixed point, $\|\bar{p} - \bar{p}'\| = (\bar{p} - \bar{p}')P\| \leq \theta\|p - \bar{p}'\|$ follows, whence $\|\bar{p} - \bar{p}'\| = 0$, i.e. $\bar{p} = \bar{p}' : \bar{p}$ is the only fixed point of P in V. For any $q \in V$ we now obtain

$$\|qP^n - \bar{p}\| = \|qP^n - \bar{p}P^n\| \leq \theta^n\|q - \bar{p}\| \leq 2\theta^n \ (n = 1, 2, \ldots).$$

Thus $V \supseteq VP \supseteq VP^2 \supseteq \ldots$ contracts into $\bar{p}$.

Applying this to the unit vectors $e^{(j)}$ $(j \in D)$, we see from $e^{(j)}P^n \to \bar{p}$, and the fact that $e^{(j)}P^n$ is the jth row of P^n, that

$$P^n \longrightarrow \bar{P}$$

where $\bar{P}$ is the stochastic matrix with all rows equal to $\bar{p}$. We even have shown exponential speed of convergence: norm distances go to 0 like θ^n, with $0 \leq \theta = (1 - |D|\epsilon) < 1$.

We conclude this subsection with a general

Theorem 3.2. Let P a stochastic matrix over D. Then $\lim_{n\to\infty} P^n = \bar{P}$, where $\bar{P}$ is a matrix with all rows equal iff P has an *attractive cell*, i.e. some $k_0 \in D$ such that there is an $n_0 \geq 1$ with

(3) $P^{(n_0)}_{jk_0} > 0 \quad (j \in D)$

PROOF. I) Necessity: if a stochastic Matrix $\bar{P}$ has all rows equal to some

$\bar{p} \in V$, there is some $k \in D$ such that $\bar{p}_{k_0} > 0$, as $\sum_k \bar{p}_k = 1$. It follows that

$$\bar{P}_{jk_0} > 0 \quad (j \in D),$$

and if $P^n \to \bar{P}$, there must be some n_0 such that (3) holds.
II) Sufficiency: if there is an attractive cell k_0, essentially the same argument as in special case II, above leads to $\|e^{(j)}P^{n_0} - e^{(k)}P^{n_0}\| \leq (1-\epsilon)\|e^{(j)} - e^{(k)}\|$ $(j \neq k)$ form some $\epsilon > 0$ (we don't get $|D|\epsilon$ because we have assumed only *one* attractive cell). But then we can continue as before and establish that $P^{n_0} : V \to V$ is a strict contraction of V. This proves contraction of $V \supseteq VP^{n_0} \supseteq VP^{2n_0} \supseteq \ldots$ to some $\bar{p} \in V$. But as $V \supseteq VP \supseteq VP^2 \supseteq \ldots$, this sequence contracts to $\bar{p}$ as well. □

In general, the iterates $I = P^0, p = P^1, P^2, P^3, \ldots$ of a given stochastic matrix P will not form a convergent sequence of matrices:

Example 3.3. Let τ be a permutation of D and $P = \Delta_\tau$. Then $P^n = \Delta_{\tau^n}$ and P^n behaves, as $n \to \infty$, like the sequence $\tau^0, \tau^1, \tau^2, \ldots$. If r is the least common multiple of all lengths of cycles into which τ decomposes, the sequence $\tau^0, \tau^1, \tau^2, \ldots$ has the period r, which may well nigh be ≥ 2. A similar argument works if $\tau : D \to D$ is an arbitrary mapping; here eventual periodicity is the result.

Exercise 3.4. Let $0 < \alpha < 1$ and $D = \{1, 2, 3\}$,

$$P = \begin{pmatrix} \alpha & \frac{1-\alpha}{2} & \frac{1-\alpha}{2} \\ 0 & 0 & 1 \\ 0 & 1 & 0 \end{pmatrix}$$

Show that

$$P^n = \begin{pmatrix} \alpha^n & \frac{1-\alpha^n}{2} & \frac{1-\alpha^n}{2} \\ 0 & 1 & 0 \\ 0 & 0 & 1 \end{pmatrix} \quad (n \text{ even})$$

$$P^n = \begin{pmatrix} \alpha^n & \frac{1-\alpha^n}{2} & \frac{1-\alpha^n}{2} \\ 0 & 0 & 1 \\ 0 & 1 & 0 \end{pmatrix} \quad (n \text{ odd})$$

Prove that the sequence $P^n \quad (n = 0, 1, \ldots)$ is exponentially asymptotic periodic.

These examples display asymptotic periodicity of $P^0, P^1, P^2, \ldots$, and in fact this will be the general result for arbitrary stochastic matrices P, as we will see in §4. But asymptotic periodicity clearly entails Cesáro convergence to the average over one period. This corollary to our later results can, however, easily be proved directly:

Theorem 3.5. (ergodic theorem for stochastic matrices). Let P be any stochastic matrix over a finite set D. Then there is a stochastic $D \times D$-matrix $\bar{P}$ such that

$$P\bar{P} = \bar{P}$$

$$\lim_{n\to\infty} \tfrac{1}{n} \textstyle\sum_{u=0}^{n-1} P^u = \bar{P} \quad \text{(componentwise, of course).}$$

PROOF. As W is compact, there is a subsequence $n_1 < n_2 < n_3 < \dots$ of $\mathbb{N}$ such that

$$\text{(4)} \quad \lim_{\nu\to\infty} \frac{1}{n_\nu} \sum_{u=0}^{n_\nu - 1} P^u = \bar{P} \in W$$

exists. Clearly

$$\lim_{\nu\to\infty} P \left[\frac{1}{n_\nu} \sum_{u=0}^{n_\nu - 1} \right] = P\bar{P},$$

but as $P\left[\frac{1}{n_\nu} \sum_{u=0}^{n_\nu-1} P^u\right] = \frac{1}{n_\nu} \sum_{u=1}^{n_\nu} P^u$ differs from $\frac{1}{n_\nu} \sum_{u=0}^{n_\nu-1} P^u$ by $\frac{1}{n_\nu}$ $[P - P^{n_\nu}] \to 0$ only,

$$\text{(5)} \quad P\bar{P} = \bar{P}$$

follows for any $\bar{P}$ obtained this way. We will sometimes call this idea "the cancellation argument". In order to prove the theorem, it will be sufficient to show that

$$\lim_{n\to\infty} \frac{1}{n} \sum_{n=0}^{n-1} P^u = \bar{P}$$

holds without passage to subsequences n_ν. We will here conveniently employ the matrix norm

$$\|M\| = \max_{\|x\| \le 1} \|xM\|.$$

Convergence with respect to this norm is tantamount to componentwise convergence (exercise), and for two $D \times D$-matrices M, M'

$$\begin{aligned} \text{(6)} \quad \|MM'\| &= \max_{\|x\|\le 1} \|(xM)M'\| \\ &\le \max_{\|y\| \le \|M\|} \|yM'\| \\ &= \|M\| \cdot \|M'\|. \end{aligned}$$

We have in particular

(7) $\|P\| = 1 \quad (P \in W)$

(exercise). Now (4) entails, for any $\epsilon > 0$, the existence of a matrix $Q \in W$ of the form

$$Q = \sum_{\rho=1}^{r} \alpha_\rho P^{t_\rho}$$

with $\alpha_1, \ldots, \alpha_r \geq 0$, $\alpha_1 + \ldots + \alpha_r = 1$, $t_1, \ldots, t_r \in \mathbf{Z}_+$ such that

$$\|Q - \bar{P}\| < \epsilon$$

Now (5) implies $\frac{1}{n}\sum_{u=0}^{n-1} P^u \bar{P} = \bar{P}$ for all $n \geq 1$; since $\frac{1}{n}\sum_{u=0}^{n-1} P^u \in W$, we conclude (by (6), (7))

$$\|\frac{1}{n}\sum_{u=0}^{n-1} P^u Q - \bar{P}\| < \epsilon \quad (n = 1, 2, \ldots)$$

It remains to show that

$$\|\frac{1}{n}\sum_{u=o}^{n-1} P^u Q - \frac{1}{n}\sum_{u=0}^{n-1} P^u\| \to 0$$

as $u \to \infty$. But

$$\frac{1}{n}\sum_{u=0}^{n-1} P^u Q = \sum_{\rho=1}^{r} \alpha_r \left[\frac{1}{n}\sum_{u=t_\rho}^{t_\rho+n-1} P^u\right]$$

follows from the above representation of Q_u, and $\frac{1}{n}\sum_{u=t_\rho}^{t_\rho+n-1} P^u$ differs from $\frac{1}{n}\sum_{u=0}^{n-1} P^u$ by $\frac{1}{n}[P^0 + \ldots + P^{t_\rho - 1} - P^n - P^{n+1} - \ldots - P^{n+t_\rho-1}]$ which tends to 0 as $n \to \infty$ because the number of terms in the square brackets remains constant (=$2r$). This "extended cancellation argument" proves the theorem □

Corollary 3.6. For every stochastic $D \times D$-matrix P there is at least one eigenvector for eigenvalue 1, i.e. a fixed point $\bar{p} = \bar{p}P$.

PROOF. Choose any $p \in V$, put $\bar{p} = p\bar{P}$ with $\bar{P}$ from theorem 3.2. Then $\bar{p}P = p\bar{P}P = p\bar{P} = \bar{p}$. – In fact, only part of the ideas used in the proof of theorem 3.2. is needed in order to establish this corollary: every limiting point of $\frac{1}{n}\sum_{u=0}^{n-1} pP^u$ is fixed under P, by the cancellation argument. □

4. Asymptotic behavior of $P, P^2, \ldots$: the method of invariant sets

Let $P = (P_{jk})_{j,k \in D}$ be an arbitrary stochastic matrix over our finite state space $D \neq \emptyset$. We will keep P fixed throughout this section, and some of the notions to be introduced in the sequel will depend upon our choice of P. We will prove that the sequence $I, P, P^2, \ldots$ is (exponentially) asymptotic periodic (theorem 4.16). We will achieve this by a method which we call the ***method of invariant sets***. In Appendix D the same result, and even a lot more, will be obtained by a different method.

Definition 4.1. Let $x = (x_j)_{j \in D} \in \mathbf{R}^D$ be arbitrary.

1. The set
$$supp(x) = \{j \mid x_j \neq 0\}$$
is called the ***support*** (or ***carrier***) of x.
2. The set
$$trk(x) = \bigcup_{n=0}^{\infty} supp(xP^n)$$
($P^0 = I$) is called the ***track*** of x (under our chosen P).
3. A set $M \subseteq D$ is called (P-) ***invariant*** if
$$supp(x) \subseteq M \Longrightarrow trk(x) \subseteq M$$
4. A set $M \subseteq D$ is called ***minimally invariant*** (under P) if
$M \neq \emptyset$
M is P-invariant
no proper subset $\neq \emptyset$ of M is P-invariant.

If $k \in D$ and $e^{(k)} = (\delta_{jk})_{j \in D}$ is the k-th unit vector then $trk(e^{(k)})$ is also denoted by $trk(k)$: the track of k.

Obviously
$$supp(x + y) \subseteq supp(x) \cup supp(y)$$
$$trk(x + y) \subseteq trk(x) \cup trk(y)$$
holds for any $x, y \in \mathbf{R}^D$. Applying this to $x = x_+ - x_-$, we conclude that it will always suffice to verify $supp(x) \subseteq M \Longrightarrow trk(x) \subseteq M$ for ***nonnegative*** $x \in \mathbf{R}^D$ in order to prove the P-invariance of a subset M of D.

Propositon 4.2. Every P-invariant set $\emptyset \neq M \subseteq D$ contains at least one minimally P-invariant subset.

As D is finite, the proof is obvious and left to the reader.

Proposition 4.3. There is at least one minimally P-invariant subset of D. Any two such sets are either identical or disjoint. More generally, if $M \subseteq D$ is invariant and $M' \subseteq D$ is minimally invariant then

$$\begin{array}{ll} \text{either} & M' \subseteq M \\ \text{or} & M' \cap M = \emptyset \end{array}$$

Any intersection of invariant sets is an invariant set.

PROOF. As D is invariant, the first statement follows from proposition 4.2. The second statement is a special case of the forelast, and the forelast one follows from the last, whose proof is obvious: if we have any family of invariant sets M_ι and define $M = \bigcap_\iota M_\iota$, then

$$\begin{aligned} supp(x) \subseteq M & \implies supp(x) \subseteq M_\iota \\ & \implies trk(x) \subseteq M_\iota \end{aligned}$$

holds for any ι, and $trk(x) \subseteq \bigcap_\iota M_\iota$ follows. □

Next we look into the interplay between $supp$, trk and the componentwise order of vectors. Recall that $x \leq y$ means $x_j \leq y_j \ \ (j \in D)$ for $x = (x_j)_{j \in D}$, $y = (y_j)_{j \in D} \in \mathbf{R}^D$. As all entries of P are ≥ 0, we have

$$0 \leq x \leq y \Longrightarrow 0 \leq xP \leq yP \quad (x, y \in \mathbf{R}^D)$$

Proposition 4.4. Let $x, y \in \mathbf{R}^D$. Then
1) $0 \leq x \leq y$ implies

$$\begin{aligned} supp(x) & \subseteq supp(y) \\ trk(x) & \subseteq trk(y) \end{aligned}$$

2) $0 < \alpha \in \mathbf{R}, 0 \leq x$ implies

$$\begin{aligned} supp(\alpha x) & = supp(x) \\ trk(\alpha x) & = trk(x) \end{aligned}$$

The proof is obvious and left to the reader.

Proposition 4.5. If $0 \leq x \in \mathbf{R}^D$, $0 < \alpha \in \mathbf{R}$, $0 \leq y \leq \alpha x P^{n_0}$ for some integer $n_0 \geq 0$, then

$$supp(yP^n) \subseteq supp(xP^{n_0+n}) \qquad (n \geq 0)$$

$$trk(y) \subseteq trk(xP^{n_0}) \subseteq trk(x).$$

We have

$$trk(x) = \bigcup_{j \in supp(x)} trk(j) = \bigcup_{j \in trk(x)} trk(j)$$

PROOF. By proposition 4.4. 2) we may assume $\alpha = 1$. Now

$$\begin{aligned} 0 \le y \le xP^{n_0} &\implies 0 \le yP^n \le P^{n_0+n} \quad (n \le 0) \\ &\implies supp(yP^n) \subseteq supp(xP^{n_0+n}) \qquad (n \ge 0) \end{aligned}$$

Taking the union over all $n \ge 0, trk(y) \subseteq trk(xP^{n_0}) \subseteq trk(x)$ follows. Applying this to $e^{(j)}$, $j \in supp(x)$, we find $trk(x) \supseteq \bigcup_{j \in supp(x)} trk(j)$. $\subseteq$ is read from $x \le \beta \sum_{j \in supp(x)} e^{(j)}$ ($\beta > 0$ suitable). The last equality of our proposition is obtained in a similar way (exercise). □

As an easy consequence, we obtain

Proposition 4.6. For every $0 < x \in \mathbf{R}^D$, $trk(x)$ is a P-invariant subset of D; it thus contains at least one minimally P-invariant subset.

PROOF. Whenever $y \ge 0, supp(y) \subseteq trk(x)$, then for every $j \in supp(y)$ we have $j \in trk(x)$, and $trk(y) = \bigcup_{j \in supp(y)} trk(j) \subseteq \bigcup_{j \in trk(y)} trk(j) \subseteq trk(x)$ follows in case $y \ge 0$. But as observed previously, this suffices in order to prove the P-invariance of $trk(x)$. □

Definition 4.7. A subset $M \subseteq D$ is called (P-)*absorbing* if

M is P-invariant
$M \cap trk(j) \ne \emptyset$ for all $j \in D$.

Proposition 4.8. The union of all minimally P-invariant subsets of D – there is, as we know, at least one such set, and these sets form a finite disjoint family – is absorbing.

PROOF. For any $j \in E, trk(j)$ is invariant, hence contains a minimally invariant M', hence nontrivially intersects the said union, which obviously is an invariant set. □

The second statement of the following theorem will tell us that any absorbing set "evacuates" its complement with exponential speed.

Theorem 4.9. Let $M \subseteq D$ be a P-absorbing set. Then the following holds:

1) M contains all minimally P-invariant sets: the union of the latter is the smallest absorbing set.
2) For any $J \subseteq D$ and any $x \in \mathbf{R}^D$ define

$$x_J = \sum_{j \in J} x_j.$$

There are constants $A > 0, 0 < \theta < 1$ such that

$$(pP^n)_{D\setminus M} \leq A\theta^n \quad (p \in V, n = 1, 2, \ldots).$$

PROOF. 1) Let M_0 be minimally invariant. By proposition 4.3. either $M_0 \subseteq M$ or $M_0 \cap M = \emptyset$. The latter alternative would imply $trk(j) \in M_0$ and hence $trk(j) \cap M = \emptyset$ for every $j \in M_0$, which is impossible for an absorbing set M. Thus $M_0 \subseteq M$ follows.
2) As M in invariant (definition 4.1.), for every $p \in V$ the sequence $(pP^n)_M$ is nondecreasing; in fact, for every $q \in V$ we have

$$\begin{aligned}(qP)_M &= \sum_{k\in M}(qP)_k = \sum_{j\in D}\sum_{k\in M} q_j P_{jk} \\ &\geq \sum_{j\in M}\sum_{k\in M} q_j P_{jk} \\ &= q_M\end{aligned}$$

because the invariance of M implies $P_{jk} = 0 (j \in M, k \notin M)$ – look at $e^{(j)}P$ whose support is contained in M if $j \in M$. Applying this to $q = pP^n$, we arrive at $(pP^{n+1})_M \geq (pP^n)_M (n = 0, 1, \ldots)$. As the sum of *all* components of pP^n is always 1, we now infer, by passage to the complement,

$$(1) \quad p_{D\setminus M} \geq (pP)_{D\setminus M} \geq (pP^2)_{D\setminus M} \geq \cdots$$

Let us now go into quantitative details: as M is absorbing, there is, for every $j \in D\setminus M$, some n_j such that $supp(e^{(j)}P^{n_j}) \cap M \neq 0$, hence $(e^{(j)}P^{n_j})_M > 0$. Put $\alpha = \min_{j\in D\setminus M}(e^{(j)}P^{n_j}), n_0 = \max_{j\in D\setminus M} n_j$. Then, making use of the abovementioned monotonicity, we obtain

$$(e^{(j)}P^n)_M \geq \alpha \quad (j \in D\setminus M, n \geq n_0).$$

For any $p \in V, n \geq n_0$, we now conclude

$$\begin{aligned}(pP^n)_{D\setminus M} &= \left(\sum_{j\in D} p_j e^{(j)} P^n\right)_{D\setminus M} \\ &= \sum_{j\in D} p_j (e^{(j)}P^n)_{D\setminus M} \\ &= \sum_{j\in M} p_j (e^{(j)}P^n)_{D\setminus M} + \sum_{j\in D\setminus M} p_j (e^{(j)}P^n)_{D\setminus M}\end{aligned}$$

As M is invariant, $j \in M$ implies $supp(e^{(j)}P^n) \subseteq M$. Thus the first sum is $= 0$, and we may continue

$$= \sum_{j\in D\setminus M} p_j (e^{(j)}P^n)_{D\setminus M}$$

$$\begin{aligned} &= \sum_{j \in D\setminus M} p_j(1-(e^{(j)}P^n)_M) \\ &\leq (1-\alpha) \sum_{j \in D\setminus M} p_j \\ &= (1-\alpha)p_{D\setminus M} \end{aligned}$$

This holds for any $n \geq n_0$, hence for $n = n_0$ in particular, and we see: applying P^{n_0} to any $p \in V$ we diminish $p_{D\setminus M}$ at least by a factor $1-\alpha < 1$. Thus

$$(pP^{\nu n_0})_{D\setminus M} \leq (1-\alpha)^\nu p_{D\setminus M} \quad (\nu = 0, 1, \ldots)$$

follows. Combining this with (1), we arrive at the desired result by standard arguments – please solve the following □

Exercise 4.10. Let $\mathbf{R} \ni a_0 \geq a_1 \geq \ldots \geq 0$ and assume the existence of $n_0 > 0$, $0 < q < 1$ such that

$$a_{\nu n_0} \leq q^\nu \quad (\nu = 0, 1, \ldots).$$

Then there is an $A > 0$ and some $0 < \theta < 1$ such that

$$a_n \leq A\theta^n \quad (n = 0, 1, \ldots).$$

We may now draw the following general picture:

> For a given stochastic matrix P over our finite sete D, we may split D into
>
> > M = the union of all minimally P-invariant subsets of D, and $D\setminus M$.
>
> M is absorbing and iterated application of P empties $D\setminus M$ with exponential speed.

We will now concentrate on what is going on inside M and since M is a disjoint union of minimal (invariant) sets, we may and shall w.l.o.g. assume

$$D \text{ is minimally } P\text{-invariant}$$

for the time being. It is our aim to establish certain periodicity phenomena within $D = M$ – periodicity which may turn out as constancy in special cases. To be more specific: we shall split D into a finite cyclus $D_0, D_1, \ldots, D_{d-1}$ of mutually disjoint nonempty subsets such that D_0 empties its content entirely into D_1, D_1 into $D_2, \ldots, D_{d-1}$ into D_0. That is

$$j \in D_0 \Longrightarrow supp(e^{(j)}P) \subseteq D_1 \quad \text{etc.}$$

We will handle indices mod d in this context, of course.

To this end we define Z_{jk} to be the set of all time moments where mass from j arrives at k, i.e.

$$Z_{jk} = \{n \mid n \geq 0, k \in supp(e^{(j)}P^n)\} \quad (j, k \in D)$$

Proposition 4.11. For any $i, j, k \in D$ the following holds:

$$Z_{ij} + Z_{jk} \subseteq Z_{ik}$$

$$(i.e.\ m \in Z_{ij},\ n \in Z_{jk} \Longrightarrow m + n \in Z_{ik})$$

In particular, Z_{jj} is, for any $j \in D$, an additive subsemigroup of $\mathbb{Z}_+ : 0 \in Z_{jj}$ and

$$Z_{jj} + Z_{jj} \subseteq Z_{jj}.$$

PROOF. Let $m \in Z_{ij}$, $n \in Z_{jk}$. From $j \in supp(e^{(i)}P^m)$ we conclude $\alpha e^{(j)} \leq e^{(i)}P^m$ for some $\alpha > 0$, whence $\alpha e^{(j)}P^n \leq e^{(i)}P^{m+n}$. From $k \in supp\ e^{(j)}P^n$ we now infer $k \in supp(e^{(i)}P^{m+n})$, *i.e.* $n + m \in Z_{ik}$. – The second statement of our proposition is an obvious consequence hereof ($0 \in Z_{jj}$ follows from $P^0 = I$). □

Lemma 4.12. let Z be any additive subsemigroup of $\mathbb{Z}_+$. Then there is some integer $d \geq 1$ such that every number in Z is a multiple of d:

(2) $\quad Z \subseteq \{0, d, 2d, \ldots\}$

There is some $\nu_0 \geq 0$ such that

(3) $\quad Z \supseteq \{\nu_0 d, (\nu_0 + 1)d, \ldots\}$

PROOF. Let $n_0, n_1, \ldots$ be an enumeration of Z and d_μ the greatest common divisor of $n_0, n_1, \ldots, n_\mu$. Clearly $d_1 \geq d_2 \geq \ldots \geq 1$, hence this sequence of integers is constant, say, from $\mu_0 = r$ onward: we define $d = d_r = d_{r+1} = \ldots$ and get (2). By Euclid's algorithm we obtain a representation

$$d = a_0 n_0 + \ldots + a_r n_r$$

with suitable integers $a_0, \ldots, a_r$. We may and shall assume them all to be $\neq 0$. If they are all > 0, $d \in Z$ follows, with equality in (3) as an obvious consequence. If some (not all!) a_ϱ are < 0, we may argue as follows: After a suitable renumbering we may assume $a_0, \ldots, a_s < 0 < a_{s+1}, \ldots, a_r$ for some $0 \leq s < r$. We thus obtain

$$a_{s+1}\, n_{s+1} + \ldots + a_r n_r = (-a_0)n_0 + \ldots + (-a_s)n_s + d,$$

that is, we have two members $n, n+d$ of Z. By suitable additions we find

$$\varrho n, \varrho n + d, \varrho n + 2d, \ldots, \varrho n + \varrho d \in Z \quad (\varrho = 1, 2, \ldots)$$

As soon as $\varrho n + \varrho d \geq (\varrho + 1)n$ we see that all these finite arithmetic progressions of step width d overlap, and (3) follows. □

Lemma 4.13. With our previous notations, the same d results for all semigroups Z_{jj}.

PROOF. Let d_j result for Z_{jj}. From proposition 2.23. we get $Z_{jj} \supseteq Z_{jk} + Z_{kk} + Z_{kj}$, from which we conclude $d_j \leq d_k$. By symmetry, equality follows. □

Having established the fact that there is some integer $d \geq 1$ such that every Z_{jj} consists of multiples of d only, but contains all sufficiently large ones of them, we may draw a related conclusion for the other Z_{jk}'s, too:

$$Z_{jk} + Z_{kj} \subseteq Z_{jj}$$

shows

$$m \in Z_{jk},\ n \in Z_{kj} \Longrightarrow m + n \equiv 0 \bmod d.$$

Keeping, say, n fixed we see: every Z_{jk} is contained in some residue class $\varrho \bmod d$. Looking at

$$Z_{jk} + Z_{kk} \subseteq Z_{jk}$$

we see: all sufficiently large $\varrho + \nu d$ belong to Z_{jk}.

We now choose any $j_0 \in D$ and define

$$D_\varrho = \{k \mid k \in D,\ Z_{j_0 k} \subseteq \varrho \bmod d\}, \quad (\varrho = 0, \ldots, d-1)$$

that is, every $k \in D$ gets mass from j_0 only at time moments of the form $\varrho + \nu d$, but at all of these that are sufficiently large.

Clearly $j_0 \in D_0$. Every $j \in D_0$ has mass from j_0 (i.e. fulfils $j \in \operatorname{supp}(e^{(j_0)} P^n)$) only at time moments of the form $n = \nu d$, and certainly at all sufficiently large ones of this form.

For arbitrary $i, k \in D_0$ we see:

> $Z_{j_0 i}$ consists of multiples of d only, by the definition of D_0; the same holds for $Z_{j_0 k}$.
>
> $Z_{j_0 k} + Z_{k j_0} \subseteq Z_{j_0 j_0}$ shows that $Z_{k j_0}$ consists of multiples of d only

$Z_{j_0 i} + Z_{ik} + Z_{k j_0} \subseteq Z_{j_0 j_0}$ shows that Z_{ik} consists of multiples of d only; but it contains all sufficiently large ones of them. Consequently, k gets mass from i only at times νd, but at all sufficiently large ones of this form.

For vectors $p \in V$ with $\text{supp}(p) \subseteq D_0$, i.e. $p = \sum_{j \in D_0} p_i \, e^{(j)}$ we now conclude by superposition:

$$\begin{aligned} supp(pP^n) &\subseteq D_0 \text{ only for } n = \nu d \\ supp(pP^n) &= D_0 \text{ for } n = \nu d \text{ sufficiently large} \end{aligned}$$

Similar arguments lead to

$$\begin{aligned} supp(pP^n) &\subseteq D_\varrho \text{ only for } n = \varrho + \nu d \\ supp(pP^n) &= D_\varrho \text{ for } n = \varrho + \nu d \text{ sufficiently large} \end{aligned}$$

We leave the details to the reader and thus arrive at

Theorem 4.14. Let D be minimally P-invariant and $D_0, D_1, \ldots, D_{d-1}$ be defined as before. Then for $p \in V$, $\text{supp}(p) \subseteq D_\varrho$ we have

$$supp(pP^n) \subseteq D_{\varrho+n} \quad \text{(indices mod } d\text{)}$$

with equality for n sufficiently large.

Obviously, D_0 is minimally P^d-invariant here. Replacing P by P^d for the moment, or, equivalently, assuming $d = 1$, we obtain

Theorem 4.15. Let D be minimally P-invariant and $\text{supp}(pP^n) = D$ for all $p \in V$ and sufficiently large n. Then there is exactly one $\bar{P} \in W$ such that

$$\lim_{n \to \infty} P^n = \bar{P} = \bar{P}^2 = \bar{P}P$$

The speed of this convergence is exponential if we employ the matrix norm

$$\|M\| = \sup_{\|x\| \leq 1} \|xM\|.$$

PROOF. The hypotheses of this theorem imply that every cell in D is attractive. We thus have only to apply theorem 3.2. □

We are now able to clear up the asymptotic behavior of $I = P^0, P^1, \ldots$ for any stochastic matrix in full generality.

Theorem 4.16. Let P be any stochastic matrix over a finite set D. Then there is a disjoint decomposition

$$D = S \cup M_1 \cup \ldots \cup M_\ell,$$

and for every $\lambda = 1, \ldots, \ell$ an integer $d_\lambda \geq 1$ and a disjoint decomposition

$$M_\lambda = D_{\lambda 0} \cup \ldots \cup D_{\lambda, d_\lambda - 1}$$

such that the following holds:

1. M_λ is minimally P-invariant $(\lambda = 1, \ldots, \ell)$
2. $p \in V, supp(p) \in D_{\lambda,\varrho}$ implies
   $supp(pP^n) \subseteq D_{\lambda,\varrho+n}$ (count $\varrho + n \bmod d_\lambda$), with equality for n sufficiently large.
3. For every $D_{\lambda\varrho}$ there is exactly one $q^{(\lambda,\varrho)} \in V$ such that

$$\begin{aligned} supp\, q^{(\lambda,\varrho)} &= D_{\lambda\varrho} \\ q^{(\lambda,\varrho)} P^n &= q^{(\lambda,\varrho+n)} \text{ (count } \varrho + n \bmod d_\lambda) \\ lim_{n\to\infty} \|pP^n - q^{(\lambda,\varrho+n)}\| &= 0 \end{aligned}$$

   exponentially fast for every $p \in V$ with $\text{supp}(p) \subseteq D_{\lambda\varrho}$
4. For every $p \in V$ there is a unique convex combination q of the vectors $q^{(\lambda,\varrho)}$ such that

$$lim_{n\to\infty} \|pP^n - qP^n\| = 0 \qquad \text{(exponentially fast)}$$

   As the sequence $qP^n (n = 0, 1, \ldots)$ is periodic, the sequence $pP^n (n = 0, 1, \ldots)$ is (exponentially) ***asymptotic periodic***, and thus so is the sequence $P^n (n = 0, 1, \ldots)$.

The reader is invited to verify that this theorem contains all informations obtained step by step in this subsections. It could be used in order to establish a decomposition of P into "boxes" which reflect the way how P acts on the various unit vectors $e^{(j)}$; we will not carry out here such a program but encourage the reader to do that.

III. Discrete Probability Spaces

In the preceding chapters we have dealt with finite probability spaces:

> let D be a nonempty finite set. A D-vector $p = (P_j)_{j \in D}$ is called a ***probability vector*** over D if
>
> $$p_j \geq 0 \qquad (j \in D)$$
>
> (1) $$\sum_{k \in D} p_k = 1$$
>
> If p is a probability vector over D, then (D, p) is called a ***finite probability space.***

This concept carries over to countable sets D without difficulty – we only have to replace the finite sum in (1) by an infinite series. Convergence and divergence of infinite series are easily handled if all terms are ≥ 0, and the order of the terms is of no importance in this case. Thus everything sounds simple and easy. In fact, large parts of this book may be read without establishing a more elaborate theory of probability spaces beforehand.

It seems nevertheless appropriate to have such a theory at one's disposal. Some parts of discrete probability theory can only be rigorously formulated on such a basis. The present rather technical chapter has the purpose to fill this need. We recommend the reader to have a look on it before going ahead to the subsequent chapters, and to use it as an arsenal of methods whenever necessary.

1. The Notion of a Discrete Probability Space (DPS)

If (D, p) is a finite probability space, we may define

(1) $$p(E) = \sum_{j \in E} p_j \qquad (E \subseteq D)$$

and then make the following simple observations:

(2) $$p(E) \geq 0 \qquad (E \subseteq D)$$

$$p(\emptyset) = 0$$

(3) $p(D) = 1$

isotony:	$p(E) \leq p(F)$	$(E \subseteq F \subseteq D)$
	$0 \leq p(E) \leq 1$	$(E \subseteq D)$
additivity:	$p(E \cup F) = p(E) + p(F)$	$(E, F \subseteq D, E \cap F = \emptyset)$

The latter property easily generalizes to

additivity: $p(E_1 \cup \ldots \cup E_n) = p(E_1) + \ldots + p(E_n)$ $(E_1, \ldots, E_n \subseteq D$ pairwise disjoint)

Actually, the probability vector

$$p = (p_j)_{j \in D}$$

and the additive set function $p : \mathcal{P}(D) \to [0, 1]$ are nothing but two aspects of the same thing: by (1) we pass from the former to the latter, and by

(4) $p_j = p(\{j\}) \qquad (j \in D)$

– we will also occasionally write $p(j)$ for p_j – from the latter to the former; every additive non-negative function $p : \mathcal{P}(D) \to [0, 1]$ fulfilling the *normalization* (3) defines, via (4), a probability vector p which, upon application of (1), leads back to that same set function; in fact, additivity implies

(5) $p(E) = p(\bigcup_{j \in E} \{j\}) = \sum_{j \in E} p(\{j\}) = \sum_{j \in E} p_j$

A reader with an analytical mind will immediately apply (1), (4) and (5) to arbitrary real-valued vectors and additive set functions, seeing them thus linked in a one-one linear fashion, while (2) and (3) effectuate the restriction to probability vectors or distributions.

All this easily generalizes to countable basic sets D as long as we restrict ourselves to nonnegative D-vectors resp. set functions, thus avoiding all difficulties concerning convergence, divergence and re-ordering of the infinite series which then most naturally come into play. Additivity generalizes to

σ-additivity: $p(\bigcup_{j=1}^{\infty} E_\nu) = \sum_{\nu=1}^{\infty} p(E_r)$

$(E_1, E_2, \ldots \subseteq D$ pairwise disjoint)

and it is this σ-additivity which makes (5) also work in the countable case. Additivity appears as a special case of σ-additivity if we choose all but finitely many of the E_ν to be $\emptyset$.

We should also observe that isotony is a consequence of additivity and non-negativity

$$E \subseteq F \Longrightarrow p(F) = p(E) + p(F \backslash E) \geq p(E).$$

We mention in passing that for a countably infinite D, the existence of additive but not σ-additive set functions ≥ 0 on $P(D)$ can be established via Banach limits (that is, by the Hahn-Banach theorem, and thus on the basis of Zorn's Lemma ($\Longleftrightarrow$ Axiom of Choice (AC))): see e.g. Jacobs [1978] pp.87. Such peculiar set functions are only of marginal significance in probability theory; see, however, Dubins-Savage [1965].

But we have still to generalize farther. In various important sections of this book we have to consider probability distributions in even non-denumberable sets Ω. $\Omega = \mathbf{R}$ is the really important example.

It is very easy to achieve such a generalization: let $D \neq \emptyset$ be an at most countable subset of an arbitrary set Ω, and let p be a probability vector over D. Extend p to Ω by setting $p_\omega = 0$ for all $\omega \in \Omega \backslash D$, and call the result $p = (p_\omega)_{\omega \in \Omega}$ by now. Clearly

$$p_\omega \geq 0 \quad (\omega \in \Omega),$$

and, if we adopt the here obvious definition

$$\sum_{\omega \in \Omega} p_\omega = \sup_{\substack{E \subseteq \Omega \\ E \text{ finite}}} \sum_{\omega \in E} p_\omega,$$

we clearly obtain

$$\sum_{\omega \in \Omega} p_\omega = \sum_{j \in D} p_j = 1.$$

Exercise 1.1. Let $\Omega \neq 0$ be an arbitrary set and $q : \Omega \to \mathbf{R}_+$, written as $q = (q_\omega)_{\omega \in \Omega}$ be a nonnegative real function of Ω. Show that

$$\sup_{\substack{E \subseteq \Omega \\ E \text{ finite}}} \sum_{\omega \in E} q_\omega < \infty$$

implies that there is an at most countable subset D of Ω such that

$$\omega \in \Omega \backslash D \quad \Longrightarrow \quad q_\omega = 0$$

$$\sum_{j \in D} q_j = \sup_{E \subseteq \Omega, |E| < \infty} \sum_{\omega \in E} q_\omega.$$

So much for a generalization of the notion of a probability *vector*. What about the corresponding σ-additive set functions?

Let again $D \neq \emptyset$ be an at most denumerable subset of Ω, p a probability vector over D and $p : \mathcal{P}(D) \to E[0,1]$ the corresponding σ-additive set function. If we define

$$\bar{p}(E) = p(E \cap D) \qquad (E \subseteq \Omega),$$

the new set function $\bar{p}$ coincides with p on all subsets of D, and is certainly ≥ 0, normalized to $\bar{p}(\Omega) = p(\Omega \cap D) = p(D) = 1$, and σ-additive: for any pairwise disjoint subsets $E_1, E_2, \ldots$ of Ω, their intersections $E_1 \cap D,\ E_2 \cap D, \ldots$ with D are pairwise disjoint again, and we conclude

$$\begin{aligned}\sum_{\nu=1}^{\infty} \bar{p}(E_\nu) &= \sum_{\nu=1}^{\infty} p(E_\nu \cap D) = p(\bigcup_{\nu=1}^{\infty}(E_\nu \cap D)) \\ &= p((\bigcup_{\nu=1}^{\infty} E_\nu) \cap D) = \bar{p}(\bigcup_{\nu=1}^{\infty} E_\nu)\end{aligned}$$

Thus $\bar{p}$ is nothing but an obvious σ-additive extension of p from $P(D)$ to $P(\Omega)$ (and could justly be denoted by p again).

How can we characterize the σ-additive normalized set functions ≥ 0 on $P(\Omega)$ obtained this way from σ-additive set functions on denumerable subsets of D? The answer is easy from exercise 1.1.:

Exercise 1.2. Let Ω be any nonempty set and $\bar{p} : \mathcal{P}(\Omega) \to \mathbf{R}_+$ be σ-additive. Prove that the following two statements are equivalent:

1. There is an at most countable subset $D \neq \emptyset$ of Ω, and a σ-additive set function $p : \mathcal{P}(D) \to \mathbf{R}_+$ (no values ∞ allowed!) such that
$$\bar{p}(E) = p(E \cap D) \qquad (E \subseteq \Omega)$$
2. $sup_{E \subseteq \Omega, |E| < \infty}\ p(E) < \infty$

The results of our discussion lead to the following

Definition 1.3. Let Ω be an arbitrary non-empty set.

1. A nonnegative function p on Ω, written as $p = (p_\omega)_{\omega \in \omega}$, is called a *probability vector* (over the basic set Ω) if there is an at most countable subset $D \neq \emptyset$ of Ω such that $p_\omega = 0\ (\omega \in \Omega \backslash D)$ and
$$\sum_{j \in D} p_j = 1 \quad (\text{ sum resp. series!})$$
2. A nonnegative σ-additive function p, defined for all subsets of Ω, is called a *discrete probability distribution (DPD)* on Ω if $p(\Omega) = 1$ and

there is at at most countable subset $D \neq \emptyset$ of Ω such that

$$p(E) = \sum_{j \in E \cap D} p(\{j\}) \quad \text{(sum resp. series!)}$$

holds for all $E \subseteq \Omega$.

3. If $E \subseteq \Omega$ fulfils $p(E) = 1$, E is called a *carrier* or *support* of p, and p is said to *live* or to *sit* on E resp. on the points belonging to E.
4. If $N \subseteq \Omega$ fulfils $p(N) = 0$ (i.e. if $E = \Omega \backslash N$ is a carrier of p), N is called a *p-nullset.* A statement about points $\omega \in \Omega$ which holds true for all ω except those from a certain p-nullset (the "exceptional set" for that statement) is said to hold *p-almost-everywhere (p.-a.e.)*
5. If p is a discrete probability distribution on Ω, we call (Ω, p) a *discrete probability space (DPS).* If Ω is finite, a DPS (Ω, p) is also called a *finite probability space (FPS)* , and if Ω is denumerable, we speak of a *countable probability space (CPS).*

Summing up previous results, we obtain

Theorem 1.4.. Let Ω be any nonempty set. Any probability vector $p = (p_\omega)_{\omega \in \Omega}$ over Ω defines a DPD p on Ω by

$$(6) \quad p(E) = \sup_{F \subseteq E, |F| < \infty} \sum_{\omega \in F} p_\omega,$$

and different probability vectors define different DPDs that way. Conversely, a nonnegative σ-additive set function p defined on all subsets of Ω is a DPD iff

$$\sup_{E \subseteq \Omega, |E| < \infty} p(E) = 1.$$

If this is the case

$$p_\omega = p(\{\omega\}) \quad (\omega \in \Omega)$$

defines a probability vector on Ω (the only one) which yields the set function p via (6); the set

$$D = \{j | j \in \Omega,\ p_j > 0\}$$

is then at most countable, and we have

$$(7) \quad p(E) = \sum_{j \in E \cap D} p_j \quad (E \subseteq \Omega)$$

the right hand side here being a sum or a series with nonnegative terms, which makes its value independent of the order in which the summation is carried out. In short: *every* DPD is obtained from a unique probability vector by (6) resp. (7); we shall say that this probability vector *defines* or *represents* that DPD.

The simplest imaginable DPDs in an arbitrary basic set Ω are those with a one-element support: the so-called *Dirac* DPDs (named after Dirac (1902-1984)) or *one-point masses*. For an arbitrary $\omega_0 \in \Omega$ the Dirac DPD or one-point mass ϵ_{ω_0} sitting on ω_0 is defined by

$$\epsilon_{\omega_0}(E) = 1_E(\omega_0) = \begin{cases} 1 & \text{if} \quad \omega_0 \in E \\ 0 & \text{if} \quad \omega_0 \notin E \end{cases}$$

We sometimes shortly call a DPD "Dirac" if it is of this form.

2. Obtaining New Probability Spaces from Given Ones

An obvious (and clearly universal) technique of constructing a probability vector p over an arbitrary set $\Omega \neq \emptyset$ runs as follows:

select a sequence $\omega_1, , \omega_2, \ldots$ in Ω

take any convergent series with non-negative entries $a_1, a_2, \ldots$, not all 0. Let $a_1 + a_2 + \ldots = a$; clearly $a > 0$;

put $p_{\omega_j} = \frac{a_j}{a}$ on ω_j $(j = 1, 2, \ldots)$ and $p_\omega = 0$ otherwise.

In this section we present some techniques producing new discrete probability spaces from given ones.

2.1. Convex Combination.

If p, q are two probability vectors over some $\Omega \neq \emptyset$, then for every $0 \leq \alpha \leq 1$ the convex combination

$$\alpha p + (1 - \alpha) q$$

is a probability vector over Ω again. This generalizes easily to finitely many, and even to countably many probability vectors given in advance.

Exercise 2.1. Let $\alpha_1, \alpha_2, \ldots \geq 0$, $\alpha_1 + \alpha_2 + \ldots = 1$, and let $p^{(1)}, p^{(2)}, \ldots$ be discrete probability distributions on the same basic set $\Omega \neq \emptyset$. Show that

$$p(E) = \alpha_1 p^{(1)}(E) + \alpha_2 p^{(2)}(E) + \ldots \qquad (E \subseteq \Omega)$$

defines a discrete probability distribution p in Ω. Prove that every DPD on Ω can be represented in this form, with all $p^{(k)}$'s "Dirac".

2.2. Restricting and Renorming.

If (Ω, p) in a DPS and $\Omega_0 \subseteq \Omega$ is a set with $p(\Omega_0) > 0$, then

$$p_0(E) = \frac{p(E \cap \Omega_0)}{p(\Omega_0)} \qquad (E \subseteq \Omega)$$

defines a DPD on Ω with

$$p_0(\Omega_0) = 1, \;\; p(\Omega \backslash \Omega_0) = 1.$$

We may call this procedure "restricting and renorming". We will encounter it again under the headlines "conditioning", "conditional probability" in ch.VI. §1.

2.3. Random Variables and their Distributions.

Let (Ω, p) be a discrete probability space. A mapping f from Ω into some other set X is also called a *random variable* (RV) with *state space* X. The idea behind the words "random variable" is that $f(\omega) \in X$ varies with $\omega \in \Omega$, and ω is the outcome of a random experiment: $f(\omega)$ is "randomly variable".

Now if, for every $\omega \in \Omega$, p_ω is the probability to obtain the outcome ω in the said random experiment, what is the probability of observing some specified $x \in X$ as the value of f? x comes out precisely for all these realizations $\omega \in \Omega$ which yield $f(\omega) = x$. Thus

$$(8) \quad q_x = \sum_{f(\omega)=x} p_\omega$$

is the correct answer to our question. The facts behind (1) may be visualized as *mass transport* by f : p has nonzero mass only on at most countably many points $\omega \in \Omega$; f throws these masses into the set X like a hail; let them stick where they hit; at most countably many can hit the same specified point $x \in X$ adding up there according to (1).

We may look at (1) still a bit differently. A mapping $f : \Omega \to X$ constitutes a decomposition of Ω into "sets of constancy" of f: defining

$$\Omega_x = \{\omega | f(\omega) = x\} \quad (x \in X)$$

we obtain a family $(\Omega_x)_{x \in X}$ of pairwise disjoint (and possibly empty) subsets of Ω which constitute a *disjoint decomposition*

$$\Omega = \bigcup_{x \in X} \Omega_x$$

Clearly

$$p(\Omega_x) = \sum_{p(\omega)=x} p_\omega$$

is precisely the part of the total mass of p which is thrown into point x by f. Only at most countably many of the Ω_x can have $p(\Omega_x) > 0$, and the σ-additivity formula for those countably many set $\Omega_x \subseteq \Omega$ shows that these $p(\Omega_x)$ sum up to 1. Thus (1) defines a probability vector $q = (q_x)_{x \in X}$ over X.

We may look at this situation also under the set function aspect; formally, things get even simpler then:

If p is a discrete probability distribution in Ω and $f : \Omega \to X$ a mapping, the "pullback formula"

$$q(F) = p(f^{-1}F) \qquad\qquad (F \subseteq X)$$

(where $f^{-1}F = \{\omega | f(\omega) \in F\}$ is the "pullback" of F from X to Ω by means of f) defines a discrete probability distribution q in X. The proof is easy and essentially left to the reader as an exercise. We only remark that $f^{-1}X = \Omega$, whence $q(X) = p(f^{-1}X) = p(\Omega) = 1$, and that the pullback of disjoint subsets of X are disjoint of Ω, which yields the proof that q is σ-additive again. Discreteness is easily obtained via theorem 1.4. For convenient quotation we concentrate our results into

Theorem 2.2.. Let (Ω, p) be a PPS and $f : \Omega \to X$ a random variable with state space X. Then

$$q(F) = p(f^{-1}F) \qquad (F \subseteq X)$$

defines a discrete probability distribution q in X. It is also denoted by fp or by pf^{-1}, and is called *the distribution of f under p*. If $D \subseteq \Omega$ is at most countable with $p(D) = 1$, then $fD \subseteq X$ is at most countable with $q(fD) = 1$.

If $f_1 : \Omega \to X_1, \ldots, f_n : \Omega \to X_n$ are RVs on Ω with state spaces $X_1, \ldots, X_n$, their (cartesian) *join* $f = f_1 \times \ldots \times f_n$ maps Ω into the cartesian product $X = X_1 \times \ldots \times X_n$. The distribution fp of this join is also called the *joint distribution* of the RVs $f_1, \ldots, f_n$.

Exercise 2.3. Let $f_1, \ldots, f_n$ etc. as above and, for every $j = 1, \ldots, n$, $\varphi_j : X \to X_j$ the j-th component mapping: $\varphi_j(x) = \varphi_j((x_1, \ldots, x_n)) = x_j$ $(x = (x_1, \ldots, x_n) \in X = X_1 \times \ldots \times X_n)$. Show that φ_j sends fp into the distribution of f_j $(j = 1, \ldots, n)$.

2.4. Product Probabilities.

Let $(\Omega_0,\ p^{(0)})$, $(\Omega_1,\ p^{(1)})$ be DPSs. Then there is a unique probability distribution p in

$$\Omega = \Omega_0 \times \Omega_1$$

such that the *product formula*

$$p(E_0 \times E_1) = p^{(0)}(E_0)p^{(1)}(E_1) \qquad (E_0 \subseteq \Omega_0,\ E_1 \subseteq \Omega_1)$$

holds. If $(p^{(u)}_{\omega_u})_{\omega_u \in \Omega_u}$ is the probability vector representing $p^{(u)}$ $(u = 0, 1,)$, then the product formula

$$(9) \quad p_{(\omega_0,\omega_1)} = p^{(0)}_{\omega_0}\, p^{(1)}_{\omega_1}$$

defines the probability vector

$$p = (p_\omega)_{\omega\in\Omega} = (p_{(\omega_0,\omega_1)})_{\omega_0\in\Omega_0,\ \omega_1\in\Omega_1}$$

representing the above discrete probability distribution p in Ω. We call p the *product* probability distribution of $p^{(0)}, p^{(1)}$ and denote it also by $p = p^{(0)} \times p^{(1)}$. The DPS $(\Omega, p) = (\Omega_0 \times \Omega_1,\ p^{(0)} \times p^{(1)})$ is called the *product probability space* of $(\Omega_0, p^{(0)})$ and $(\Omega_1, p^{(1)})$ The proof is practically obvious and left as an exercise to the reader.

Exercise 2.4. Generalize the result of theorem 2.3. to an arbitrary finite number of DPSs $(\Omega_1, p^{(1)}), \ldots, (\Omega_t, p^{(t)})$, thus obtaining a discrete probability distribution $p = p^{(1)} \times \ldots \times p^{(t)}$ on $\Omega = \Omega_1 \times \ldots \times \Omega_t = \{\omega = (\omega_1, \ldots, \omega_t) | \omega_1 \in \Omega_1, \ldots, \omega_t \in \Omega_t\}$. Show that, for every $u = 1, 2, \ldots, t$, the random variable $\varphi_u : \Omega \to \Omega_u$ defined by $\varphi_u(\omega) = \varphi_u((\omega_1, \ldots, \omega_t)) = \omega_u$ $(\omega \in \Omega)$ – the u-th *component mapping* – has the distribution $\varphi_u p = p^{(u)}$.

It is a natural question whether this simple *product construction* couldn't be carried over even to countably many given DPSs $(\Omega_1,\ p^{(1)})$, $(\Omega_2,\ p^{(2)}), \ldots$. This is actually possible in the framework of measure theory (see e.g. Jacobs [1978]), but doesn't lead to a *discrete* probability space as a rule. In fact, the countable analogon of the product formula (2) leads to infinite products diverging down to 0 for every ω unless very special conditions are imposed. This indicates one of the fundamental technical limitations of discrete probability theory, and motivates the study of non-discrete probability distributions, which will, however, not be treated in this book, up to a few exceptions.

2.5. Markovian Constructions.

Let $D \neq \emptyset$ be a finite set, $p = (p_j)_{j \in D}$ a probability vector over D, and $P = (P_{jk})_{j,k \in D}$ a stochastic $D \times D$-matrix. For any integer $t > 0$, consider the product space

$$\Omega = D^{t+1} = \{j = (j_0, j_1, \dots, j_t) | j_0, j_1, \dots, j_t \in D\}$$

and define

$$m_j = m_{(j_0, j_1, \dots, j_t)} = p_{j_0} P_{j_0 j_1} \dots P_{j_{t-1} j_t} \qquad (j_0, j_1, \dots, j_t \in D)$$

Then $m_j \geq 0$ $(j \in \Omega)$ and

$$\begin{aligned}
\sum_{j \in \Omega} m_j &= \sum_{j_o, j_1, \dots, j_t \in D} p_{j_0} P_{j_0 j_1} \dots P_{j_{t-1} j_t} \\
&= \left(\sum_{j_0, j_1, \dots, j_{t-1} \in D} p_{j_0} P_{j_0 j_1} \dots P_{j_{t-2} j_{t-1}} \right) \sum_{j_t \in D} P_{j_{t-1} j_t} \\
&= \sum_{j_0, j_1, \dots, j_{t-1} \in D} p_{j_0} P_{j_0 j_1} \dots P_{j_{t-1} j_t} \\
&= \dots = \sum_{j_0 \in D} p_{j_0} = 1.
\end{aligned}$$

Thus $m = (m_j)_{j \in \Omega}$ is a probability vector over D. The corresponding probability distribution in Ω is called the *Markov probability distribution* in Ω, with *initial distribution* p and *transition matrix* P. We may also denote it by $p \times P \times \dots \times P$. If all rows of P are $= p$, then $m = p \times \dots \times p$ $(t+1$ factors) comes out as a special case. This *Markovian construction* is easily generalized to the case of matrices $P^{(1)}, \dots, P^{(t)}$ instead of all the same $P, \dots, P$, and even to varying basic sets $D_0, D_1, \dots, D_t$ instead of $D, D, \dots, D$. Generalization to countable D is more or less obvious as well. Further generalization is requested by

Exercise 2.5. Defining the concept of a discrete stochastic $X \times X$-matrix $(P_{x,y})_{x,y \in X}$ for an arbitrary $X \neq \emptyset$ in a suitable fashion, generalize the above Markovian construction such as to obtain discrete Markovian probability distribution in arbitrary finite product spaces $\Omega = X \times \dots \times X$ $(t+1$ factors), for any given DPD p in X as an initial distribution. Calculate the distributions of the random variables $\varphi_0, \varphi_1, \dots, \varphi_t$ defined as component mappings (exercise 2.4.).

Remark 2.6. For reasons which will be explained in ch. VI. §1, the DPDs $\varphi_0 m, \dots, \varphi_t m$ will also be called the *marginal distributions* or the *marginals* of m, for any DPD m in $\Omega = X^{t+1}$.

3. Independence

The pupils of a highschool can be swimmers (= able to swim) or non-swimmers. They can also be catholic or non-catholic. If

> the proportion of swimmers is roughly the same among the catholics as in the whole high school

we will, of course, conclude that being catholic has no influence on the ability to swim. Or, more symmetrically: the distinction swimmer/non-swimmer is *independent* of the distinction catholic/non-catholic. Actually, if the proportions in question are not too extreme, i.e. not near 0 or 1, and if the approximation indicated by the word "roughly" is not too bad, the above statement entails corresponding statements about non-swimmers, catholics etc. in any of the four meaningful combinations. Independence is a fundamental phenomenon in stochastics. Its rigorous definition is with probabilities rather than frequencies. We will treat it here, in a somewhat technical fashion (in order to save technicalities in later chapters), within the framework of discrete probability spaces. We will encounter here, however, concepts and techniques originally developed for measure-theoretical probability theory: they make things smoother even on our more elementary level.

3.1. Independence of sets and set systems.

Definition 3.1. Let (Ω, p) be a DPS.

1. For a finite sequence $E_1, \ldots, E_n$ of subsets of Ω, the *product formula* is
$$p(E_1 \cap \ldots \cap E_n) = p(E_1) \cdot \ldots \cdot p(E_n).$$
2. $E_1, \ldots, E_n \subseteq \Omega$ are said to be *independent (for p)*, if for any $1 \leq r \leq n$ and any choice of pairwise distinct indexes $1 \leq k < \ldots < k_r \leq r$, the product formula holds for $E_{k_1}, \ldots, E_{k_r}$:
$$p(E_{k_1} \cap \ldots \cap E_{k_r}) = p(E_{k_1} \cdot \ldots \cdot p(E_{k_r})$$
3. Let $\mathcal{S}, \ldots, \mathcal{S} \subseteq \mathcal{P}(\Omega)$ be n systems of subsets of Ω. They are called *independent for p*, if for any choice $E_1 \in \mathcal{S}_1, \ldots, E_n \in \mathcal{S}_n$ the sets $E_1, \ldots, E_n$ are independent (for p).

Exercise 3.2. Let (Ω, p) be a DPS

1. Show that any finite sequence from $\{\emptyset, \Omega\}$ is independent
2. Show that $E_1, \ldots, E_n$ are independent (for p) iff any sequence $F_1, \ldots,$ F_n with $F_k = E_k$ or $= \Omega$ $(k = 1, \ldots, n)$ fulfils the product formula.
3. Show that $\mathcal{S}_1, \ldots, \mathcal{S}_n \in \mathcal{P}(\Omega)$ are independent iff for any choice of $F_1 \in \mathcal{S}_1 \cup \{\Omega\}, \ldots, F_n \in \mathcal{S}_n \cup \{\Omega\}$ the product formula holds.

Exercise 3.3. Let $\Omega = \{1, 2, 3, 4\}$ and $p_j = \frac{1}{4}$ $(j = 1, 2, 3, 4)$. Show that any two different two-element subsets of Ω containing 4 are independent, but any three of them are not.

3.2. Extending independence from smaller to larger set systems.

Exercise 3.4. let (Ω, p) be a DPS.

1. Let $E, F \subseteq \Omega$ fulfil the product formula, i.e. let
$$p(E \cap F) = p(E) \cdot p(F).$$
Show that every one of the following pairs of sets also fulfils the product formula
$$E, \Omega$$
$$\Omega, F$$
$$\Omega \backslash E, F$$
$$E, \Omega \backslash F$$
$$\Omega \backslash E, \Omega \backslash F$$
2. Let $E_1, \ldots, E_n \subseteq \Omega$ be independent for p. Show that for any choice of $F_1 \in \{E_1, \Omega \backslash E_1\}, \ldots, F_n \in \{E_n, \Omega \backslash E_n\}$ the sets $F_1, \ldots, F_n$ are independent as well.

This exercise displays extension of independence statements from smaller set systems to larger ones. The next exercise exhibits such a possibility once again.

Exercise 3.5. Let (Ω, p) be a DPS and every $\mathcal{D}_1, \ldots, \mathcal{D}_n \subseteq \mathcal{P}(\Omega)$ be a disjoint decomposition of Ω. Assume them independent (for p). For every $k = 1, \ldots, n$, let $\bar{\mathcal{D}}_k$ be the system of all those subsets of Ω which can be represented as an at most countable union of sets from $\mathcal{D}_k$. Show that $\bar{\mathcal{D}}_1, \ldots, \bar{\mathcal{D}}_n$ are independent again.

Let us now investigate such possibilities in a more systematic fashion. We shall use ideas from measure theory without going too far into that direction.

Definition 3.6. Let $\Omega \neq \emptyset$. A system $\mathcal{B} \subseteq \mathcal{P}(\Omega)$ of subsets of Ω is called a *σ-algebra* (in Ω) if

1. $\Omega \in \mathcal{B}$
2. $E, F \in \mathcal{B} \Longrightarrow E \backslash F \in \mathcal{B}$ (stability under the formation of differences)
3. $E_1, E_2, \ldots \in \mathcal{B} \Longrightarrow E_1 \cup E_2 \cup \ldots \in \mathcal{B}$ (stability under the formation of countable unions)

Clearly 1) and 2) together imply $\emptyset = \Omega\backslash\Omega \in \mathcal{B}$ plus stability under complementation: $E \in \mathcal{B} \Longrightarrow \Omega\backslash E \in \mathcal{B}$. By passage to complements stability against passage to countable intersections results: $E_1 \cap E_2 \cap \ldots = \Omega\backslash[(\Omega\backslash E_1) \cup (\Omega\backslash E_2) \cup \ldots] \in \mathcal{B}$ if $E_1, E_2, \ldots \in \mathcal{B}$. Finite unions and intersections are special cases of countable ones: insert countably many copies of $\emptyset$ resp. Ω. Observations like these should become the daily routine of the reader.

The easiest examples of σ-algebras in a set $\Omega \neq \emptyset$ are, of course

$\{\emptyset, \Omega\}$ (the minimal σ-algebra)

$\mathcal{P}(\Omega)$ (the maximal σ-algebra)

The one-element sets $\{\omega\}$ ($\omega \in \Omega$) are, in a way, the "atoms" of $\mathcal{P}(\Omega)$. They form a disjoint decomposition of Ω, and every set $E \subseteq \Omega$ is a union of some of them. Blowing up these "singletons" into subsets of Ω we arrive at

Exercise 3.7. Let any disjoint decomposition of a set $\Omega \neq \emptyset$ be given, and let $\mathcal{B}$ be the system of all subsets which can be represented as a union of some members of that decomposition. Show that $\mathcal{B}$ is a σ-algebra.

Exercise 3.8. In exercise 3.7., replace $\mathcal{B}$ by $\mathcal{B}_0$ = the system of all sets which can be represented as a union of countably many, or of *all but* countably many members of the given disjoint decomposition of Ω. Show that $\mathcal{B}_0$ is a σ-algebra.

In general probability theory, probability distributions are σ-additive functions of σ-algebras, and the theory of such functions – measure theory – is therefore the proper tool of full-fledged probability theory. Here we restrict ourselves to discrete probability spaces. As we have seen in §1, a DPD in an arbitrary set Ω lives on an at most countable subset of Ω. It is therefore natural to focus attention to the case of a *countable* Ω for a while.

Let thus Ω be *countable* and $\mathcal{B} \subseteq \mathcal{P}(\Omega)$ a σ-algebra. We call two points $\omega, \eta \in \Omega$ ***separated*** by a set $E \subseteq \Omega$ if

$$\begin{array}{lll} \text{either} & \omega \in E, & \eta \in \Omega\backslash E \\ \text{or} & \omega \in \Omega\backslash E, & \eta \in E \end{array}$$

Define a binary relation $\sim$ in Ω by

$$\omega \sim \eta \text{ iff no } E \in \mathcal{B} \text{ separates } \omega \text{ and } \eta$$

Exercise 3.9. Show that this $\sim$ is an equivalence relation on Ω.

Thus we may decompose Ω into equivalence classes for $\sim$, and since Ω is assumed to be *countable*, this decompositon is at most countable. If $\mathcal{B} = \{\emptyset, \Omega\}$

there is only one equivalence class, namely Ω, and if $\mathcal{B} = \mathcal{P}(\Omega)$, the equivalence classes are the singletons $\{\omega\}$ $(\omega \in \Omega)$. We have now all ingredients for a proof of

Theorem 3.10. Let Ω be at most countable and $\mathcal{B} \subseteq \mathcal{P}(\Omega)$ a σ-algebra. Then there is at at most countable decomposition of Ω such that $E \in \mathcal{B}$ iff E can be represented as a union of some members of that decomposition. The relationship thus established between σ-algebras in Ω and decompositions of Ω is one-to-one.

The reader is invited to complete the proof of this theorem.

If Ω is uncountable, the above theorem fails, as can e.g. be seen from exercise 3.8. where two different σ-algebras (prove it!) for the same decomposition arise if the latter has uncountably many nonempty members.

In view of theorem 3.10., exercise 3.5. establishes the independence of certain σ-algebras. It is our purpose to do this under fairly general conditions. For this purpose, it is convenient here to introduce the notion of a *Dynkin algebra* (after E.B. Dynkin, *1924).

Definition 3.11. Let $\Omega \neq \emptyset$. A system $\mathcal{D} \subseteq \mathcal{P}(\Omega)$ of subsets of Ω is called a *Dynkin algebra* (in Ω) if

1. $\Omega \in \mathcal{D}$
2. $E, F \in \mathcal{D},\ E \supseteq F \Longrightarrow E \backslash F \in \mathcal{B}$ (stability under *proper* differences)
3. $E_1, E_2, \ldots \in \mathcal{D},\ E_j \cap E_k = \emptyset \quad (j \neq k) \Longrightarrow E_1 \cup E_2 \cup \ldots \in \mathcal{D}$ (stability under countable *disjoint* unions)

The reader is invited to make analogous routine observations about Dynkin algebras as we did for σ-algebras after definition 3.6. The sharper requirement in 2) and 3) forestall, of course, some of the conclusions made there.

Exercise 3.12. Show that a Dynkin algebra $\mathcal{D} \subseteq \mathcal{P}(\Omega)$ is a σ-algebra iff it is stable under finite intersections, i.e. iff $E, F \in \mathcal{D} \Longrightarrow E \cap F \in \mathcal{D}$ (hint: intersection stability allows to represent every difference $E \backslash F$ of sets $E, F \in \mathcal{D}$ as a *proper* difference $E \backslash (E \cap F)$ with $E \cap F \in \mathcal{D}$, and every countable union $E_1 \cup E_2 \cup \ldots$ of sets from $\mathcal{D}$ as a *disjoint* union $E_1 \cup (E_2 \backslash (E_1 \cap E_2)) \cup \ldots$ (continue this appropriately!) of sets from $\mathcal{D}$).

Now the properties 2) and 3) of Dynkin algebras are just the right tools for deriving new product formulas, and hence independencies, from given ones:

Exercise 3.13. Let (Ω, p) be a DPS and $E \subseteq \Omega$ arbitrary. Show that $\mathcal{D}_E = \{F | F \subseteq \Omega,\ p(E \cap F) = p(E)p(F)\}$ is a Dynkin algebra in Ω.

According to this exercise we expect Dynkin algebras to occur most naturally whenever we investigate independence phenomena. It is rather inconvenient to deal with a large variety of such Dynkin algebras, e.g. one $\mathcal{D}_E$ for each $E \subseteq \Omega$. In order to avoid this, we borrow one more idea from measure theory, actually an idea which is ubiquitous in mathematics: *generation.*

Since the concept of a Dynkin algebra is defined by certain stability properties like stability under (the formation of) proper differences, it follows that the intersection of an arbitrary system of Dynkin algebras in a given basic set Ω is a Dynkin algebra in Ω again (that's obvious, yes?). If we apply this to the system of all Dynkin algebras containing a certain given set system $\mathcal{S} \subseteq \mathcal{P}(\Omega)$ – there is at least one such Dynkin algebra, namely, $\mathcal{P}(\Omega)$ itself – we clearly obtain the *smallest* Dynkin algebra $\supseteq \mathcal{S}$: it is called the *Dynkin algebra generated by* $\mathcal{S}$, and denoted by $\mathcal{D}(\mathcal{S})$. This scheme of definition is the customary way to introduce generation in all mathematical subdisciplines. In many cases there is an explicit generation procedure which implements such an abstract definition; e.g. in linear algebra, the linear subspace generated by a certain subset of a given vector space can be obtained as the set of all finite linear combinations of vectors from that subset. It is a characteristic feature of measure-theoretical techniques that such explicit constructions often fail to exist. One works very elegantly with generated σ-algebras and generated Dynkin algebras, but the results are non-constructive. It is this elegance which I take as a motive of using such techniques, although, in the framework of discrete stochastics, they could, in principle be replaced by more elementary albeit clumsy devices.

First remark: applying the above scheme of definition to the notion of a σ-algebra, we arrive at the concept of *the* σ*-algebra* $\mathcal{B}(\mathcal{S})$ *generated* by a given $\mathcal{S} \subseteq \mathcal{P}(\Omega)$.

Second, we observe that $\mathcal{B}(\mathcal{S}) \supseteq \mathcal{D}(\mathcal{S})$ holds generally, since every σ-algebra is also a Dynkin algebra. Now, when does equality hold here?

Proposition 3.14. If $\Omega \neq \emptyset$ and $\mathcal{S} \subseteq \mathcal{P}(\Omega)$ is intersection stable (i.e. if $E, F \in \mathcal{S} \Longrightarrow E \cap F \in \mathcal{S}$), then

$$\mathcal{D}(\mathcal{S}) = \mathcal{B}(\mathcal{S}).$$

PROOF. By exercise 3.12., every intersection stable Dynkin algebra is a σ-algebra. Thus it is clearly sufficient to infer the intersection stability of $\mathcal{D}(\mathcal{S})$ from the intersection stability of $\mathcal{S}$. We do this in two similar steps:

Step I: Let $E \in \mathcal{S}$ be arbitrary. Then $\mathcal{D}' = \{F | F \in \mathcal{D}(\mathcal{S}),\ E \cap F \in \mathcal{D}(\mathcal{S})\}$ is a Dynkin subalgebra of $\mathcal{D}(\mathcal{S})$ — an easy exercise for the reader. As $\mathcal{S}$ is intersection stable, $\mathcal{S} \subseteq \mathcal{D}'$ follows. From the minimality of $\mathcal{D}(\mathcal{S})$ we now conclude $\mathcal{D}' = \mathcal{D}(\mathcal{S})$. Thus we have shown $E \in \mathcal{S},\ F \in \mathcal{D}(\mathcal{S}) \Longrightarrow E \cap F \in \mathcal{D}(\mathcal{S})$.

Step II: Let now $E \in \mathcal{D}(\mathcal{S})$ be arbitrary and define $\mathcal{D}'' = \{F | F \in \mathcal{D}(\mathcal{S}),\ E \cap F \in \mathcal{D}(\mathcal{S})\}$. By the result of step I, we have $\mathcal{D}'' \supseteq \mathcal{S}$. By the same easy exercise as in step I, we find that $\mathcal{D}''$ is a Dynkin subalgebra of $\mathcal{D}(\mathcal{S})$. By the minimality of $\mathcal{D}(\mathcal{S})$, $\mathcal{D}'' = \mathcal{D}(\mathcal{S})$ follows. Thus we have shown $E, F \in \mathcal{D}(\mathcal{S}) \Longrightarrow E \cap F \in \mathcal{D}(\mathcal{S})$, i.e. the intersection stability of $\mathcal{D}(\mathcal{S})$. □

One traditionally favorite case of application for this result is $\Omega = \mathbf{R}$, $\mathcal{S} = \{]-\infty, \alpha] | \alpha \in \mathbf{R}\}$.

We are now in the position to prove

Theorem 3.15. Let (Ω, p) be a DPS and $\mathcal{S}_1, \ldots, \mathcal{S}_n \subseteq \mathcal{P}(\Omega)$ independent for p. Assume that, for every $j = 1, \ldots, n$, $\mathcal{S}_j$ is intersection stable. Then the generated σ-algebras $\mathcal{B}_1 = \mathcal{B}(\mathcal{S}_1), \ldots, \mathcal{B}_n = \mathcal{B}(\mathcal{S}_n)$ are independent as well.

PROOF. Choose $F_2 \in \mathcal{S}_2 \cup \{\Omega\}, \ldots, F_n \in \mathcal{S}_n \cup \{\Omega\}$. By exercise 3.13, $\{F_1 | F_1 \subseteq \Omega,\ p(F_1 \cap (F_2 \cap \ldots \cap F_n)) = p(F_1)p(F_2 \cap \ldots \cap F_n)\}$ is a Dynkin algebra which clearly contains $\mathcal{S}_1$, and thus also $\mathcal{D}(\mathcal{S}_1)$, the smallest Dynkin algebra $\supseteq \mathcal{S}_1$. By proposition 3.14, $\mathcal{D}(\mathcal{S}_1) = \mathcal{B}(\mathcal{S}_1)$. Thus $F_1 \in \mathcal{B}(\mathcal{S}_1)$, $F_2 \in \mathcal{S}_2 \cup \{\Omega\}, \ldots, F_n \in \mathcal{S}_n \cup \{\Omega\}$ entails $p(F_1 \cap F_2 \cap \ldots \cap F_n) = p(F_1)p(F_2 \cap \ldots \cap F_n) = p(F_1)p(F_2)\ldots p(F_n)$ that is, $\mathcal{B}(\mathcal{S}_1), \mathcal{S}_2, \ldots, \mathcal{S}_n$ are independent. Applying the same argument to $\mathcal{S}_2, \ldots, \mathcal{S}_n$ successively, we complete the proof of the theorem. □

3.3. Independent Random Variables.

In exercise 3.5 we have seen how easy it is to extend independence from disjoint decompositions to sets representable as unions of the components ("atoms") of such a decomposition.

The reader is certainly familiar with the basic scheme which identifies disjoint decompositions of a set Ω with mappings f of Ω into other sets X:

> The "sets of constancy" $\{\omega | f(\omega) = x\}$ $(x \in X)$ form a disjoint decomposition of Ω.
>
> Every disjoint decomposition of Ω can be obtained that way, that is via a suitable mapping $f : \Omega \to X$; one may e.g. choose $X =$ the set of all atoms of the decomposition and $f(\omega) =$ the atom to which ω belongs $(\omega \in \Omega)$.

It is thus an obvious idea to *define* the independence of mappings (= random variables) by the independence of their attached sets-of-constancy decompositions. Mathematical (and in particular stochastic) experience motivates the following generalization of this idea.

Recall, for an arbitrary mapping $f : \Omega \to X$, the definition of "pullback" for sets resp. set systems in X:

$$\begin{aligned} f^{-1}F &= \{\omega | f(\omega) \in F\} \quad (F \subseteq X) \\ f^{-1}\mathcal{D} &= \{f^{-1}F | F \in \mathcal{D}\} \quad (\mathcal{D} \subseteq \mathcal{P}(X)) \end{aligned}$$

The reader should be aware (exercise!) that $F \to f^{-1}F$ commutes with set operations like union, difference etc. and sends $X \to \Omega$, $\emptyset \to \emptyset$.

Definition 3.16. Let (Ω, p) be be a DPS and $f_1 : \Omega \to X_1, \ldots, f_n : \Omega \to X_n$ random variables on Ω with state spaces $X_1, \ldots, X_n$. We call them *independent* (for p), if the set systems

$$\begin{aligned} \mathcal{B}_1 &= f_1^{-1}\mathcal{P}(X_1) \\ \mathcal{B}_2 &= f_2^{-1}\mathcal{P}(X_2) \\ &\cdots\cdots\cdots \\ \mathcal{B}_n &= f_n^{-1}\mathcal{P}(X_n) \end{aligned}$$

are independent (for p).

As we are working within *discrete* stochastics here, we observe that among the possibly uncountable many pairwise disjoint "sets of constancy" $f_j^{-1}\{x\}$ ($x \in X_j$) of one of our RVs $f_j : \Omega \to X_j$, only at most countably many ones can have a probability > 0; it is only these sets that play a nontrivial rôle in our independence definition for RVs. The reader should realize how much this fact simplifies the reality underlying our rather general set theoretical considerations:

Exercise 3.17. Let (Ω, p) a DPS and $f_1 : \Omega \to X_1, \ldots, f_n : \Omega \to X_n$ RVs with state spaces $X_1, \ldots, X_n$. For $j = 1, \ldots, n$, let D_j be the at most countable disjoint decomposition of X_j all but possibly one of whose members are one-element sets having probability > 0 under the distribution $f_j p$ of f_j. Show that $f_1, \ldots, f_n$ are independent (for p) iff the set system $f_1^{-1}\mathcal{D}_1, \ldots, f_n^{-1}\mathcal{D}_n \subseteq \mathcal{P}(\Omega)$ are independent.

The next few exercises offer the reader an opportunity to study set-theoretical-minded independence theory in more detail. Their results will, however, not be used in this book.

Exercise 3.18. Let $f : \Omega \to X$ be a mapping and $f^{-1} : \mathcal{P}(\Omega) \leftarrow \mathcal{P}(X)$ be the corresponding set pullback. Prove:

1. If $\mathcal{B} \subseteq \mathcal{P}(X)$ is a σ-algebra, then $f^{-1}\mathcal{B} \subseteq \mathcal{P}(\Omega)$ is also a σ-algebra
2. If $E, F \subseteq X$ $f^{-1}E \cap f^{-1}F = \emptyset$, then $f^{-1}F = f^{-1}(F \backslash E)$.
3. If $E_1, E_2, \ldots \subseteq X$ and if $f^{-1}E_1, f^{-1}E_2, \ldots$ are pairwise disjoint, then there exist $E_1' \subseteq E_1$, $E_2' \subseteq E_2, \ldots$ such that $E_1', E_2', \ldots$ are pairwise disjoint and $f_1^{-1}E_1' = f_1^{-1}E_1, \ldots$
4. If $\mathcal{D} \subseteq \mathcal{P}(X)$ is a Dynkin algebra, then $f^{-1}\mathcal{D} \subseteq \mathcal{P}(\Omega)$ is a Dynkin algebra
5. If $\mathcal{B} \subseteq \mathcal{P}(\Omega)$ is a σ-algebra , then $\{E | E \subseteq X,\ f^{-1}E \in \mathcal{B}\} \subseteq \mathcal{P}(X)$ is a σ-algebra.
6. Same for Dynkin algebras.
7. Pullback commutes with the generation of σ-algebras:

   If $\mathcal{S} \subseteq \mathcal{P}(X)$, then
   $$f^{-1}\mathcal{B}(\mathcal{S}) = \mathcal{B}(f^{-1}\mathcal{S}),$$
   and the same holds for generated Dynkin algebras.

Exercise 3.19. Let (Ω, p) be a DPS and $f_1 : \Omega \to X_1, \ldots, f_n : \Omega \to X_n$ RVs with state spaces $X_1, \ldots, X_n$. For every $j = 1, \ldots, n$ let $\mathcal{S}_j \subseteq \mathcal{P}(X)$ be an intersection-stable set system separating all points of X_j which have probability > 0 under $f_j p$. Prove that $f_1, \ldots, f_n$ are independent (for p) iff the set systems $f_1^{-1}\mathcal{S}_1, \ldots, f_n^{-1}\mathcal{S}_n$ are independent. —
The result of this exercise may be applied to the case $X_1 = \ldots = X_n = \mathbf{R}$ and $\mathcal{S}_1 = \ldots = \mathcal{S}_n = \{\,]-\infty, \alpha] | \alpha \in \mathbf{R}\}$ = the system of all closed half-lines bounded from above, justifying a definition of independence of real-valued RVs which pervades the literature before, say, 1945.

Proposition 3.20. Let (Ω, p) be a DPS and $f_1 : \Omega \to X_1, \ldots, f_n : \Omega \to X_n$ RVs with state spaces $X_1, \ldots, X_n$ and distributions $p_1 = f_1 p, \ldots, p_n = f_n p$. Then $f_1, \ldots, f_n$ are independent (under p) iff their *joint* distribution q is $p_1 \times \ldots \times p_n$.

PROOF. $p_1 \times \ldots \times p_n$ is the only DPD q in $X_1 \times \ldots \times X_n$ which fulfils

$$q(E_1 \times \ldots \times E_n) = p_1(E_1) \ldots p_n(E_n) \quad (E_1 \subseteq X_1, \ldots, E_n \subseteq X_n)$$

But this is $= p(f_1^{-1}E_1) \ldots p(f_n^{-1}E_n)$
and this is $= p(f_1^{-1}E_1 \cap \ldots \cap F_n^{-1}E_n)$
$= q(E_1 \times \ldots \times E_n)$

for all choices of $E_1 \subseteq X_1, \ldots, E_n \subseteq X_n$ iff $f_1, \ldots, f_n$ are independent. □

Exercise 3.21. (Disjoint grouping). Let (Ω, p) be a DPS and $f_1 : \Omega \to X_1, \ldots, f_n : \Omega \to X_n$ be independent RV's with state spaces $X_1, \ldots, X_n$. Prove:

1. For any choice of mappings $h_1 : X_1 \to Y_1, \ldots, h_n : X_n \to Y_n$ into sets $Y_1, \ldots, Y_n$, the RVs $h_1(f_1) : \Omega \to Y_1, \ldots, h_n(f_n) : \Omega \to Y_n$ (we might write $h_j \circ f_j$ ("first f_j then h_j") instead $(j = 1, \ldots, n)$) are independent.
2. For any disjoint decomposition $\{1, \ldots, n\} = J_1 \cup \ldots \cup J_r$ with all $J_\rho \neq \emptyset$, the RV's
$$\begin{aligned} f_{J_1} &= \text{join of all } f_j \text{ with } j \in J_1 \\ &\ldots \\ f_{J_r} &= \text{join of all } f_j \text{ with } j \in J_n \end{aligned}$$
are independent.
3. Combine 1. and 2.

Our remark at the end of 2.4 clearly implies (via proposition 3.20) that it is impossible, in discrete stochastics, to have an infinity of IID random variables, except for trivial cases.

IV. Independent Identically Distributed (IID) Random Variables

In this chapter, the reader is to get acquainted with the core of classical probability

independent identically distributed (IID) random variables,

mostly real-valued: the "IID world". The notions of *expectation* and *variance* are central, and

limit theorems

are the main theme. The classical triad of limit theorems

law of large numbers (LLN)
central limit theorem (CLT)
laws of iterated logarithm (loglog theorem, LIL)

is only partly accessible within the framework of discrete stochastics.

In §1 we treat the convolution of DPDs on $\mathbf{R}$. In §2 we deal with expectation and variance. In §3 we prove the weak law of large numbers (WLLN), within the bounds of discrete probability theory, and in §4 we present de Moivre-Laplace's special case of the central limit theorem (CLT), introducing a non-discrete probability distribution, the normal distribution ("bell-shaped curve") $N(0,1)$ on $\mathbf{R}$ for that purpose. In §5 we prove a general CLT using a "convolution operator" method (after Trotter [1959]) which would work in a general measure-theoretical setting as well. In §6 we give an outlook on the classical canon LLN /CLT/LIL, sketching a landscape beyond discrete stochastics.

1. Addition of independent RVs → convolution of their distributions

We begin our excursion into the IID world with the following question

if we add independent real RVs – what happens to their distributions?

The answer is given by

Theorem 1.1. Let f, g be independent real-valued RVs on the DPS (Ω, m), and let $p = fm$, $q = gm$ be their distributions. Then the distribution of the RV $f + g$ is the ***convolution*** $p * q$ of p and q, defined by

$$(1) \quad (p*q)_x = \sum_{\substack{y,z \in \mathbf{R} \\ y+z=x}} p_y\, q_z = \sum_{y \in \mathbf{R}} p_{x-y}\, q_y = \sum_{z \in \mathbf{R}} p_z q_{x-z} \quad (x \in \mathbf{R})$$

(the sums have only at most countably many nonzero terms).

PROOF. By proposition III.3.20, the joint distribution of f, g is the product DPD $p \times q$ on $\mathbf{R}^2$. The mapping $a : (y, z) \to y + z$ of $\mathbf{R}^2 \to \mathbf{R}$ sends the joint distribution of f, g into the distribution of $f + g$, since $f + g = a(f, g)$. But for every $x \in \mathbf{R}$, $a^{-1}(\{x\}) = \{(y, z) | y + z = x\}$, which proves the theorem. □

Since the addition of real RVs is commutative, so is the convolution of DPD's. $p \times q = q \times p$ can as well be read from the symmetry of terms in (1).

Theorem 1.1. can, of course, be extended to arbitrary finite sums of independent RVs:

> if $f_1, \ldots, f_t$ are independent real RVs with distributions $p^{(1)}, \ldots, p^{(t)}$, the distribution of $f_1 + \ldots + f_t$ is the convolution $p^{(1)} * \ldots * p^{(t)}$ of $p^{(1)}, \ldots, p^{(t)}$, and this DPD in $\mathbf{R}$ may be calculated e.g. by
>
> $$(p^{(1)} * \ldots * p^{(t)})_x = \sum_{\substack{x_1, \ldots, x_t \in \mathbf{R} \\ x_1 + \ldots + x_t = x}} p^{(1)}_{x_1} \cdots p^{(t)}_{x_t} \quad (x \in \mathbf{R})$$
>
> (only at most countably many terms of this sum are nonzero).

In the IID case, we have $p^{(1)} = \ldots = p^{(t)} = p$ and will write p^{*t} for a convolution of t equal copies of p. So we may say that the IID world is, distributionwise, pervaded by ***convolution powers*** p^{*t}. We will come back to this in §5 again.

Exercise 1.2. Calculate p^{*n} for $p = \alpha\epsilon_1 + (1 - \alpha)\epsilon_0$ $(0 \le \alpha \le 1;\ n = 1, 2, \ldots)$, where generally ϵ_x denotes the "Dirac" DPD "point mass 1 sitting at x" ($x \in \mathbf{R}$ arbitrary).

The reader is invited to carry over the considerations of this section to $\mathbf{R}^d$-valued RVs, $d \ge 1$ arbitrary. He should also realize that generalizations to RVs with values in arbitrary groups are easy to achieve; they actually play a considerable rôle in research, but will not be a theme of this book.

2. Expectation and Variance

Let (Ω, m) be a DPS; we shall keep it fixed throughout this section.

Let $f : \Omega \to \mathbf{R}$ be a real RV and $p = fm$ its distribution. p is a DPD in $\mathbf{R}$, sitting on pairwise distinct points $x_1, x_2, \ldots \in \mathbf{R} : p_x = m(f = x) = m(\{\omega|)f(\omega) = x\})$ $(x \in \mathbf{R})$, $p_{x_1}, p_{x_2}, \ldots \geq 0$, $p_{x_1} + p_{x_2} \cdots = 1$.

2.1. Expectation.

If we want to tell to someone where roughly p is situated, the physical notion of *barycenter* offers itself: the barycenter of p is the real number $\bar{x}$ defined by

$$(1) \quad \bar{x} = x_1 p_{x_1} + x_2 p_{x_2} + \ldots = \sum_{k=1}^{\infty} x_k p_{x_k}$$

– provided this series converges. There is no question about that if we have only finitely many x_k's with $p_{x_k} > 0$, or if the set $\{x_1, x_2, \ldots\}$ is bounded. In particular, if p is unit mass sitting on x_1, then $\bar{x} = x_1$. But in general convergence is a nontrivial hypothesis here.

Exercise 2.1. Put $x_k = 2^k$, $p_{x_k} = \frac{1}{2^k}$ $(k = 1, 2, \ldots)$ and show that (1) diverges. Invent two more essentially distinct examples of divergence of (1).

If (1) converges, we may calculate

$$\begin{aligned} \bar{x} &= \textstyle\sum_{k=1}^{\infty} x_k p_{x_k} \\ &= \textstyle\sum_{k=1}^{\infty} x_k m(f = x_k) \\ &= \textstyle\sum_{\omega \in \Omega} f(\omega) m_\omega. \end{aligned}$$

In fact, the forelast sum arises from the last one by grouping the points ω (only the at most countably many ones with $m_\omega > 0$ really count) according to the values attained by $f(\omega)$ (only countably many ones really count). This should be enough to motivate the

Definition 2.2. Let f be a real RV on the DPS (Ω, m). We define the *expected value* or the *expectation* of f (under m) as

$$(2) \quad E_m(f) = \sum_{\omega \in \Omega} f(\omega) m_\omega$$

provided the right hand term is is meaningful.

That is, we say that the expectation of f *exists* or that f *has an expectation* if the – in general infinite – sum on the right hand side of (2) converges unconditionally. If m sits on $\omega_1, \omega_2, \ldots$, that is, if $m_{\omega_1} + m_{\omega_2} + \ldots = 1$, then $E_m(f)$ exists iff the series $\sum_{j=1}^{\infty} f(\omega_j)m_{\omega_j}$ is absolutely convergent. In particular, $E_m(f)$ exists iff $E_m(|f|)$ exists. Finally, if two real RVs coincide m-a.e., and if one of them has an expectation, so has the other, and if they both have expectations, these coincide.

Exercise 2.3. Let f be a real RV on the DPS (Ω, m), and let $\Omega_0, \Omega_1, \Omega_2, \ldots$ be a disjoint decomposition of Ω such that $m(\Omega_0) = 0$ and f is constant – say, $= y_i$ – on every Ω_i $(i = 1, 2, \ldots)$. Prove: $E_m(f)$ exists iff the series $\sum_{i=1}^{\infty} y_i m(\Omega_i)$ is absolutely convergent; if this is the case, then $E_m(f) = \sum_{i=1}^{\infty} y_i\, m(\Omega_i)$.

Exercise 2.4. Prove: if (Ω, m) is a DPS, then $E_m(1_E) = m(E)$ for $1_E =$ indicator function of a set $E \subseteq \Omega$.

Proposition 2.5. Let $f, f_1, \ldots, f_t$ be real RVs on the DPS (Ω, m).

1) Assume that $E_m(f_1), \ldots, E_m(f_t)$ exist. Then for any reals $\alpha_1, \ldots, \alpha_t$, $E_m(\alpha_1 f_1 + \ldots \alpha_t f_t)$ also exists, and

$$E_m(\alpha_1 f_1 + \ldots + \alpha_t f_t) = \alpha_1 E_m(f_1) + \ldots + \alpha_t E_m(f_t) \tag{1}$$

2) If $E_m(f)$ exists and $f \geq 0$ m-a.e., then

$$E_m(f) \geq 0 \tag{2}$$

In short: the set – usually denoted by $\mathcal{L}^1_m$ – of all real RVs whose expectation exists – is a real vector space and $E_m : \mathcal{L}^1_m \to \mathbf{R}$ is a *positive linear form* on it.

3) If f is constant $= \alpha$ m-a.e., then $E_m(f) = \alpha$.

PROOF.

1) We only have to list the points $\omega_1, \omega_2, \ldots \in \Omega$ on which m sits (i.e. $m_{\omega_1} + m_{\omega_2} + \ldots = 1$) and read the formula

$$\sum_{k=1}^{\infty} \left(\sum_{u=1}^{t} \alpha_u\, f(\omega_k) \right) m_{\omega_2} = \sum_{u=1}^{t} \alpha_u \sum_{k=1}^{\infty} f(\omega_k) m_{\omega_k}$$

appropriately, in order to arrive at the linearity formula (3). (4) is even simpler, and

2) and 3) is trivial.

□

For technical reasons, we still prove

Proposition 2.6. Let (Ω, m) be a DPS and $\varphi : \Omega \to X$ a RV with state space X and distribution p – a DPD in X. Let $f : X \to \mathbf{R}$ be a RV on the DPS (X, p). Then $E_m(f(\varphi))$ exists iff $E_p(f)$ exists, and if both exist, then

$$E_m(f(\varphi)) = E_p(f).$$

PROOF. Let p sit on $x_1, x_2, \ldots \in X$ and

$$E_j = \{\omega | \varphi(\omega) = x_j\} \quad (j = 1, 2, \ldots).$$

Clearly $p_{x_j} = m(E_j)$, and $\sum_j p_{x_j} = \sum_j m(E_j) = 1$. The proposition now follows upon reading

$$\begin{aligned} E_m(f(\varphi)) &= \sum_{\omega \in \Omega} f(\varphi(\omega)) m_\omega \\ &= \sum_j f(x_j) m(E_j) \\ &= \sum_j f(x_j) p_{x_j} \end{aligned}$$

intelligently. □

Our introduction of $E_m(f)$ as the barycenter of the distribution $p = fm$ of f reappears as a particular case of this proposition (replace φ by f, and f by $id_{\mathbf{R}}$).

2.2. Variance.

Let again p be the distribution of a real-valued RV f on the DPS (Ω, m); thus p is a DPD sitting on points $x_1, x_2, \ldots \in \mathbf{R}$, i.e. fulfilling $p_{x_1} + p_{x_2} + \ldots = 1$. Assume that p has a barycenter; that is, $E_m(f)$ exists and tells us, where roughly p lies, or, in other words, where roughly the values of f, directed by random, that is, by m, tend to lie.

We might now like to tell to someone, how much p is concentrated near $\bar{x} = E_m(f)$, that is, how much the values of f tend to be concentrated near $E_m(f)$. One way of making such a statement in a quantitative form is *mean quadratic deviation* or *variance*:

$$\sigma_p^2 = \sum_{k=1}^{\infty} (x_k - \bar{x})^2 \, p_{x_k}$$

– provided this series converges to a finite limit.

Exercise 2.7. Prove, with the above notation: if $\sum x_k^2 p_{x_k} < \infty$, then the series $\sum_{k=1}^{\infty} x_k p_{x_k}$ is absolutely convergent, and $\sum_{k=1}^{\infty} (x_k - a)^2 p_{x_k}$ is, for every real a, a convergent series whose limit we denote by $\varphi(a)$. Show that $\varphi(a)$ attains its minimum for $a = \bar{x} = \sum_{k=1}^{\infty} x_k p_{x_k}$ ("characterization of the barycenter (expectation) by the least squares method").

As in the case of the expectation, we may easily calculate

$$\sigma_p^2 = E_m((f - E_m(f))^2)$$

This motivates the

Definition 2.8. Let f be a real-valued RV on the DPS (Ω, m). We define the *variance* $\sigma_m^2(f)$ of f by

$$(3) \quad \sigma_m^2(f) = E_m((f - E_m(f))^2) = \sum_{\omega \in \Omega} (f(\omega) - E_m(f))^2 m_\omega,$$

provided the right hand form is meaningful and a finite real number.

This definition needs, like definition 2.2. a few explanations. In any case, we shall say that f *has a variance* or that the *variance* of f *exists* if the right hand term in (3) are meaningful and finite. Further details can be obtained from

Exercise 2.9 . Let f be a real RV on the DPS (Ω, m).

a) Prove: if f^2 has an expectation, then f has both an expectation and a variance and
$$\sigma_m^2(f) = E_m(f^2) - (E_m(f))^2.$$

b) Prove: if $E_m(f)$ exists, then f has a variance iff $E_m(f^2)$ also exists.

c) Prove: if $E_m(f^2)$ exists, then $\varphi(a) = E_m((f-a)^2)$ exists for all reals a and attains its minimum $\sigma_m^2(f)$ for $a = E_m(f)$.

d) Prove that for every real α the variance $\sigma_m^2(f - \alpha)$ exists iff $\sigma_m^2(f)$ exists, and that both, if existent, are equal.

e) Prove that for every real $\alpha \neq 0$ the variance $\sigma^2(\alpha f)$ exists iff $\sigma_m^2(f)$ exists, and $\sigma_m^2(\alpha f) = \alpha^2 \sigma_m^2(f)$ holds in this case (and even (trivially) for $\alpha = 0$).

(Hint: have a look on exercise 2.7.).

Exercise 2.10. Let (Ω, m) be a DPS and $E \subseteq \Omega$. Show that $\sigma_m^2(1_E) = p(E)(1 - p(E))$.

Let us now see how the variance behaves if we add and multiply real RVs.

Proposition 2.11. Let f, g be *independent* real-valued RVs on the DPS (Ω, m) and assume that $E_m(f)$, $E_m(g)$ exist. Then $E_m(fg)$ exists and we have

$$E_m(fg) = E_m(f) \cdot E_m(g).$$

PROOF. Let p be the distribution of f, sitting on $x_1, x_2, \ldots \in \mathbf{R}$ (i.e. $p_{x_1} + p_{x_2} + \ldots = 1$); let likewise q be the distribution of g, sitting on $y_1, y_2, \ldots \in \mathbf{R}$. By proposition III.3.20 $p \times q$ is the joint distribution of f and g, sitting on $\{(x_j, y_j) | j, k = 1, 2, \ldots\} \subseteq \mathbf{R}^2$. From proposition 2.6 (replace φ by the join $f \times g : \Omega \to \mathbf{R}^2 = X$, and f by $(x, y) \to xy$ there) we see that $E_m(fg)$ exists iff

$$(4) \quad \sum_{j,k} x_j y_k p_{x_j} q_{y_k}$$

converges absolutely, and $E_m(fg)$ is given by (4) if this convergence takes place. Now this is obviously the case since the two series $\sum_j x_j p_{x_j}$, $\sum_k y_k q_{y_k}$ are absolutely convergent by hypothesis, and clearly we have

$$\begin{aligned} \textstyle\sum_{jk} x_j y_k p_{x_j} q_{y_k} &= (\textstyle\sum_j x_j p_{x_j}) \cdot (\textstyle\sum_k y_k q_{y_k}) \\ &= E_m(f) E_m(g). \end{aligned}$$

□

Proposition 2.12. Let $f_1, \ldots, f_t$ be independent real-valued RVs on the DPS (Ω, m), and assume that $\sigma_m^2(f_1), \ldots, \sigma_m^2(f_t)$ exist. Then $\sigma_m^2(f_1 + \ldots + f_t)$ exists, and

$$\sigma_m^2(f_1 + \ldots + f_t) = \sigma_m^2(f_1 + \ldots + \sigma_m^2(f_t)$$

PROOF. Subtracting expectations without affecting existence and values of variances (exercise 2.9.), we may and will assume $E_m(f_1) = \ldots = E_m(f_t) = 0 = E_m(f_1 + \ldots f_t)$. From proposition 2.11. we now derive

$$\begin{aligned} \sigma_m^2(f_1 + \ldots f_t) &= E_m\left((\textstyle\sum_{u=1}^t f_u)^2\right) \\ (***) \qquad &= E_m(\textstyle\sum_{u,v=1}^t f_u f_v) \\ &= E_m(\textstyle\sum_{u=1}^t f_u^2 + \sum_{u \neq v} f_u f_v) \end{aligned}$$

$$\begin{aligned}
&= \textstyle\sum_{u=1}^{t} E_m(f_u^2) + \sum_{u \neq v} E_m(f_u)E_m(f_v) \\
&= \textstyle\sum_{u=1}^{t} E_m(f_u^2) \\
&= \textstyle\sum_{u=1}^{t} \sigma_m^2(f_u)
\end{aligned}$$

The existence of $(***)$ proves the existence statement of our proposition. □

It is certainly in a way surprising that the *quadratic* functional $\sigma_m^2(\cdot)$ behaves *additively*, as we have just shown.

Exercise 2.13. Let $f_1, \ldots, f_t$ be IID real-valued RVs on the DPS (Ω, m), and assume that $\sigma_m^2(f_1) = \sigma^2$ exists and is > 0. Prove the existence of the following variances, and the formulas

$$\begin{aligned}
\sigma_m^2\left(\tfrac{f_1 - E_m(f_1)}{\sqrt{\sigma^2}}\right) &= 1 \\
\sigma_m^2\left(\tfrac{1}{t}(f_1 + \ldots + f_t)\right) &= \tfrac{\sigma^2}{t}
\end{aligned}$$

3. The Weak Law of Large Numbers (WLLN)

In this section we present a result which may be, roughly and verbally, restated follows:

> The arithmetical mean (average) of a large number of real IID random variables is a good estimate of their expected value: it is highly improbable that it deviates from the latter significantly.

In a special case, this law, nowadays called "weak law of large numbers (WLLN)", had been observed already by Jacob Bernoulli (1654-1705) in 1685 (Bernoulli [1975] p. 76 ff). For the proof of the WLLN in its nowadays usual form, we need an inequality involving variances and usually named after Irénée Jules Bienaymé (1796-1878) and Pavnuti Tschebyshev (1821-1894). See also Heyde-Seneta [1977].

3.1. The Biénaymé-Tschebyshev Inequality.

in its simplest form is contained in

Proposition 3.1. Let f be a *nonnegative* real RV on the DPS (Ω, m). Assume that $E_m(f)$ exists. Then for every $\epsilon > 0$ we have the *Biénaymé-Tschebyshev inequality*.

$$(1) \quad m(f \geq \epsilon) \leq \frac{1}{\epsilon} E_m(f)$$

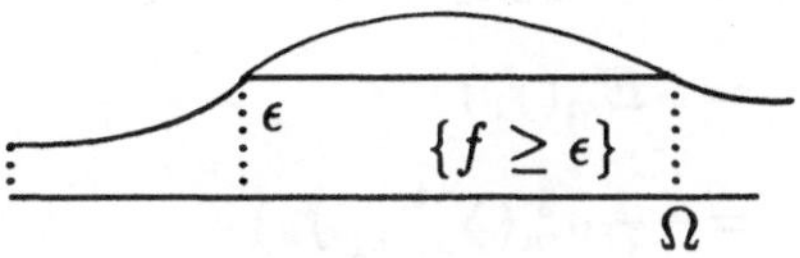

PROOF. Let $E = \{\omega | f(\omega) \geq \epsilon\}$. Clearly

$$\epsilon 1_E \leq f,$$

and hence

$$\epsilon m(E) \leq E_m(f),$$

which yields (1) upon division by ϵ. □

A variant of this inequality sounds

Proposition 3.2. Let f be a real RV on the DPS (Ω, m) and assume that $E_m(f)$ and $\sigma_m^2(f)$ exist. The we have for every $\epsilon > 0$, the *Biénaymé- Tschebyshev inequality*

$$(2) \quad m(|f - E_m(f)| \geq \epsilon) \leq \frac{\sigma_m^2(f)}{\epsilon^2}$$

PROOF. Replace f by $|f - E_m(f)|^2$ and $\epsilon > 0$ by ϵ^2 in proposition 2.1. □

3.2. The WLLN in the Form of an Inequality of Biénaymé- Tchebyshev Type.

is contained in

Theorem 3.3.

Let $f_1, \ldots, f_t$ be IID real RVs on the DPS (Ω, m). Assume that $E_m(f_1)$, $\sigma_m^2(f_1)$ exist. Then the (again so called) *Biénaymé-Tchebyshev inequality*

$$m(|\frac{1}{t}\sum_{u=1}^{t} f_u - E_m(f_1)| \geq \epsilon) \leq \frac{\sigma_m^2(f_1)}{t\epsilon^2}$$

holds.

PROOF. We calculate, making use of the additivity of the variance for independent RVs (proposition 2.12.),

$$\begin{aligned}
E_m(\tfrac{1}{t}\textstyle\sum_{u=1}^t f_u) &= \tfrac{1}{t}\textstyle\sum_{u=1}^t E_m(f_u) = \tfrac{1}{t}\cdot t\cdot E_m(f_1) \\
&= E_m(f_1) \\
\sigma_m^2(\tfrac{1}{t}\textstyle\sum_{u=1}^t f_u) &= \tfrac{1}{t^2}\sigma_m^2(\textstyle\sum_{u=1}^t f_u) \\
&= \tfrac{1}{t^2}\textstyle\sum_{u=1}^t \sigma_m^2(f_u) \\
&= \tfrac{1}{t^2}\cdot t\cdot \sigma_m^2(f_1) \\
&= \tfrac{1}{t}\sigma_m^2(f_1).
\end{aligned}$$

Replacing f by $\frac{1}{t}\sum_{u-1}^n f_u$ in proposition 2.3., we obtain the desired result. □

Exercise 3.4. Let $f_1,\ldots,f_n$ be independent real-valued RVs on the DPS (Ω, m). Assume that all $E_m(f_k)$, $\sigma_m^2(f_k)$ $(k=1,\ldots,n)$ exist, and that $\sup_{1\le k\le n}\sigma_m^2(f_k) = M < \infty$. Prove, for

$$g = \frac{1}{n}\sum_{k=1}^n f_k,$$

the inequality

$$m(|g - E_m(g)|) \le \frac{M}{n\epsilon^2}$$

Theorem 3.3. resp. the result of exericse 3.4. are what is nowadays often called the *weak law of large numbers (WLLN)* in its finite form, or in the form of a Biénaymé-Tschebyshev type inequality. If we conduct a large number of independent repetitions of a random experiment with a real-valued outcome with expected value a and variance $\sigma^2 < \infty$, then theorem 3.3. (resp. exercise 3.4.) tells us that, upon replacing a by the average of the values observed in t repetitions, we commit an error $\ge \epsilon$ with a probability $\le \frac{\sigma^2}{t\epsilon^2}$, which becomes as small as we want, if we only repeat our experiment sufficiently often.

As a corollary, we may treat the case where our experiment is of yes-no type: symbolize "yes" by 1, and "no" by 0. This defines a 0-1-valued RV whose expectation we may again denote by

$$a = \text{probability of "yes"}.$$

The variance then is $a(1-a)$, and the addition of the values of this RV in independently repeated experiments means simply counting the number of outcomes "yes" while averaging means calculating the *relative frequency* of "yes" in, say, t independent repetitions. Theorem 3.3. tells us that, upon replacing a by the relative frequency of "yes", we commit an error $\geq \epsilon > 0$ with a probability $\leq \frac{a(1-a)}{t\epsilon^2}$, which becomes as small as we want if we repeat our experiment often enough.

In §6 we shall sketch more results of LLN type. These, as well as WLLN, may serve as a good reason to replace or estimate expectations by averages, and probabilities by relative frequencies, for practical purposes. This is often the only way of getting an idea about the value of an unknown expectation, and it is daily practice in administrative or insurance statistics – we all e.g. watch the averages which give us an idea about our expected life span, or about the probability to get cured from a certain disease.

4. The Central Limit Theorem (CLT) I: de Moivre-Laplace's Version

The WLLN told us that the distribution of the arithmetic mean of IID real RVs with existing variance concentrates more and more on one single point: the expectation. The reason: the variance of the average goes to zero as $\frac{1}{t}$, where t is the number of RVs over which the average is taken. This result is coarse, and its proof is extremely simple. It doesn't tell us anything about the *shape* which the said distribution takes while concentrating more and more near the expected value.

Maybe we can see sort of a limit shape if we look at the said distribution through a magnifying lens which endows it with variance 1: magnification factor $\sqrt{t}$. This idea is indeed realizable, and the result is the so-called *central limit theorem* (CLT; this name seems to go back to Pólya [1920]). We shall prove it here first in a very special case which had been settled already by Abraham de Moivre (1667-1754) [1718] resp. Pierre Simon de Laplace (1749-1829) [1814]. A much more general result will be proved in §5.

The limiting procedure which is the central feature of the CLT, leads us out of the realm of DPDs. Thus we begin by introducing

4.1. The Normal Distribution ("The Bell-Shaped Curve").

Proposition 4.1. The function ϱ defined on $\mathbf{R}$ by

$$\varrho(x) = \frac{1}{\sqrt{2\pi}} e^{-\frac{x^2}{2}}$$

fulfils

$$\int_{-\infty}^{\infty} \varrho(x)dx = 1 \tag{3}$$

PROOF. The integral over all of $\mathbf{R}^2$, of the function

$$\varrho(x)\varrho(y) = \frac{1}{2\pi} e^{-\frac{x^2+y^2}{2}} \quad ((x,y) \in \mathbf{R}^2)$$

can be calculated as

$$\int_{\mathbf{R}^2} \varrho(x)\varrho(y)dxdy = \frac{2\pi}{2\pi} \int_0^{\infty} re^{-\frac{r^2}{2}} dr,$$

as $\varrho(x)\varrho(y)$ is constant $= e^{-\frac{r^2}{2}}$ on the circle line of radius r around (0,0), the length of which is $2\pi r$. We obviously may continue with

$$= \int_0^{\infty} e^{-s} ds = 1$$

On the other hand, we have

$$\int_{\mathbf{R}^2} \varrho(x)\varrho(y)dxdy = \left[\int_{-\infty}^{\infty} \varrho(x)dx\right]^2,$$

which concludes the proof. □

Viewing the "normalization" equation (1) as an analogue to the normalization property $m(\mathbf{R}) = 1$ of an arbitrary DPD m on $\mathbf{R}$, we may consider ϱ as the "density" function of a probability distribution on $\mathbf{R}$ which is not discrete but "continuous" and can be defined from ϱ as the set function

$$p_\varrho(E) = \int_E \varrho(x)dx = \left(= \int_{\mathbf{R}} 1_E(x)\varrho(x)dx\right),$$

not for all E, but for those E for which we are able to reasonably define the integral written in brackets. So-called measure theory is required in order to make this sketch really precise. We will however, only have to consider intervals E here. They suffice already to demonstrate that p_ϱ is "continuous" at least in the sense that it cannot ascribe a value > 0 to any one-point set

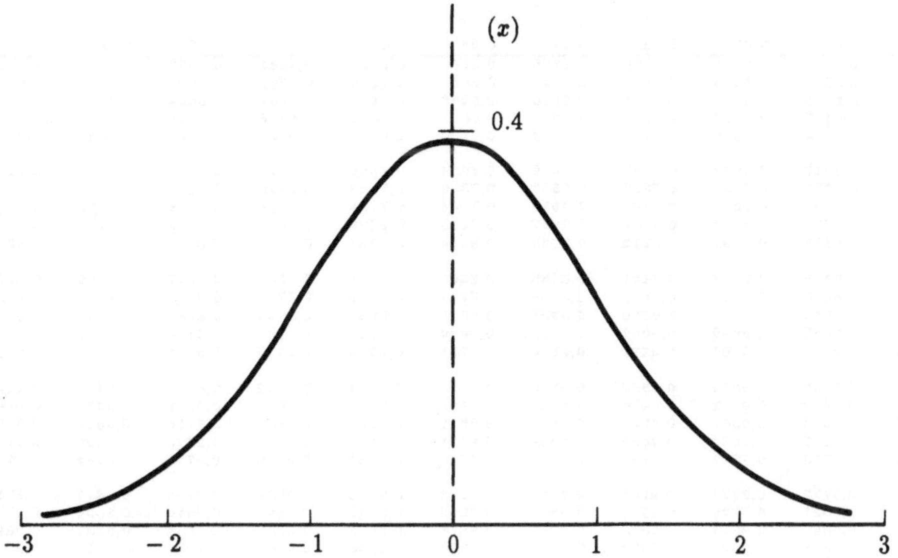

fig. IV.4.1

$\{x\}$, because $p_\varrho(\{x\}) \leq \int_E \varrho(x)dx$ would have to hold for every interval E containing x, and by chosing E sufficiently short, we can make $\int_E \varrho(x)dx$ as small as we want.

Exercise 4.2. For any $a \in \mathbf{R}$ and $\sigma^2 > 0$, let

$$(4) \quad \varrho_{a,\sigma^2}(x) = \frac{1}{\sqrt{2\pi\sigma^2}}\, e^{-\frac{(x-a)^2}{2\sigma^2}} \quad (x \in \mathbf{R})$$

Prove $\int_{-\infty}^{\infty} \varrho_{a,\sigma^2}(x)dx = 1$.

By tradition, the "continuous" probability distribution p_ϱ is called the

standard normal (or Gaussian) distribution $N(0,1)$

It has been considered long before Gauss, e.g. by de Moivre [1718]. Its density $\varrho(x)$ is also called

the Gaussian bell-shaped curve;

in fact, its graph looks as in figure IV.4.1.

We enclose also a table with values of $\phi(x) = \int_{-\infty}^{x} \varrho(y)dy$ (next page).

They are of constant use in practical statistics, for reasons which we explain at the end of §5 and in ch. V.

For any $a \in \mathbf{R}$, $\sigma^2 > 0$, the density ϱ_{a,σ^2} defines the so-called "normal distribution $N(a, \sigma^2)$ with expected value a and variance σ^2". In fact, the

x	0,00	0,02	0,03	0,04	0,05	0,06	0,06	0,07	0,08	0,09
0,0	0,5000	0,5040	0,5080	0,5120	0,5160	0,5199	0,5239	0,5279	0,5319	0,5359
0,1	0,5398	0,5438	0,5478	0,5517	0,5557	0,5596	0,5636	0,5675	0,5714	0,5753
0,2	0,5793	0,5832	0,5871	0,5910	0,5948	0,5987	0,6026	0,6064	0,6103	0,6141
0,3	0,6179	0,6217	0,6255	0,6293	0,6331	0,6368	0,6406	0,6443	0,6480	0,6517
0,4	0,6554	0,6591	0,6628	0,6664	0,6700	0,6736	0,6772	0,6808	0,6844	0,6879
0,5	0,6915	0,6950	0,6985	0,7019	0,7054	0,7088	0,7123	0,7157	0,7180	0,7224
0,6	0,7257	0,7291	0,7324	0,7357	0,7389	0,7422	0,7454	0,7486	0,7517	0,7549
0,7	0,7580	0,7611	0,7642	0,7673	0,7704	0,7734	0,7764	0,7794	0,7823	0,7852
0,8	0,7881	0,7910	0,7939	0,7967	0,7995	0,8023	0,8051	0,8078	0,8106	0,8133
0,9	0,8159	0,8186	0,8212	0,8238	0,8264	0,8289	0,8315	0,8340	0,8365	0,8389
1,0	0,8413	0,8438	0,8461	0,8485	0,8508	0,8531	0,8554	0,8577	0,8599	0,8621
1,1	0,8643	0,8665	0,8686	0,8708	0,8729	0,8749	0,8770	0,8790	0,8810	0,8830
1,2	0,8849	0,8869	0,8888	0,8907	0,8925	0,8944	0,8962	0,8980	0,8997	0,9015
1,3	0,9032	0,9049	0,9066	0,9082	0,9099	0,9115	0,9131	0,9147	0,9162	0,9177
1,4	0,9192	0,9207	0,9222	0,9236	0,9251	0,9265	0,9279	0,9292	0,9306	0,9319
1,5	0,9332	0,9345	0,9357	0,9370	0,9382	0,9394	0,9406	0,9418	0,9429	0,9441
1,6	0,9452	0,9463	0,9474	0,9484	0,9495	0,9505	0,9515	0,9525	0,9535	0,9545
1,7	0,9554	0,9564	0,9573	0,9582	0,9591	0,9599	0,9608	0,9616	0,9625	0,9633
1,8	0,9641	0,9649	0,9656	0,9664	0,9671	0,9678	0,9686	0,9693	0,9699	0,9706
1,9	0,9713	0,9719	0,9726	0,9732	0,9738	0,9744	0,9750	0,9756	0,9761	0,9767
2,0	0,9772	0,9778	0,9783	0,9788	0,9793	0,9798	0,9803	0,9808	0,9812	0,9817
2,1	0,9821	0,9826	0,9830	0,9834	0,9838	0,9842	0,9846	0,9850	0,9854	0,9857
2,2	0,9861	0,9864	0,9868	0,9871	0,9875	0,9878	0,9881	0,9884	0,9887	0,9890
2,3	0,9893	0,9896	0,9898	0,9901	0,9904	0,9906	0,9909	0,9911	0,9913	0,9916
2,4	0,9918	0,9920	0,9922	0,9925	0,9927	0,9929	0,9931	0,9932	0,9934	0,9936
2,5	0,9938	0,9940	0,9941	0,9943	0,9945	0,9946	0,9948	0,9949	0,9951	0,9952
2,6	0,9953	0,9955	0,9956	0,9957	0,9959	0,9960	0,9961	0,9962	0,9963	0,9964
2,7	0,9965	0,9966	0,9967	0,9968	0,9969	0,9970	0,9971	0,9972	0,9973	0,9974
2,8	0,9974	0,9975	0,9976	0,9977	0,9977	0,9978	0,9979	0,9979	0,9980	0,9981
2,9	0,9981	0,9982	0,9982	0,9983	0,9984	0,9984	0,9985	0,9985	0,9986	0,9986
3,0	0,9987	0,9987	0,9987	0,9988	0,9988	0,9989	0,9989	0,9989	0,9990	0,9990

continuous analogues of barycenter and variance, as considered in §2, are $\int_{-\infty}^{\infty} x\varrho_{a,\sigma^2}(x)dx$ and $\int_{-\infty}^{\infty}\left[x-\int y\varrho_{a,\sigma^2}(y)dy\right]^2 \varrho_{a,\sigma^2}(x)dx$. The reader is invited to evaluate these two integrals

Exercise 4.3. Prove, for any $a \in \mathbf{R}$, $\sigma^2 > 0$

$$\int_{-\infty}^{\infty} x\varrho_{a,\sigma^2}(x)dx = a$$

$$\int_{-\infty}^{\infty} (x-a)^2 \varrho_{a,\sigma^2}(x)dx = \sigma^2$$

4.2. de Moivre-Laplace's Theorem.

can now be stated as

Theorem 4.4. Let $b^{(n)}$ be the binomial distribution no. n with parameter value $p \in]0,1[$, set on the points $x_{nk} = \frac{k-np}{\sqrt{np(1-p)}}$ $(k = 0,1,\ldots,n)$. That is, $b^{(n)}$ is the DPD defined on $\mathbf{R}$ by

$$b_x^{(n)} = \begin{cases} \binom{n}{k} p^k (1-p)^{n-k} & \text{if } x = \frac{k-np}{\sqrt{np(1-p)}} \quad (k = 0,1,\ldots,n) \\ 0 & \text{otherwise.} \end{cases}$$

Then for any $-\infty < a < b < \infty$

$$\lim_{n\to\infty} b^{(n)}([a,b]) = \frac{1}{\sqrt{2\pi}} \int_a^b \varrho(x)dx.$$

This theorem states sort of a convergence of the sequence $b^{(1)}, b^{(2)}, \ldots$ of DPDs towards the continuous probability distribution $N(0,1)$: for any finite closed interval $[a,b]$ the values

$$b^{(n)}([a,b,]) = \sum_{a \le \frac{k-np}{\sqrt{np(1-p)}} \le b} \binom{n}{k} p^k (1-p)^{n-k}$$

tend to the value which is given by $N(0,1)$ to that same interval. If we visualize $b^{(n)}$ by plotting a vertical strip of width $\frac{1}{\sqrt{np(1-p)}}$ with base interval

$$\left[\frac{k-np}{\sqrt{np(1-p)}} - \frac{1}{2\sqrt{np(1-p)}}, \frac{k-np}{\sqrt{np(1-p)}} + \frac{1}{2\sqrt{np(1-p)}}\right]$$

and height $\sqrt{np(1-p)}\binom{n}{k}p^k(1-p)^{n-k}$ (so that the area of that strip is $\binom{n}{k}p^k(1-p)^{n-k}$ then the resulting roughly bell-shaped plane figure which is the union of all those strips, tends in an obvious geometric sense to the plane set under the bell-shaped curve $\varrho(x) = \frac{1}{\sqrt{2\pi}} e^{-\frac{x^2}{2}}$. Now, this $b^{(n)}$ is nothing than the distribution of a suitable modification of a sum of n IID 0-1-valued RVs $f_1, \ldots, f_n$ (on some DPS) which attain value 1 with probability p and value 0 with probability $1-p$ $(0 < p < 1)$; the modification is such that $f_1 + \ldots + f_n$, which has expectation np and variance $np(1-p)$ (see e.g. exercise 2.10) is normalized to expectation 0 and variance 1:

$$\frac{f_1 + \ldots + f_n - np}{\sqrt{np(1-p)}}$$

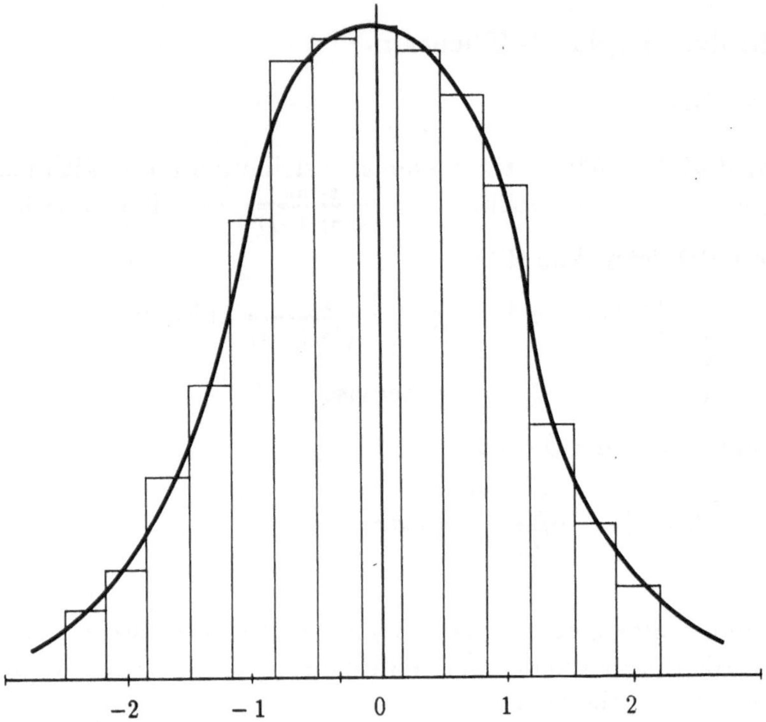

fig. IV.4.2

Clearly this RV has the distribution $b^{(n)}$, and the normalization is nothing but the precise realization of what we had, in the introduction to this section, called intuitively "looking through a magnifying glass". Our theorem states that the "shape" which we see becomes $N(0,1)$ in the limit $n \to \infty$.

PROOF. **of theorem 4.4.**

1) We employ Stirling's formula in the version

$$n! = \sqrt{2\pi n}\left(\frac{n}{e}\right)^n Q(n) \quad (n = 1, 2, \ldots)$$

with $\lim_{n\to\infty} Q(n) = 1$

(one could e.g. prove $Q(n) = e^{q(n)}$ with $\frac{1}{12n+1} < q(n) < \frac{1}{12n}$). We calculate (setting $\bar{p} = 1 - p$)

$$\begin{aligned} \binom{n}{k} p^k \bar{p}^{n-k} &= \frac{n!}{k!(n-k)!} p^k \bar{p}^{n-k} \\ &= \frac{1}{\sqrt{2\pi}} \sqrt{\frac{n}{k(n-k)}} \left(\frac{np}{k}\right)^k \left(\frac{n\bar{p}}{n-k}\right)^{n-k} \frac{Q(n)}{Q(k)Q(n-k)} \end{aligned}$$

2) In order to get a smoother notation, we shall write, for any two sequences a_n, b_n $(n = 1, 2, \ldots)$ of reals > 0,

$$a_n \sim b_n \text{ iff } \frac{a_n}{b_n} \to 1.$$

3) Let us now restrict attention to values $0 \leq k = k_n \leq n$ such that the sequence $k_1, k_2, \ldots$ fulfils $\frac{k_n}{n} \to p$ and hence $\frac{n-k_n}{n} \to \bar{p}$. As $0 < p < 1$, this is tantamount to

$$k_n \sim np,\ n - k_n \sim n\bar{p},$$

and entails

$$\sqrt{\frac{n}{k_n(n-k_n)}} \sim \sqrt{\frac{n}{npn\bar{p}}} = \frac{1}{\sqrt{n\sigma^2}}$$

where

$$\sigma^2 = p\bar{p} = p(1-p)$$

denotes the variance of f_1.

4) The result of 1) may now be reformulated as

$$\binom{n}{k} p^k \bar{p}^{n-k} \sim \frac{1}{\sqrt{2\pi}} \frac{1}{\sqrt{n\sigma^2}} \left(\frac{np}{k}\right)^k \left(\frac{n\bar{p}}{n-k}\right)^{n-k} \qquad (k = k_n)$$

The terms inside the brackets are ~ 1 each, but this need not hold any more after passage to the powers under consideration. We will investigate the behavior of these terms by passage to logarithms:

5)
$$\log\left[\left(\frac{np}{k}\right)^k \left(\frac{n\bar{p}}{n-k}\right)^{n-k}\right]$$

$$= -n\left[\frac{k}{n}\log\frac{k}{np} + \left(1 - \frac{k}{n}\right)\log\left(\frac{1-\frac{k}{n}}{\bar{p}}\right)\right]$$

With the understanding $k = k_n$, $\frac{k}{n} \to p$ $(\Longrightarrow 1 - \frac{k}{n} \to \bar{p})$, we see the term inside the bracket tending to 0, but this may be outweighed by the factor $-n$ outside the bracket. Let us therefore have a closer look on the function

$$g(x) = x\log\frac{x}{p} + (1-x)\log\frac{1-x}{\bar{p}}$$

near $x = p$:

$$\begin{aligned} g(p) &= 0 \\ g'(p) &= 0 \\ g''(p) &= \tfrac{1}{p} + \tfrac{1}{\bar{p}} = \tfrac{1}{p\bar{p}} \end{aligned}$$

(prove it!). Taylor's formula yields

$$(5) \quad g(x) = \frac{(x-p)^2}{2p\bar{p}} + r(x-p)$$

with $|r(x-p)| \leq C|x-p|^3$ for a suitable constant $C > 0$. If we sharpen our hypothesis

$$\frac{k_n}{n} \to p$$

by

(6) $$n|\frac{k_n}{n} - p|^3 \to 0,$$

we obtain, putting $x = \frac{k}{n} = \frac{k_n}{n}$ in (3)

$$\lim_n \left(-n \left[g\left(\frac{k}{n}\right) - \frac{(\frac{k}{n}-p)^2}{2p\bar{p}} \right] \right) = 0$$

i.e., as $p\bar{p} = \sigma^2$

$$\lim_n \left\{ \log \left[\left(\frac{np}{k}\right)^k \left(\frac{n\bar{p}}{n-k}\right)^{n-k} \right] + n\frac{(\frac{k}{n}-p)^2}{2\sigma^2} \right\} = 0$$

Taking exp of this, we arrive at

$$\left(\frac{np}{k}\right)^k \left(\frac{n\bar{p}}{n-k}\right)^{n-k} \sim e^{-\frac{\left(\frac{k-np}{\sqrt{n\sigma^2}}\right)^2}{2}}$$

(assuming (4))

6) If we put this together with 3), we finally get

(7) $$\binom{n}{k} p^k \bar{p}^{n-k} \sim \frac{1}{\sqrt{2\pi}} \frac{1}{\sqrt{n\sigma^2}} e^{-\frac{\left(\frac{k-np}{\sqrt{n\sigma^2}}\right)^2}{2}}$$

(assuming (2))

This statement is often called the "local limit theorem", de Moivre-Laplace's version.

7) We have now all ingredients for a proof of the conclusion of our theorem: the right hand side of (5) is the area of a vertical strip of width $\frac{1}{\sqrt{n\sigma^2}} = \frac{1}{\sqrt{np(1-p)}}$ centered at $x = x_{nk} = \frac{k-np}{\sqrt{n\sigma^2}}$, and with height $\frac{1}{\sqrt{2\pi}} e^{-\frac{x^2}{2}}$. Summing these areas for all x_{nk} with $a \leq x_{nk} \leq b$, we obtain, up to negligible (for $n \to \infty$) "border effects" at a and b, a Riemann sum for $\frac{1}{\sqrt{2\pi}} \int_a^b e^{-\frac{x^2}{2}} dx$. The only thing about which we still have to make sure is

$$a \leq \frac{k_n - np}{\sqrt{n\sigma^2}} \leq b \Longrightarrow n\left(\frac{k_n}{n} - p\right)^3 \to 0,$$

which is, however, obviously true as $n|\frac{k_n}{n} - p|^3 = \frac{1}{\sqrt{n}} |\frac{k_n - np}{\sqrt{n\sigma^2}}|^3 \cdot \sigma^3$

□

In view of the practical importance of theorem 4.4., we display a few numerical results here, after Meyer [1971].

k	$n=8,\ p=0{,}2$		$n=8,\ p=0{,}5$		$n=25,\ p=0{,}2$	
	Approximation	Exakt	Approximation	Exakt	Approximation	Exakt
0	0,130	0,168	0,005	0,004	0,009	0,004
1	0,306	0,336	0,030	0,031	0,027	0,024
2	0,331	0,294	0,104	0,109	0,065	0,071
3	0,164	0,147	0,220	0,219	0,121	0,136
4	0,037	0,046	0,282	0,273	0,176	0,187
5	0,004	0,009	0,220	0,219	0,199	0,196
6	0+	0,001	0,104	0,109	0,176	0,163
7	0+	0+	0,030	0,031	0,121	0,111
8	0+	0+	0,005	0,004	0,065	0,062
9					0,027	0,029
10					0,009	0,012
11					0,002	0,004

Exercise 4.5. Prove that the convergence in theorem 4.4. holds uniformly for all $-\infty < a < b < \infty$.

4.3. A parallel: the Poisson approximation.

De Moivre's CLT smoothes the whole of the binomial distribution $b^{(n)}$ into Gauss' bell-shaped curve, for $n \to \infty$. There is another approximation, named after Siméon Denis Poisson (1781-1840) which displays an increasing similarity of the upper (i.e. right) tail of $b^{(n)}$ to the Poisson distribution with parameter $\lambda > 0$ as $n \to \infty$ and $p = p_n \to 0$ such that $np_n \to \lambda$. The proof is much simpler than the proof of the Moivre's theorem.

Theorem 4.4 (Poisson approximation). Let $0 < p_1, p_2, \ldots < 1$ be such that

$$\lim_{n\to\infty} np_n = \lambda > 0$$

Then

$$\lim_{n\to\infty} \binom{n}{k} p_n^k (1-p_n)^k = \frac{\lambda^k}{k!} e^{-\lambda} \quad (k = 0, 1, \ldots)$$

PROOF. Writing p instead of p_n for short, we have

$$\binom{n}{k} p^k (1-p)^{n-k} = \frac{(np)^k}{k!} \cdot (1 - \frac{np}{n})^n \cdot \frac{n(n-1)\ldots(n-k+1)}{n^k} (1 - \frac{np}{n})^{-k}$$
$$\to \frac{\lambda^k}{k!} e^{-\lambda}$$

as $n \to \infty$, $np \to \lambda$. □

A numerical comparison (after Krickeberg-Ziezold [1979]) for $n = 600$, $p = \frac{1}{365}$, $\lambda = np \approx 1.64384$ shows the following values:

k	0	1	2	3	4	5	6	7
$\binom{n}{k}p^k(1-p)^{n-k}$	0.1928	0.3178	0.2616	0.1432	0.0587	0.0192	0.0052	0.0012
$\frac{\lambda^k}{k!}e^{-\lambda}$	0.1932	0.3177	0.2611	0.1431	0.0588	0.0193	0.0053	0.0012

5. The Central Limit Theorem (CLT) II: Lindeberg-Trotter's Version

Our proof of theorem 4.4. makes use of tools closely adapted to the special situation envisaged there. The reader might therefore surmise that generalizations to other situations be utterly difficult if not impossible. This is, however, only true for the method of proof, not for the final result. In this section, we shall prove a very general CLT which largely goes back to Lindeberg [1922]; our proof, due to Trotter [1959], would easily carry over to general measure-theoretical stochastics, but is meaningful and interesting also in our DPS framework. Before formulating the theorem, we set the stage in

5.1. Normalized schemes.

formulate the intuitive idea of a sequence of sums of independent real RVs normalized to expectation 0 and variance 1:

Definition 5.1. Let (Ω, m) be a DPS and $f_1, \ldots, f_r$ independent real RVs on Ω such that

$$E_m(f_j) = 0 \qquad (j = 1, \ldots, r)$$

$$\sum_{j=1}^{r} \sigma_m^2(f_j) = 1$$

Then we call the r-tuple of RVs $f_1, \ldots, f_r$ a *normalized row* of real RVs.

We will later formulate the CLT in a way involving DPDs in $\mathbf{R}$ only. So let us see what a normalized row of real RVs yields distributionswise:

let $p^{(j)} = f_j m$ be the distribution of f_j under m. Then, by theorem 1.1, the distribution p of $f_1 + \ldots + f_r$ is

$$p = p^{(1)} * \ldots * p^{(r)},$$

where $*$ means convolution. All of the DPDs $p, p^{(1)}, \ldots, p^{(r)}$ in $\mathbf{R}$ have

$$\begin{aligned} E_{p^{(j)}}(x) &= E_m(f_j) &= 0 \qquad (j = 1, \ldots, r) \\ E_p(x) &= E_m(\textstyle\sum_j f_j) &= 0 \end{aligned}$$

and their variances

$$\sigma_j^2 = E_{p(j)}(x^2) = E_m(f_j^2)$$

sum up to 1. Let us therefore make the

Definition 5.2. 1) Let $p^{(1)}, \ldots, p^{(r)}$ be DPDs in $\mathbf{R}$ such that

$$E_{p^{(j)}}(x) = \sum_{x \in \mathbf{R}} x p_x^{(j)} = 0$$

holds and the variances

$$\sigma_j^2 = \sigma_{p^{(j)}}^2 = \sum_{x \in \mathbf{R}} x^2 p_x^{(j)}$$

fulfil

$$\sigma_1^2 + \ldots + \sigma_r^2 = 1.$$

Then $p^{(1)}, \ldots, p^{(r)}$ is called a *normalized row* of DPDs in $\mathbf{R}$.
2) A family

$$\left(p^{(n,j)}\right)_{\substack{n=1,2,\ldots \\ j=1,\ldots,r_n}}$$

of DPDs in $\mathbf{R}$ such that for every $n = 1, 2, \ldots$

$$p^{(n,1)}, \ldots, p^{(n,r_n)}$$

is a normalized row, is called a *normalized scheme* of DPDs in $\mathbf{R}$.

This definition fits perfectly into the framework of discrete stochastics. If we would try to define normalized schemes of random *variables* along the lines of definition 5.2., we would encounter difficulties with discreteness similar to those hinted at in ch. III §3. We will thus, in the sequel, work with normalized schemes of DPDs only.

The CLT which we shall formulate and prove later on, will tell us that for any normalized scheme fulfilling a certain additional condition, the DPDs $p^{(n)} = p^{(n,1)} * \ldots * p^{(n,r_n)}$ tend to $N(0,1)$ as $n \to \infty$, in a sense which we shall specify in the next subsection.

Before geoing ahead, let us focus attention, for a moment, to the following special situation:

If $f_1, \ldots, f_n$ are IID real RVs such that the distribution p of f_1 (which is also the distribution of $f_2, \ldots, f_n$) has

(1) expectation $\sum_{x \in \mathbf{R}} x p_x = 0$

(2) variance $\sum_{x \in \mathbf{R}} x^2 p_x = 1,$

we get a normalized row of RVs by passing to $\frac{1}{\sqrt{n}} f_1, \ldots, \frac{1}{\sqrt{n}} f_n$. For the corresponding distributions, this means transport, within $\mathbf{R}$, by the mapping $T_n : x \to \frac{x}{\sqrt{n}}$. Thus, however we give a DPD p in $\mathbf{R}$ fulfilling (1), (2), we get a normalized scheme of DPDs by taking

$$\begin{aligned} r_n &= n \\ T_n : x &\to \frac{x}{\sqrt{n}} \\ p^{(n,k)} &= T_n p \qquad (k = 1, \ldots, n) \end{aligned}$$

$(n = 1, 2, \ldots)$. We shall call this the *standard situation* for short.

For technical reasons we will also form "normalized schemes" of normal distributions later on.

5.2. Weak Convergence of DPDs in R.

For every DPD p in $\mathbf{R}$ and for every bounded real function f on $\mathbf{R}$ the expectation $E_p(f) = \sum_{x \in \mathbf{R}} f(x) p_x$ makes sense.

Definition 5.3. Let $p, p^{(1)}, p^{(2)}$, be DPDs in $\mathbf{R}$. We shall say that the sequence $p^{(1)}, p^{(n)}, \ldots$ tends or converges to p *weakly*, in symbols

$$\lim_{n \to \infty} p^{(n)} = p \quad \text{(weakly)}$$

if

$$\lim_{n \to \infty} E_{p^{(n)}}(f) = E_p(f)$$

holds for every *bounded continuous* real function f on $\mathbf{R}$.

Let us display a few criteria for weak convergence:

Proposition 5.4. Let $p, p^{(1)}, p^{(2)}, \ldots$ be DPDs on $\mathbf{R}$. Then the following statements are equivalent:

1) $\lim_{n\to\infty} p^{(n)}([a,b[) = p([a,b[) \quad (-\infty < a < b < \infty, p_a = 0 = p_b)$
2) $\lim_{n\to\infty} p^{(n)} = p \quad$ weakly
3) $\lim_{n\to\infty} E_{p^{(n)}}(f) = E_p(f)$ for every bounded real function f on **R** which has a bounded continuous derivative of order 2 everywhere on **R**.

PROOF. 1) $\Longrightarrow$ 2): let f be bounded and continuous on **R**. Let $K > 0$ be such that $|f(x)| \le K \quad (x \in \mathbf{R})$. As $\lim_{N\to\infty} p([-N,N]) = p(\mathbf{R}) = 1$, we may choose for any given $\epsilon > 0$, a real (not necessarily integer) number N in such a fashion that $p_{-N} = 0 = p_N$

$$Kp(]-\infty,-N[) + Kp(]N,\infty[) < \frac{\epsilon}{8}.$$

As f is uniformly continuous on $[-N,N]$, we may choose $-N = a_0 < a_1 < \ldots < a_s = N$ in such a fashion that, $p_{a_0} = \ldots = p_{a_s} = 0$ and, with $b_k = \inf\{f(x) | a_{k-1} \le x \le a_k\} \quad (k = 1,\ldots,s)$ the step function

$$g(x) = \sum_{k=1}^{s} b_k 1_{[a_{k-1},a_k[}$$

fulfils

$$|f(x) - g(x)| < \frac{\epsilon}{16KN} \quad (-N \le x < N)$$

and hence

$$|E_p(f) - E_p(g)| < \frac{\epsilon}{4}$$

From 1) we now see

$$\text{(3)} \quad \lim_{n\to\infty} p^{(n)}([a_{k-1},a_k[) = p([a_{k-1},a_k[)$$

$$\lim_{n\to\infty} p^{(n)}([-N,N[= p([-N,N[) > 1 - \frac{\epsilon}{8K}$$

Fix n_0 such that $n \ge n_0 \Longrightarrow p^{(n)}([-N,N[) > 1 - \frac{\epsilon}{8K}$. Then $n \ge n_0$ also implies

$$|E_{p^{(n)}}(f) - E_{p^{(n)}}(g)| < \frac{\epsilon}{4}$$

From (3) we infer

$$\lim_{n\to\infty} E_{p^{(n)}}(g) = E_p(g)$$

Combining this with the above approximations we obtain

$$|E_{p^{(n)}}(f) - E_p(f)| < \epsilon$$

for $n \geq n_0$ sufficiently large. As $\epsilon > 0$ was arbitrary, 2) follows.
2) $\Rightarrow$ 3) is trivial.

3) $\Rightarrow$ 1). For any $-\infty < a < b < \infty$ such that $p_a = 0 = p_b$, and for any $\epsilon > 0$ we may find two functions f, g with the properties listed in 3) such that

$$f \leq 1_{[a,b[} \leq g$$

and $(0 \leq) E_p(g) - E_p(f) < \frac{\epsilon}{2}$. By 3) we may find $n_0 > 0$ such that $n \geq n_0$ implies

$$|E_{p^{(n)}}(f) - E_p(f)| < \frac{\epsilon}{4}$$

$$|E_{p^{(n)}}(g) - E_p(g)| < \frac{\epsilon}{4}$$

From this and

$$E_p(f) \leq p([a, b[\leq E_{p^{(n)}}(g) \leq E_p(g) + \frac{\epsilon}{4}$$

$$E_{p^{(n)}}(f) \leq p^{(n)}([a, b[) \leq E_{p^{(n)}}(g) \leq E_p(g) + \frac{\epsilon}{4}$$

we now infer

$$|p^{(n)}([a, b[) - p([a, b[)| < \epsilon \quad (n \geq n_0),$$

which was to be proved. □

We still need the analogues of all this, with p replaced by $N(0, 1)$.

Definition 5.5. A sequence $p^{(1)}, p^{(2)}, \ldots$ of DPDs in $\mathbf{R}$ is said to *converge weakly* towards $N(0, 1)$ in symbols

$$\lim_{n \to \infty} p^{(n)} = N(0, 1) \quad \text{(weakly)}$$

if for every bounded continuous real function f on $\mathbf{R}$

$$\lim_{n \to \infty} E_{p^{(n)}}(f) = \frac{1}{\sqrt{2\pi}} \int_{-\infty}^{\infty} f(x) e^{-\frac{x^2}{2}} dx$$

holds.

Observe that the integral makes sense as f is bounded and continuous.

Exercise 5.6. Formulate and prove the analogon to proposition 5.4., for weak convergence to $N(0, 1)$ (observe that assumptions like $p_a = 0$ are obsolete here).

5.3. Convolution of Normal Distributions.

For the proof of our CLT we will need analogues, for normal distributions, of certain operations which we have already introduced for DPDs. Let $N(0,\sigma^2)$ denote the normal distribution with expectation 0 and variance $\sigma^2 > 0$. It is given by the "density" function

$$\varrho_{0,\sigma^2}(x) = \frac{1}{\sqrt{2\pi\sigma^2}} e^{-\frac{x^2}{2\sigma^2}}.$$

In exercise 4.3 the reader has shown

$$\begin{aligned}
\int_{-\infty}^{\infty} \varrho_{0,\sigma^2}(x)dx &= 1 \\
\int_{-\infty}^{\infty} x\, \varrho_{0,\sigma^2}(x)dx &= 0 \\
\int_{-\infty}^{\infty} x^2\, \varrho_{0,\sigma^2}(x)dx &= \sigma^2
\end{aligned}$$

We will now prove that convolutions – here defined by integrals – of densities of the form $\varrho_{0,\sigma^2}(x)$ result in densities of the same class, with variances added:

Proposition 5.7. For any $\sigma^2, \tau^2 > 0$

$$\begin{aligned}
&\int_{-\infty}^{\infty} \varrho_{0,\sigma^2}(x-u)\varrho_{0,\tau^2}(u)du \\
= &\int_{-\infty}^{\infty} \varrho_{0,\sigma^2}(x+u)\varrho_{0,\tau^2}(u)du \\
= &\ \varrho_{0,\sigma^2+\tau^2}(x) \qquad\qquad (x \in \mathbf{R})
\end{aligned}$$

PROOF. As soon as the second equality is established, commutativity follows for the convolution considered here, and the first equality results by this, by substitution $u \to u + x$, and by the symmetry $\varrho_{0,\sigma^2}(-u) = \varrho_{0,\sigma^2}(u)$. Let us now prove the second equality:

$$\begin{aligned}
&\int_{-\infty}^{\infty} \varrho_{0,\sigma^2}(x+u)\varrho_{0,\tau^2}(u)du \\
= &\ \frac{1}{2\pi\sqrt{\sigma^2\tau^2}} \int_{-\infty}^{\infty} e^{-\frac{(x+u)^2}{2\sigma^2}}\, e^{-\frac{u^2}{2\tau^2}} du \\
= &\ \sqrt{\frac{\sigma^2+\tau^2}{2\pi\sigma^2\tau^2}} \cdot \frac{1}{\sqrt{2\pi(\sigma^2+\tau^2)}} \int_{-\infty}^{\infty} e^{-\frac{1}{2}\left[\frac{(x+u)^2}{\sigma^2}+\frac{u^2}{\tau^2}\right]} du
\end{aligned}$$

Let us recalculate the term [...] a bit differently:

$$\frac{(x+u)^2}{\sigma^2} + \frac{u^2}{\tau^2} = \frac{x^2+2xu+u^2}{\sigma^2} + \frac{u^2}{\tau^2}$$

$$
\begin{aligned}
&= \tfrac{1}{\sigma^2+\tau^2}\left\{\tfrac{\sigma^2+\tau^2}{\sigma^2}x^2 + 2\tfrac{\sigma^2+\tau^2}{\sigma^2}xu + \tfrac{(\sigma^2+\tau^2)^2}{\sigma^2\tau^2}u^2\right\}\\
&= \tfrac{x^2}{\sigma^2+\tau^2} + \tfrac{1}{(\sigma^2+\tau^2)\sigma^2\tau^2}\cdot\\
&\qquad \cdot\left\{x^2\tau^4 + 2x\tau^2u(\sigma^2+\tau^2) + (\sigma^2+\tau^2)^2u^2\right\}\\
&= \tfrac{x^2}{\sigma^2+\tau^2} + \tfrac{\sigma^2+\tau^2}{\sigma^2\tau^2}\left\{u^2 + 2u\tfrac{x\tau^2}{\sigma^2+\tau^2} + \left(\tfrac{x\tau^2}{\sigma^2+\tau^2}\right)^2\right\}\\
&= \tfrac{x^2}{\sigma^2+\tau^2} + \tfrac{1}{\frac{\sigma^2\tau^2}{\sigma^2+\tau^2}}\left(u + \tfrac{x\tau^2}{\sigma^2+\tau^2}\right)^2
\end{aligned}
$$

The above integral thus takes the form

$$
= \frac{1}{\sqrt{2\pi(\sigma^2+\tau^2)}}e^{-\frac{x^2}{2(\sigma^2+\tau^2)}}\frac{1}{\sqrt{2\pi\frac{\sigma^2\tau^2}{\sigma^2+\tau^2}}}\int_{-\infty}^{\infty} e^{-\frac{\left(u+\frac{x\tau^2}{\sigma^2+\tau^2}\right)^2}{2\frac{\sigma^2\tau^2}{\sigma^2+\tau^2}}}du
$$

Substitution $u \to u - \frac{x\tau^2}{\sigma^2+\tau^2}$ makes this

$$
\begin{aligned}
&= \varrho_{0,\sigma^2+\tau^2}(x)\int_{-\infty}^{\infty}\varrho_{0,\frac{\sigma^2\tau^2}{\sigma^2+\tau^2}}(u)du\\
&= \varrho_{0,\sigma^2+\tau^2}(x)
\end{aligned}
$$

as was to be shown. □

5.4. Convolution Operators.

Our proof of the CLT still needs one more technical preparation:

Definition 5.8.

1) For every bounded function f on $\mathbf{R}$, define
$$\|f\| = \sup_{x\in\mathbf{R}} |f(x)|$$
– the so-called *sup norm*, often also denoted by $\|\cdot\|_\infty$, defining the topology of uniform convergence of functions on $\mathbf{R}$.
2) For every DPD p on $\mathbf{R}$ and every bounded real function f on $\mathbf{R}$, define
$$(R_p f)(x) = \sum_{y\in\mathbf{R}} f(x+u)p_u \quad (x \in \mathbf{R})$$
3) For every $\sigma^2 > 0$ and every bounded continuous function f on $\mathbf{R}$, define
$$(R_{\sigma^2} f)(x) = \int_{-\infty}^{\infty} f(x+u)\varrho_{0,\sigma^2}(u)du \quad (x \in \mathbf{R})$$

Observe that all sums and integrals occurring in this definition are meaningful and yield finite values: We shall call

$$\begin{aligned} R_p : f &\rightarrow R_p f \\ R_{\sigma^2} : f &\rightarrow R_{\sigma^2} f \end{aligned}$$

the *convolution operators* associated with p resp. $N(0, \sigma^2)$.

Proposition 5.9. Let H be the real vector space of all bounded uniformly continuous real functions on $\mathbf{R}$, endowed with the sup norm $\|\cdot\| = \|\cdot\|_\infty$. Then

1) Every convolution operator $R = R_p$ or R_{σ^2} is a linear positive contraction on H, i.e.

$$\begin{aligned} &RH \subseteq H \\ &R \text{ is linear} \\ &f \geq 0 \Rightarrow Rf \geq 0 \quad (f \in H) \\ &\|Rf\| \leq \|f\| \qquad (f \in H) \end{aligned}$$

2) All our convolution operators commute:

$$\begin{aligned} R_p R_q &= R_q R_p & &(p, q \text{ DPDs on } \mathbf{R}) \\ R_{\sigma^2} R_{\tau^2} &= R_{\tau^2} T_{\sigma^2} & &(\sigma^2, \tau^2 > 0) \\ R_p R_{\sigma^2} &= R_{\sigma^2} R_p & &(p \text{ a DPD on } \mathbf{R}, \sigma^2 > 0) \end{aligned}$$

3)
$$\begin{aligned} R_p R_q &= R_{p*q} & &(p, q \text{ DPDs on } \mathbf{R}) \\ R_{\sigma^2} R_{\tau^2} &= R_{\sigma^2+\tau^2} & &(\sigma^2, \tau^2 > 0) \end{aligned}$$

PROOF.

1) follows easily from the observation that R_p and R_{σ^2} perform weighted averagings over values of f in order to produce the values of $R_p f$ resp. $R_{\sigma^2} f$.
2) For every $f \in H$ we have

$$\begin{aligned} (R_p R_q f)(x) &= \textstyle\sum_{u \in \mathbf{R}} \left[\sum_{v \in \mathbf{R}} f((x+u)+v) q_v\right] p_u \\ &= \textstyle\sum_{w \in \mathbf{R}} f(x+w) \sum_{u+v=w} p_u q_v \\ &= \textstyle\sum_{w \in \mathbf{R}} f(x+p)(p \times q)_w \\ &= (R_{p \times q} f)(x) \qquad (x \in \mathbf{R}). \end{aligned}$$

Similarly

$$\begin{aligned}(R_{\sigma^2}R_{\tau^2}f)(x) &= \int_{-\infty}^{\infty}\left[\int_{-\infty}^{\infty} f((x+u)+v)\varrho_{0,\tau^2}(v)dv\right]\varrho_{0,\sigma^2}(u)du \\ &= \int_{-\infty}^{\infty}\left[\int_{-\infty}^{\infty} f(x+w)\varrho_{0,\tau^2}(w-u)dw\right]\varrho_{0,\sigma^2}(u)du \\ &= \int_{-\infty}^{\infty} f(x+w)\left[\int \varrho_{0,\tau^2}(w-u)\varrho_{0,\sigma^2}(u)du\right]dw \\ &= \int_{-\infty}^{\infty} f(x+w)\varrho_{0,\sigma^2+\tau^2}(w)dw \\ &= (R_{\sigma^2+\tau^2}f)(x) \qquad (x\in\mathbf{R})\end{aligned}$$

follows from proposition 5.9.

3) $p*q=q*p$ for arbitrary DPDs p,q in $\mathbf{R}$ and $\sigma^2+\tau^2=\tau^2+\sigma^2$ ($\sigma^2,\tau^2>0$) now prove already the first two (the "pure") commutativity formulas. In order to also prove the last (the "mixed") one, we calculate

$$\begin{aligned}(R_{\sigma^2}R_p)(x) &= \int_{-\infty}^{\infty}\left[\sum_{v\in\mathbf{R}} f((x+u)+v)p_v\right]\varrho_{0,\sigma^2}(u)du \\ &= \sum_{v\in\mathbf{R}}\left[\int_{-\infty}^{\infty} f((x+u)+v)\varrho_{0,\sigma^2}(u)du\right]p_v \\ &= (R_pR_{\sigma^2}f)(x) \qquad (x\in\mathbf{R})\end{aligned}$$

which coincides with the above expression up to a change of notation. It should be clear that all the above exchanges of integrations and summations are perfectly legal because f is bounded.

□

5.5. The Lindeberg [1922] Condition.

Definition 5.10. A normalized scheme

$$\left(p^{(n,k)}\right)_{\substack{n=1,2,\dots\\ k=1,\dots,r_n}}$$

of DPDs in $\mathbf{R}$ is said to fulfil the *Lindeberg [1922] condition* if

$$(L) \qquad \lim_{n\to\infty}\sum_{k=1}^{n}\left[\sum_{|x|\geq\epsilon} x^2p_x^{(n,k)}\right]=0 \qquad (\epsilon>0)$$

The idea behind this condition is that the $p^{(n,k)}$ concentrate their weights more and more near 0, variancewise, as $n\to\infty$. In fact

$$(L) \qquad \Longrightarrow \lim_{n\to\infty}\left[\max_{1\leq k\leq r_n}\sigma_{nk}^2\right]=0$$

follows easily: for every n, choose $1 \le k_n \le r_n$ such that $\sigma^2_{nk_n}$ is the maximum in question. We then have, for an arbitrary $\epsilon > 0$,

$$\begin{aligned} \max_{1\le k\le r_n} \sigma^2_{nk} &= \sigma^2_{nk_n} = \sum_{x\in \mathbf{R}} x^2 p_x^{(n,k_n)} \\ &= \sum_{|x|<\epsilon} x^2 p_x^{(n,k_n)} + \sum_{|x|\ge\epsilon} x^2 p_x^{(n,k_n)} \\ &\le \epsilon^2 + \sum_{k=1}^{r_n}\left[\sum_{|x|\ge\epsilon} x^2 p_x^{(n,k_n)}\right]. \end{aligned}$$

As (L) brings the last term to 0 in the limit for $n \to \infty$, and as $\epsilon > 0$ is arbitrary, the desired conclusion follows.

In the standard situation where a DPD p with expectation 0 and variance r is given in $\mathbf{R}$, and our normalized scheme is given by

$$\begin{aligned} &r_n = n \\ &T_n : x \to \tfrac{x}{\sqrt{n}}, \\ &p^{(n,k)} = T_n p, \end{aligned}$$

the Lindeberg condition prevails: for arbitrary $\epsilon > 0$

$$\begin{aligned} \textstyle\sum_{k=1}^n \left[\sum_{x\ge\epsilon} x^2 p_x^{(n,k)}\right] &= n \textstyle\sum_{|\frac{y}{\sqrt{n}}|\ge\epsilon} (\frac{y}{\sqrt{n}})^2 p_{\frac{y}{\sqrt{n}}} \\ &= \textstyle\sum_{|z|\ge\sqrt{n}\epsilon} z^2 p_z \\ &\to 0 \qquad (n \to \infty) \end{aligned}$$

Let us also consider the analogues of all this for normal distributions: given a "normalized schema"

$$(4) \qquad (\varrho_{0,\sigma^2_{nk}})_{\substack{n=1,2,\ldots \\ k=1,\ldots,r_n}}$$

of normal distributions – here represented by their densities – fulfilling

$$\text{expectations } \int_{-\infty}^{\infty} \varrho_{0,\sigma^2_{nk}}(x)dx = 0$$

$$\text{variances } \int_{-\infty}^{\infty} x^2 \varrho_{0,\sigma^2_{nk}}(x)dx = \sigma^2_{nk}$$

$$\sum_{k=1}^{n} \sigma^2_{nk} = 1 \qquad (n = 1, 2, \ldots),$$

we will say that it fulfils the Lindeberg condition if

$$(L) \quad \lim_{n\to\infty} \sum_{k=1}^{r_n} \int_{|x|\geq\epsilon} x^2 \varrho_{0,\sigma^2_{nk}}(x)dx = 0 \qquad (\epsilon > 0)$$

Now, here things are more uniform than in our above situation with DPDs, since all $\varrho_{(0,\sigma^2_{nk})}(x)$ are derived from $\varrho_{(0,1)}(x)$ by a bare modification of scale, to the effect that

$$\int_{x\geq\epsilon} x^2 \varrho_{0,\sigma^2_{nk}}(x)dx = \sigma^2_{nk} \int_{|y|\geq\epsilon/\sqrt{\sigma^2_{nk}}} y^2 \varrho_{0,1}(y)dy$$

and thus

$$\begin{aligned} \textstyle\sum_{k=1}^{v_n} \int_{|x|\geq\epsilon} x^2 \varrho_{0,\sigma^2_{nk}}(x)dx &\leq \textstyle\sum_{k=1}^{n} \sigma^2_{nk} \cdot \\ &\quad \cdot \left[\textstyle\int_{|y|\geq\epsilon/\max_{1\leq v\leq r_n}\sigma^2_{nj}} y^2 \varrho_{0,1}(y)dy\right] \\ &= \textstyle\int_{|y|\geq\epsilon/\max_{1\leq j\leq r_n}\sigma^2_{nj}} y^2 \varrho_{0,1}(y)dy \end{aligned}$$

We thus conclude: for a normalized scheme (4) of normal distributions

$$\lim_{n\to\infty} \left[\max_{1\leq k\leq r_n} \sigma^2_{nk}\right] = 0 \quad \Rightarrow \quad (L)$$

as $\epsilon/\max_{1\leq j\leq r_n}\sigma^2_{nj} \to \infty$ in that case.

5.6. The General CLT.

Theorem (CLT) 5.11. Let

$$\left(p^{(n,k)}\right)_{\substack{n=1,2,\ldots \\ k=1,\ldots,r_n}}$$

be a normalized scheme of DPDs in $\mathbf{R}$ fulfilling the Lindeberg condition

$$(L) \quad \lim_{n\to\infty} \sum_{k=1}^{n} \left[\sum_{|x|\geq\epsilon} x^2 p_x^{(n,k)}\right] = 0 \quad (\epsilon > 0)$$

Let $p^{(n)} = p^{(n,1)} * \ldots * p^{(n,r_n)}$ (convolution). Then

$$\lim_{n\to\infty} p^{(n)} = N(0,1) \quad \text{(weakly)}$$

PROOF. (after Trotter [1959]) 1) According to proposition 5.5. we may form a family

$$\left(\varrho_{0,\sigma^2_{nk}}\right)_{\substack{n=1,2,\ldots \\ k=1,\ldots,r_n}}$$

of densities of normal distributions which have the same variances σ_{nk}^2 as the $p^{(n,k)}$ from our given normalized scheme. For any bounded uniformly continuous real function f on $\mathbf{R}$ we obtain, setting $P_{nk} = R_{p^{(n,k)}}$, $Q_{nk} = R_{\sigma_{nk}^2}$ for short, and making use of the commutativities and contractivities established in proposition 5.7.

$$\begin{aligned}
& \|P_{n1}\dots P_{nr_n}f - Q_{n1}\dots Q_{nr_n}f\| \\
= \ & \|P_{n1}\dots P_{nr_n}f - P_{n1}\dots P_{n,r_n-1}Q_{nr_n}f \\
& +P_{n1}\dots P_{n,r_n-1}Q_{nr_n}f - P_{n1}\dots P_{n,r_n-2}Q_{n,r_n-1}Q_{nr_n}f \\
& \dots \\
& +P_{n1}Q_{n2}\dots Q_{nr_n}f - Q_{n1}Q_{n2}\dots Q_{nr_n}f\| \\
\leq \ & \|P_{n1}\dots P_{n,r_n-1}(P_{nr_n} - Q_{rn_n})f\| + \dots \\
& \dots\|Q_{n2}\dots Q_{nr_n}(P_{n1} - Q_{n1})f\| \\
\leq \ & \textstyle\sum_{k=1}^{r_n}\|P_{nk}f - Q_{nk}f\|
\end{aligned}$$

We will now assume that f even has a bounded uniformly continuous second derivative $f''(x)$ everywhere on $\mathbf{R}$. Taylor expansion yields

$$(*) \qquad f(x+u) = f(u) + uf'(x) + \frac{u^2}{2}f''(x) + u^2 r(x,u)$$

where $r(x,u)$ has a representation

$$r(x,u) = \frac{1}{2}\left[f''(\xi_{x,u}) - f''(x)\right]$$

where $\xi_{x,u} \in]x, x+u[$ and hence $\lim_{0\neq u\to 0} r(x,u) = 0$ uniformly in x. Moreover, $r(x,u)$ is a bounded function of x and u: $\sup_{x,u}|r(x,u)| = K < \infty$. As all other terms in $(*)$ are continuous functions of x and u, so is $r(x,u)$. We will thus encounter no difficulties in forming the sums and integrals etc. occurring in the sequel.

Multiplying both members of $(*)$ with p and summing over $u \in \mathbf{R}$ as usual, we obtain

$$(P_{nk}f)(x) = f(x) + \frac{f''(x)}{2}\sigma_{nk}^2 + \sum_{u\in\mathbf{R}} u^2 r(x,u)p_u$$

Multiplying both members of (*) with $\varrho_{\sigma_{nk}^2}(u)$ and integrating over $\mathbf{R}$, we obtain

$$(Q_{nk}f)(x) = f(x) + \frac{f''(x)}{2}\sigma_{nk}^2 + \int_{-\infty}^{\infty} u^2 r(x,u)\varrho_{0,\sigma_{nk}^2}(u)du$$

and thus

$$(P_{nk}f)(x) - (Q_{nk}f)(x) = \sum_{u\in\mathbf{R}} u^2 r(x,u)p_u + \int_{-\infty}^{\infty} u^2 r(x,u)\varrho_{0,\sigma_{nk}^2}(u)du$$

Now the Lindeberg condition comes in. For any $\epsilon > 0$, choose $\delta > 0$ such that $|u| < \delta$ implies $|r(x,u)| < \frac{\epsilon}{4}$ ($x \in \mathbf{R}$). It follows that

$$\|P_{nk}f - Q_{nk}f\| \leq \left(\frac{\epsilon}{4} + \frac{\epsilon}{4}\right)\sigma_{nk}^2 + K\sum_{|u|\geq\delta} u^2 p_u + K\int_{|u|\geq\delta} u^2 \varrho_{0,\sigma_{nk}^2}(u)du$$

and thus

$$\sum_{k=1}^{r_1} \|P_{nk}f - Q_{nk}f\|$$

$$< \frac{\epsilon}{2}\sum_{k=1}^{r_n} \sigma_{nk}^2 + K\sum_{k=1}^{r_n}\left(\sum_{|u|\geq\delta} u^2 p_u\right) + K\sum_{k=1}^{r_n}\int_{|u|\geq\delta} u^2 \varrho_{0,\sigma_{nk}^2}(u)du$$

Since the normalized scheme $(P^{(n,k)})_{\substack{n=1,2,\ldots\\ k=1,\ldots,r_n}}$ fulfils the Lindeberg condition, and since the scheme $(\varrho_{0,\sigma_{nk}^2})$ fulfils, as previously shown, the analogous condition, the last two terms tend to 0 as $n \to \infty$.Thus we may find a $n_0 \in \mathbf{N}$ such that $n \geq n_0$ implies

$$\sum_{k=1}^{r_n} \|P_{nk}f - Q_{nk}f\| < \frac{\epsilon}{2} + \frac{\epsilon}{2} = \epsilon$$

which in turn (put $x = 0$) implies

$$|E_{p_{n1}*\ldots*p_{nr_n}} f - E_{N(0,1)} f| < \epsilon$$

By proposition 5.4, we have thus shown

$$\lim_{n\to\infty} p_{n1} * \ldots * p_{nr_n} = N(0,1) \qquad \text{(weakly)}$$

□

This CLT is an encouragement to substitute distributions of sums of independent RVs by normal distributions rather freely, namely whenever the assumption that the hypothesis of theorem 5.11. holds, seems justified. This enables the practical statistician to do his numerical work largely with $N(0,1)$ alone, thus justifying the use of the word *central* here (it goes probably back to Pólya [1920]).

6. Outlook

There is a classical triad of limit theorems in the IID world

law of large numbers (LLN)
central limit theorem (CLT)
law of the iterated logarithm (LIL)

They presuppose an infinite sequence of IID real RVs. Since models of such a situation are not available in discrete stochastics, we have treated only LLN and CLT, and these in a fashion which fits into our framework:

LLN in its weak form (WLLN)
CLT as a limit theorem for binomial distribution (§4) and
as a limit theorem for convolutions as a limit theorem of rows of normalized schemes of DPDs in $\mathbf{R}$ (§5)

LIL was completely out of reach here.

I want now to briefly report on results concerning LLN, CLT and LIL which have been obtained in measure-theoretical probability theory.

6.1. The Law of Large Numbers (LLN).

One of the most important notions in stochastics is

convergence almost everywhere (a.e.)
(= almost surely (a.s.))

with respect to a probability distribution m in a given basic set:

let $h_1, h_2, \ldots$ be real RVs on Ω; if there is a real RV h on Ω such that

$$\lim_{n\to\infty} h_1(\omega) = h(\omega)$$

except for $\omega \in N$ where the exceptional set N is a m-nullset ($m(N) = 0$), we say that the sequence $h_1, h_2, \ldots$ converges to h (m-)almost surely and write

$$\lim_{n\to\infty} h_n = h \quad (m\text{-a.s.})$$

(or(m-)almost everywhere (m-a.e.)). Clearly h is then m-a.s. uniquely determined.

The LLN in its classical *strong* version (strong law of large numbers = SLLN) tells us that for IID real RVs $f_1, f_2, \ldots$ on (Ω, m) with existing expectation $E_m f = a$

$$(1) \quad \lim_{n\to\infty} \frac{1}{n}\sum_{k=1}^{n} f_k = a \quad (m\text{-a.e.})$$

As such IID sequences don't nontrivially occur in discrete stochastics, SLLN really requires fullfledged measure theory. The first proof of a special case of SLLN goes back to Borel [1909]. One can show that (1) implies

$$\lim_{n\to\infty} m(\{\omega | \frac{1}{n}\sum_{k=1}^{n} f_k(\omega) - a| \geq \epsilon\}) = 0 \quad (\epsilon > 0)$$

that is, a version of WLLN. For a thorough treatment of the LLN theme see e.g. Bauer [1990].

6.2. The Central Limit Theorem (CLT).

Theorem 5.11. and its proof carry over to arbitrary probability distributions in $\mathbf{R}$ without difficulties. Thus it may be viewed as a fullfledged version of CLT. Important investigations about the speed of the weak convergence in CLT go back to Berry [1941], Esséen [1956][1958].

6.3. The Law of the Iterated Logarithm (Loglog-Theorem, LIL).

answers the question

> what is the speed of convergence in (1) (strong LLN)

in the following way:

> if $f_1, f_2, \ldots$ are IID real RVs with expectation 0 and variance 1, then the sequence
>
> $$(2) \quad \frac{f_1 + \ldots + f_n}{\sqrt{2n \log\log n}} \quad (n = 2, 3, \ldots)$$
>
> has $\limsup_{n\to\infty} = 1$ and $\liminf_{n\to\infty} = -1$ almost surely; actually, every $s \in [-1, 1]$ occurs as the limit value of a subsequence of (2).

Thus the speed in question is given by

$$\frac{\sqrt{2n \log\log n}}{n} = \sqrt{2}\sqrt{\frac{\log\log n}{n}}$$

This loglog theorem is certainly one of the deepest limit theorems in stochastics. It was discovered, in a special case, by Chintschin [1924]; the final result

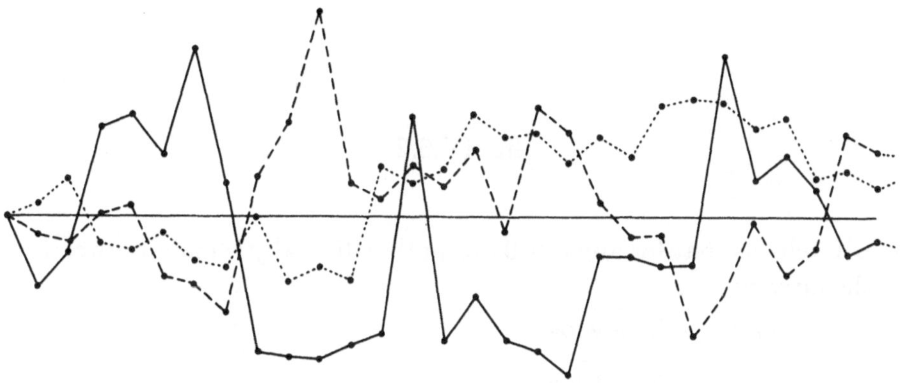

fig. IV.6.1

is due to Kolmogorov [1929]. The topic then lay practically dormant for 35 years, until Strassen [1964] revived it again, initiating a vast stream of research – see e.g. Csörgö-Revész [1975], Major [1978], Wittmann [1987]. The proof of LIL combines LLN and CLT techniques.

LLN and CLT may be visualized as follows: for any given random point ω in the underlying probability space consider the sequence

$$0, f_1(\omega), f_1(\omega) + f_2(\omega), \ldots, f_1(\omega) + \ldots f_n(\omega), \ldots$$

and draw the associated path, interpolating these reals linearly between $0, 1, \ldots, n, \ldots$. Figure IV.6.1 shows three such random paths.

Assuming that the RVs $f_1, f_2, \ldots$ have expectation 0 and variance 1,

> the strong LLN tells us that our path will finally remain between the legs of an arbitrarily small angle opening at (0,0) symmetrically to the right. We sketch this situation for two different angles, indicating the moment from which onward the path remains within that angle.

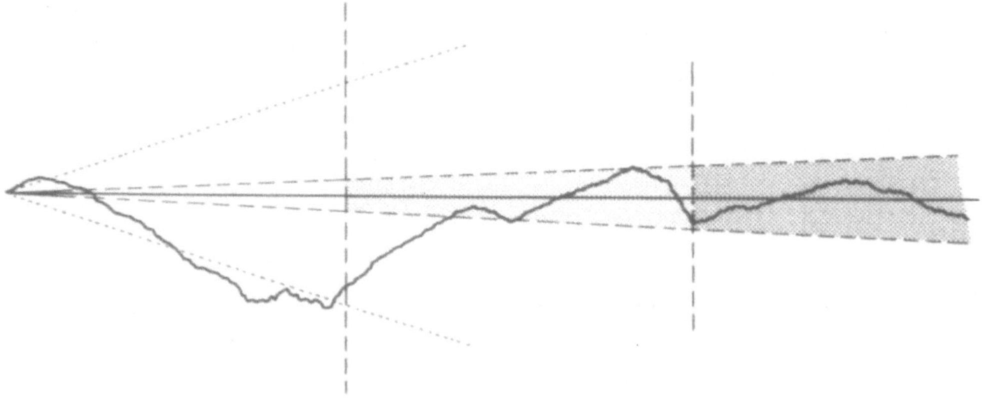

fig. IV.6.2

LIL tells us that for any $\epsilon > 0$ our path will finally remain between the curves

$$-(1+\epsilon)\sqrt{2n\log\log n}$$
$$+(1+\epsilon)\sqrt{2n\log\log n}$$

but will go

below $-(1-\epsilon)\sqrt{2n\log\log n}$

above $(1-\epsilon)\sqrt{2n\log\log n}$

over and over again. We may sketch this as shown in figure IV.6.3.

If we want to visualize the CLT in the same way, we have to plot, for arbitrary $-\infty < a < b < \infty$, the curves $a\sqrt{n}$, $b\sqrt{n}$ and to realize that for large n the probability that our path is in $[a\sqrt{n}a,\ b\sqrt{n}]$ is approximately $\frac{1}{\sqrt{2\pi}}\int_a^b e^{-\frac{x^2}{2}}\,dx$ (figure IV.6.4).

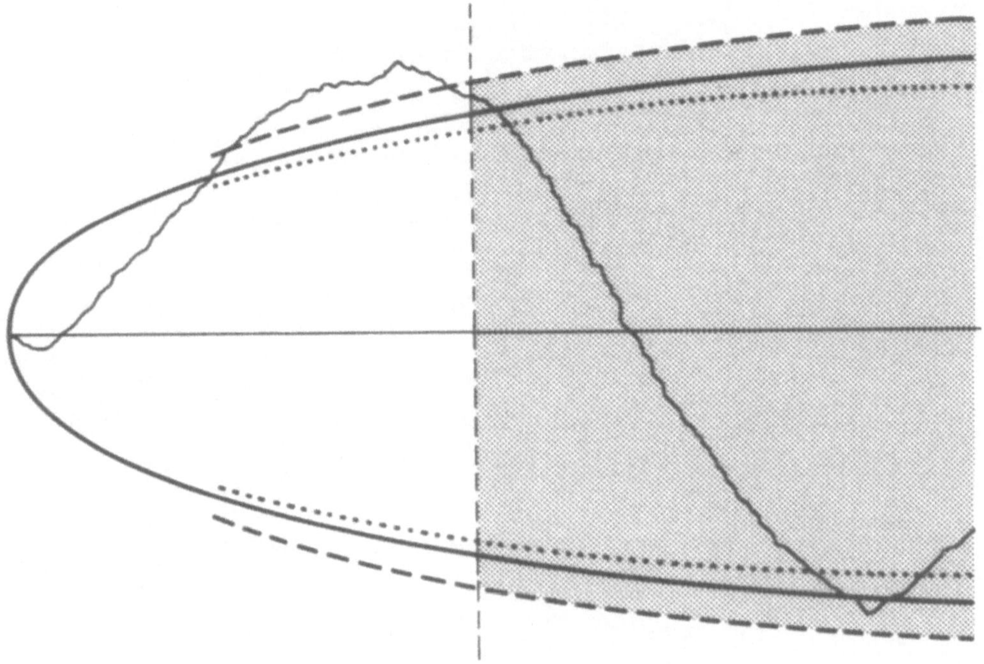

fig. IV.6.3

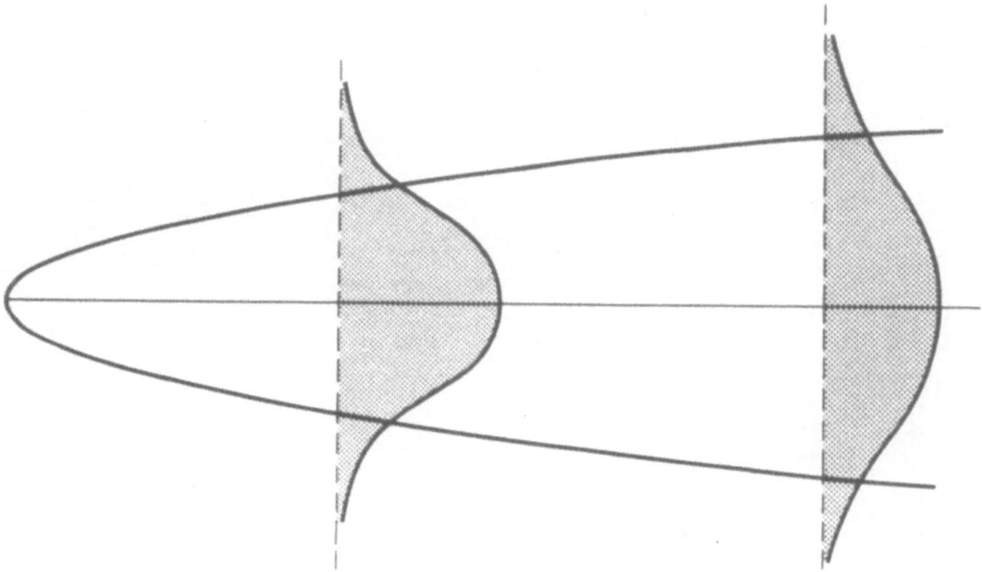

fig. IV.6.4

V. Statistics

The discipline of ***mathematical stochastics*** may be subdivided into ***probability theory*** and ***statistics***.

Probability theory usually works with one probability distribution m which is given by a few data (parameters) plus qualitative (e.g. independence) properties; it aims at calculating values $m(E)$ for vertain events E which are of particular interest, and at establishing further interesting qualitative properties of m. In ch. I §2 we considered a few examples of such calculations; the main bulk of this book is devoted to probability theory in this sense. In contrast to this,

Statistics is concerned with the problem of making ***decisions*** of probabilistic nature. One type of such decisions is to select, on grounds of observed values of certain random variables, one probability distribution out of several candidates. One might also be faced with a yes-no alternative about a certain property of an unknown probability distribution.

We begin this chapter with a series of specimens of statistical reasoning (§1). In §2 we present the so-called ***game-theoretical framework*** of statistical decision theory, after Abraham Wald (1902-1950). In §3 we elaborate on one important type of such statistical decision procedures: tests. In §4 we report about some other types of statistical decision procedures. – A good textbook is Rohatgi [1984].

1. Specimens of Statistical Reasoning

It is the basic working hypothesis of the statistician, that the world, or more modestly, the particular phenomenon which he is about to investigate, is of random type, and is governed by a certain probability distribution, about which he may know some qualitative details while he fails to know other important features, parameter values etc. A few examples:

1) The yield of an acre of arable land planted with a certain crop fruit is the sum of the yields of the individual plants on it, and the statistician might feel himself entitled to make certain independence assumptions which allow to apply some variant of the CLT (theorem IV.5.11), so that

he will presuppose the yield to be a random variable whose distribution is normal, i.e. some $N(a,\sigma^2)$ – roughly: after all yields cannot have negative values. But he doesn't know the values of the parameters a (expectation) and σ^2 (variance). He will thus make several observations of yields $Y_1,\dots,Y_n$ on the same acre in order estimate these parameters. One simple estimate of a would be

$$\bar{Y} = \frac{1}{n}(Y_1 + \ldots + Y_2)$$

– reasonable, because WLLN tells us that with high probability, this value will not be far away from a. A customary estimate of σ^2 is

$$S_n^2 = \frac{1}{n-1}[(Y_1 - \bar{Y})^2 + \ldots + (Y_n - \bar{Y})^2].$$

Why $\frac{1}{n-1}$ and not $\frac{1}{n}$? In what sense precisely are these estimates good or even optimal? Statistical theory aims at answering such questions.

2) A gambler accuses his fellow gambler to use a false die. In order to settle the argument, they consult a statistician, who asks them to throw the die 60 times and to note the outcomes. They do it and report 7 times a one, 16 times a two, 8 times a three, 17 times a four, 3 times a five, 9 times a six. The statistician mumbles "chi square, five degrees of freedom, five percent", calculates $\frac{(7-10)^2}{10} + \frac{(16-10)^2}{10} + \frac{(8-10)^2}{10} + \frac{(17-10)^2}{10} + \frac{(3-10)^2}{10} = 14.8$ and says: "that's more than 11.07 – I consider the die as false!" – What is behind the mumblings of this statistician? Statistics answers such questions.

3) The notorious "Tea Tasting Lady" claims that whenever she is being served a cup of tea with milk, she is able to taste whether the tea had been poured into the milk, or the milk into the tea. She agrees to have her ability tested by a statistician. He serves her 8 cups of tea with milk, 4 of which having been prepared by pouring the tea into the milk ("type I") and the other 4 the other way round ("type II"). A random order of tasting the 8 cups is chosen (one out of 8!) and the lady is asked to single out, after tasting, the four cups of type I. It doesn't matter whether she makes her decisions step by step or after she has tasted all 8 cups. Assume she correctly identifies 3 of the 4 cups of type I. Now the statistician has to decide whether this degree of success if a proof of her claimed ability, or a random result. He calculates: there are $\binom{8}{4}$ possibilities of choosing 4 cups out of 8. In $\binom{4}{3}\cdot\binom{4}{1}$ cases one chooses 3 of "type I". That is, this kind of success happens with probability

$$\frac{\binom{4}{3}\binom{4}{1}}{\binom{8}{4}} = \frac{4\cdot 4\cdot 4!\, 4!}{8!} = \frac{8}{35} = 0.229\ldots$$

To hit all 4 would have a probability

$$\frac{\binom{4}{4}\binom{4}{0}}{\binom{8}{4}} = \frac{4!4!}{8!} = \frac{1}{70} = 0.014\ldots$$

And to hit at least 3 has probability

$$\frac{\binom{4}{3}\binom{4}{1} + \binom{4}{4}\binom{4}{0}}{\binom{8}{4}} = \frac{16+1}{70} = \frac{17}{70} = 0.243\ldots$$

If the statistician now says "that's too improbable – the lady is not perfect, but she has a certain ability", he may be wrong, but on the basis of the above test result (3 correct) only with a probability $\frac{17}{70}$. This is typical for statistical decision procedures: you may err, but given the procedure, you may calculate the error probability.

4) A pharmaceutic firm wants to corroborate her claim that her new medicament B is better than some other medicament A. 20 couples of test persons of equal health status are chosen at random; for each couple, one person gets A, the other B. In 16 couples, B does better than A. Does this result allow to maintain the said claim? – Well if A has the same quality as B, then the results of the experiment are tantamount to flipping a coin, and the probability of getting 16 or more outcomes in favor of B is

$$\frac{1}{2^{20}}\left[\binom{20}{16} + \binom{20}{17} + \binom{20}{18} + \binom{20}{19} + \binom{20}{20}\right] = 0.006\ldots$$

"That's too improbable – I reject the hypothesis that B is not better than A", says the statistician.

5) A social psychologist wants to find out whether a certain film about juvenile delinquency would change the opinions of members of a particular community about how severely juvenile delinquents should be punished. He draws a random sample of 100 adults from the community and conducts a "before and after" study, letting each subject serve as his own control. He writes up the outcome in the following form

		"after"	
		less	more
"before"	more	59	7
	less	8	26

That is, 59 subjects favored more severe punishment before they saw the film, and less severe punishment after they saw it, etc. Did the film have any effect? 8 + 7 subjects showed no effect, and 59 + 26 = 85 showed one. The statistician ponders: if the film had no effect, about one half of those 85 who changed their opinion would have changed from "more" to "less", and the other half from "less" to "more"; the probability that ≥ 59 change from "more" to "less" would be 0.0006 only (I calculate that assuming a normal distribution with expectation $\frac{85}{2}$ and variance $85 \cdot \frac{1}{4}$). Well, that's too improbable to the "no-effect" hypothesis: I reject it: the film had an effect" (after Siegel [1956]).

These specimens hopefully allow the reader to develop certain intuitions about what is going on in statistics:

The statistician collects random data and derives *decisions* from them

He does this, in each situation, according to some methods. Each such method has its *error probabilities.*

The statistical theorist weighs one method against others – mainly by comparing error probabilities – and tries to find out *optimal methods.*

We will deal primarily with the third aspect here: optimality of statistical methods. In the next section we will develop a general theoretical frame for such investigations. A reader who wants to see more specimens of statistical methods is referred to P.G.Sachs [1984], Siegel [1956].

2. The Game-Theoretical Framework of Statistical Theory

Abraham Wald (1902-1950) proposed (in Wald [1950][1950a]) to interpret statistical methods as strategies in two-person games with "nature" (or "the god of random") as one, and the statistician as the other player. This nowadays generally adopted idea in fact accomodates all aspects of statistics which we have tried to convey to the reader in the preceding section.

We assume the reader to be informed about the general ideas of two-person-game theory:

strategy sets for each player

payoffs resulting for each player, as soon as everyone has chosen a strategy for himself

best answers, equilibria, minimax strategies etc.

(see e.g. Franklin [1981]). We will now describe how statistical theory fits into this frame. As an introduction, let us consider a simple *test problem*:

On a given basic set Ω, we consider two DPDs p_0 and p_1. We don't know which of the two is the *true* DPD governing the random experiment whose possible outcomes are the points ω of Ω. We are to make a decision between p_0 (the *null hypothesis*) and p_1 (the *counter-hypothesis*), based on an observation $\omega \in \Omega$. That is, we are to make a clean-cut 0-1-decision: for some ω's we decide for p_0, and for the other ω's we decide for p_1. A decision *procedure* in this situation is called a *test*, and such a test is fully described by the set K of all those $\omega \in \Omega$ for which we decide for p_1; this K *is* the test, it is also called the *critical region* of the test. The notation reflects a certain asymmetry of our view on p_0 and p_1. This asymmetry is the reflex of empirical preliminaries, not of mathematical properties of the model – the reader has certainly observed such an asymmetry e.g. in specimen 2) or 5) of §1.

Now comes the game theoretical interpretation:

> A strategy of "nature" consists in choosing one of the DPDs p_0 or p_1, and thereby steering the random experiment resulting in one particular $\omega \in \Omega$.
>
> A strategy of the statistician consists in choosing a test $K \subseteq \Omega$, thereby steering the decision
>
> > for 1 if $\omega \in K$
> >
> > for 0 if $\omega \in \Omega \backslash K$
>
> "0" stands for "assuming the null-hypothesis" that p_0 is true DPD, and "1" stands for "assuming the counter-hypothesis" that p_1 is the true DPD. That is, "nature" decides for one of p_0, p_1, and the statistician decides, first for his test K, and then, on the basis of the observed ω and the chosen K, for a *hypothesis about* p_0, p_1.
>
> The *payoff* for the statistician depends on the chosen strategies of "nature" and himself: it is
>
> > the *error probability* $p_0(K)$ if nature has chosen p_0: to reject p_0 on grounds of $\omega \in K$ is traditionally called an *error of first kind*
> >
> > the error probability $p_1(\Omega \backslash K) = 1 - p_1(K)$ if nature has chosen p_1: to adopt p_0 is then called an *error of second kind*
>
> In game theory one often operates with *mixed strategies*, that is, distributions of total weight 1 over several ("pure") strategies. In

our present situation, a *mixed strategy of nature* is given by two reals $\alpha_0, \alpha_1 \geq 0$ with $\alpha_0 + \alpha_1 = 1$. If nature adopts this, the payoff for the statistician who has chosen K is

$$\alpha_0 p_0(K) + \alpha_1 p_1(\Omega \backslash K)$$

As soon as he knows to calculate his payoffs, the statistician may try to develop best answers, minimax strategies in the form of tests or mixed test strategies.

This is an example. The *general* framework of game-theoretical statistical decision theory looks as follow:

on a set $\Omega \neq \emptyset$ we have a *family* $(P_\theta)_{\theta \in \Theta}$ of DPDs, parametrized with the elements θ of a parameter set Θ, which represents the set of all strategies of "nature".

we have a set Δ of possible decisions δ (of the statistician): the *decision space*

we have a *loss function* $L(\theta, \delta)$ which tells us how large the loss inflicted on the statistician (or his client) is if "nature" has chosen $\theta \in \Theta$ and the statistician has decided for $\delta \in \Delta$.

a *statistical decision function* (SDF) is a mapping $d : \Omega \to \Delta$; that is, if $\omega \in \Omega$ has been observed, d prescribes to decide for $d(\omega) \in \Delta$. The strategy set of the statistician is the set $\mathcal{D}$ of all SDFs in the given model.

once the strategies $\theta \in \Delta$ (of "nature") and $d \in \mathcal{D}$ (of the statistician) have been chosen, we may calculate the *expected loss = the risk* of the statistician as

$$R(\theta, d) = E_{p_\theta}(L(\theta, d(\cdot)))$$

It is this *risk function* which serves as payoff function or payoff matrix in the game theoretical investigations of mathematical statistics.

We will not go into more generalities here. It is enough if the reader knows this general framework and has an idea of how special methods may fit into it. Under these auspices we shall, in the next section, investigate test problems in more detail. In §4 we shall report on some more statistical procedures.

3. Tests

3.1. The General Notion of a Test.

The simple test problem considered in §2 fits into a more general context: statistical decision functions (SDFs) with a two-element decision space Δ, say $\Delta = \{0,1\}$, are called *tests*. This conforms with every-day language in some respect: in factories, tests are methods to reach yes-no-decisions – an item passes the test or is rejected. Thus a SDF is a *test* if it attains values 0 and 1 only. It is thus the indicator function of a certain subset K of the underlying basic set Ω, the ***critical region*** of that test: after observations $\omega \in K$ you decide for 1, after observation $\omega \in \Omega \setminus K$ you decide for 0. As in every statistical decision problem, we have to presuppose a parametrized family $(p_\theta)_{\theta\in\Theta}$ to be given in Ω in advance. And for a test problem, the other important ingredient is a splitting of the parameter set Θ into a subset H_0 called the ***null-hypothesis*** and its complement $H_1 = \Theta \setminus H_0$, which we will call the *counter-hypothesis*. Thirdly, in a test problem

decision 0 is interpreted as decision
for the null-hypothesis, that is for the assumption that the "true" θ lies in H_0; and
decision 1 is interpreted as decision
for H_1, the counter-hypothesis

Lastly, the customary ***loss function*** in test problems is

$$L(\theta,\delta) = \begin{cases} 0 & \text{for "correct decisions":} \\ & \quad \theta \in H_0,\ \delta = 0, \text{ or else } \theta \in H_1,\ \delta = 1 \\ 1 & \text{for "erroneous decisions" :} \\ & \quad \theta \in H_0,\ \delta = 1 \text{ ("error of first kind")} \\ & \quad \theta \in H_1,\ \delta = 0 \text{ ("error of second kind")} \end{cases}$$

If $d = 1_K$, we thus get

$$L(\theta, 1_K) = \begin{cases} 1_K & \text{if } \theta \in H_0 \\ 1_{\Omega\setminus K} & \text{if } \theta \in H_1 \end{cases}$$

Thus the calculation of the risk function $R(\theta,d)$ leads to bare probabilities:

$$\begin{aligned} R(\theta,d) &= E_{p_\theta}(L(\theta,d(\cdot))) \\ &= E_{p_\theta}(L(\theta,1_K)) \\ &= \begin{cases} E_{p_\theta}(1_K) = p_\theta(K) & (\theta \in H_0) \\ \quad \text{("error probabilities of first kind")} & \\ E_{p_\theta}(1_{\Omega\setminus K}) = 1 - p_\theta(K) & (\theta \in H_1) \\ \quad \text{("error probabilities of second kind")} & \end{cases} \end{aligned}$$

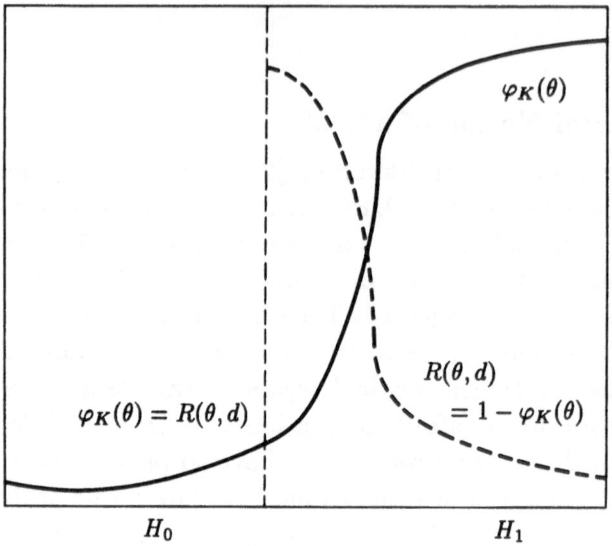

fig. V.3.1

All these probabilities can be read from a single function of θ, the *power function*

$$\varphi_K(\theta) = p_\theta(K) \quad (\theta \in \Theta)$$

of the test $d = 1_K$. We may sketch this as in figure V.3.1.

The aim of the statistician – our aim – is to get small values of $R(\theta, d)$ $(\theta \in \Theta)$, or, equivalently, to push the power function down to 0 on H_0, and up to 1 on H_1, by proper choice of d. The basic problem is: what means "down" (or "up") if we have *functions* of θ, and not single real numbers only?

Exercise 3.1. Show that in our present setup, the "ideal" power function 1_{H_1} ("zero on H_0, 1 on H_1") can be achieved if and only if the p_θ with $\theta \in H_0$ live "disjointly" from the p_θ with $\theta \in H_1$: no g_θ with $\theta \in H_0$ gives probability > 0 to an $\omega \in \Omega$ to which at least one p_θ with $\theta \in H_1$ gives probability > 0. Indicate a critical region K for which $p_\theta(K) = 1_{H_1}(\theta) \quad (\theta \in \Theta)$.

In a family of functions $R(\cdot, d)$ on Θ obtained from various SDFs d, we would certainly be glad to find one $R(\cdot, d_0)$ which minorizes all other ones pointwise ("uniformly") on H_1:

$$R(\theta, d_0) \le R(\theta, d) \quad (\theta \in H_1)$$

for all other d's. Such SDFs are called ***uniformly most powerful*** (UMP) within that set of SDFs. We may, in particular, speak of UMP tests within a given set of tests (figure V.3.2).

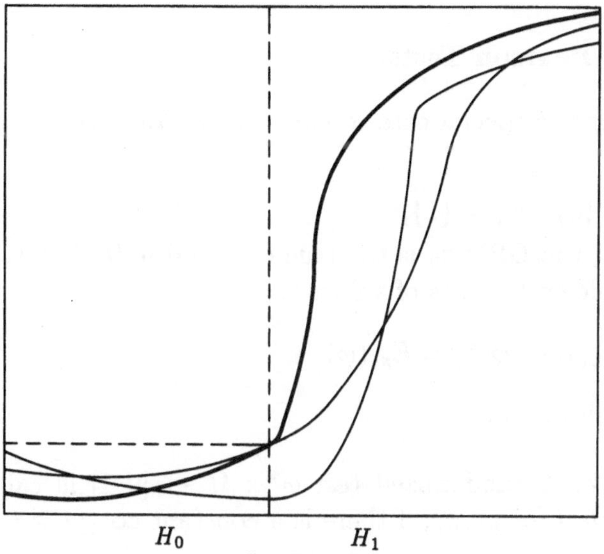

fig. V.3.2

It is indeed possible to find UMP tests, if the situation under consideration is sufficiently simple. On the other hand, we have to generalize the notion of a test slightly:

3.2. Randomized Tests.

A randomized test is a SDF with decision space not $\Delta = \{0, 1\}$ but $\Delta =$ all probability distributions on $\Delta = \{0, 1\}$, that is

$$\begin{aligned} \Delta &= \{(\varphi_0, \varphi_1) | \varphi_0, \varphi_1 \geq 0,\ \varphi_0 + \varphi_1 = 1\} \\ &= \{(1-\varphi, \varphi) | 0 \leq \varphi \leq 1\}. \end{aligned}$$

If you decide for such a $(1-\varphi, \varphi)$, you don't say "I decide for 0" or "I decide for 1" but "I decide for 1 with probability φ, and for 0 with probability $1 - \varphi$". And this means: when deciding for $(1 - \varphi, \varphi)$, you toss an, as a rule, asymmetric, coin which results in 1 with probability φ, and in 0 with probability $1 - \varphi$, and follow the decision of the coin. You may realize this, for arbitrary $0 \leq \varphi \leq 1$, by partitioning the circumference of a roulette wheel proportionally to $\varphi : 1 - \varphi$.

A *randomized test* thus assigns to every $\omega \in \Omega$ a real number $\varphi(\omega)$: after having observed ω an "$\varphi(\omega)$-coin" is thrown in order to decide for 1 or 0 finally. And the resulting power function then is

$$\varphi(\theta) = E_{p_\theta}(\varphi(\cdot)).$$

Our original, non-randomized tests given by critical regions $K \subseteq \Omega$ fit in here as the special case $\varphi(\omega) = 1_K(\omega)$: if $\omega \in K$, you decide for 1 with probability $1_K(\omega) = 1$, and if $\omega \in \Omega \backslash K$, for 0 with probability $1 = 1 - 1_K(\omega)$.

3.3. Neyman-Pearson Tests.

are designed for the special case of *one-element hypotheses*, that is,

$$\Theta = \{0,1\}$$
$$H_0 = \{0\},\ H_1 = \{1\},$$

i.e. if H_0 holds, the DPD p_0 is the true one, and if H_1 holds, it is p_1. Power functions boil down to pairs of values

$$\varphi(0) = E_{p_0}(\varphi),\ \varphi(1) = E_{p_1}(\varphi)$$

in this simple situation.

Definition 3.2. A randomized test $\varphi^* : \Omega \to [0,1]$ is called a *Neyman-Pearson (NP) test* for p_0, p_1, if there is a constant $\infty \geq c \geq 0$ such that

$$\varphi^*(\omega) = \begin{cases} 1 & \text{if} \quad p_1(\{\omega\}) > cp_0(\{\omega\}) \\ 0 & \text{if} \quad p_1(\{\omega\}) < cp_0(\{\omega\}) \end{cases}$$

(no requirements in case $p_1(\{\omega\}) = cp_0(\{\omega\})$).

Theorem 3.3. (Neyman-Pearson Lemma, part I). Let, in the above situation

φ^* be a NP test
φ any test with $E_{p_0}(\varphi) \geq E_{p_0}(\varphi^*)$.

Then

$$E_{p_1}(\varphi^*) \geq E_{p_1}(\varphi)$$

That is, NP tests are uniformly most powerful (UMP).

This lemma (along with part II, our theorem 3.4.) was first published in Neyman [1933]. Jerzy Neyman (1894-1981), who collaborated with Egon S. Pearson (1895-1980), son of Karl Pearson (1857-1936), on this and related subjects, was, along with Ronald Alymer Fisher (1890-1962) and Abraham Wald (1902-1950), one of the most important figures in modern statistical theory. For biographical details see Reid [1982], Fisher [1978]. The idea behind the definition of a NP-test is:

> try to collect as much of p_1 as you can into a critical region, while accepting as little of p_0 as possible; do this by taking the ω's with large proportions $p_1(\{\Omega\})/p_0(\{\omega\})$ before taking those with smaller proportions.

PROOF. of theorem 3.3.

$$\begin{aligned} & E_{p_1}(\varphi^*) - E_{p_1}(\varphi) \\ = \;& E_{p_1}(\varphi^* - \varphi) \\ = \;& E_{p_1}(1_{\{\varphi^* > \varphi\}}(\varphi^* - \varphi) + E_{p_1}(1_{\{\varphi^* < \varphi\}}(\varphi^* - \varphi) \end{aligned}$$

Now, if $\varphi^*(\omega) > \varphi(\omega)$, then $\varphi^*(\omega) > 0$, hence $p_1(\{\omega\}) \geq cp_0(\{\omega\})$; thus the first term above is $\geq cE_{p_0}(1_{\{\varphi^* > \varphi\}}(\varphi^* - \varphi))$. Similarly, if $\varphi^*(\omega) < \varphi(\omega)$, then $\varphi^*(\omega) < 1$, hence $p_1(\{\omega\}) \leq cp_0(\{\omega\})$, thus the second term above is $\geq cE_{p_0}(1_{\{\varphi^* < \varphi\}}(\varphi^* - \varphi))$. Putting things together, we arrive at

$$E_{p_1}(\varphi^*) - E_{p_1}(\varphi) \geq 0.$$

□

Theorem 3.4. (Neyman-Pearson-Lemma, part II). For any $0 \leq \alpha \leq 1$ there is a NP-test φ^* such that

$$E_{p_0}(\varphi^*) = \alpha.$$

PROOF. If $\alpha = 0$, we put $c = \infty$, get $p_0(\{\omega\}) > 0 \Longrightarrow p_1(\{\omega\}) < cp_0(\{\omega\})$ $\Longrightarrow \varphi^*(\omega) = 0$, and thus $E_{p_0}(\varphi^*) = 0$.

Let now $\alpha > 0$. For every $c \geq 0$ put

$$K(c) = \{\omega | p_1(\{\omega\}) > cp_0(\{\omega\})\}$$

$$K(c-0) = \{\omega | p_1(\{\omega\}) \geq cp_0(\{\omega\})\}$$

and consider the non-increasing function

$$a(c) = p_0(K(c)) \quad (c \geq 0)$$

Clearly its left sided limits are

$$a(c-0) = p_0(K(c-0)) \geq a(c) \quad (c \geq 0)$$

and

$$a(0-0) = 1, \; \lim_{c \to \infty} a(c) = \lim_{c \to \infty} a(c-0) = 0.$$

If $a(c) = \alpha$ for some c, we put

$$\varphi^* = 1_{K(c)}$$

and clearly have a NP-test φ^* with $E_{p_0}(\varphi^*) = p_0(K(c)) = a(c) = \alpha$. If $a(c-0) = \alpha$ for some c, we put

$$\varphi^* = 1_{K(c-0)}$$

and again get a NP-test φ^* with $E_{p_0}(\varphi^*) = \alpha$.

In all other cases, there is some $c > 0$ such that

$$a(c-0) > \alpha > a(c),$$

namely,

$$a(c-0) - a(c) = p_0(K(c-0)\backslash K(c)) > 0$$

We now put

$$\varphi^*(\omega) = \begin{cases} 1 & \text{for} \quad \omega \in K(c) \\ 0 & \text{for} \quad \omega \notin K(c-0) \\ \gamma & \text{for} \quad \omega \in K(c-0)\backslash K(c) \end{cases}$$

where $0 < \gamma < 1$ is chosen in such a fashion that $a(c) + \gamma p_0(K(c-0)\backslash K(c)) = \alpha$. Clearly φ^* is a NP-test with

$$\begin{aligned} E_{p_0}(\varphi^*) &= p_0(K(c)) + \gamma p_0(K(c-0)\backslash K(c)) \\ &= a(c) + \gamma p_0(K(c-0)\backslash K(c)) \\ &= \alpha. \end{aligned}$$

□

Theorems 3.3. and 3.4. (NP-Lemma) are the basis for the construction of UMP tests in the case of one-element hypotheses. Under suitable monotonicity assumptions they turn out to be UMP even for some cases of many-element hypotheses. We display this in the very simple case of

Example 3.5. (Optimality of the Binomial Test). Let $\Omega = \{0, 1, \dots, n\}$ and, for any $0 \leq \theta \leq 1$,

$$p_\theta(\{k\}) = \binom{n}{k} \theta^k (1-\theta)^{1-k}$$

be the binomial distribution with parameter θ (= "success probability"). Choose $0 < \theta_0 < \theta_1 < 1$ arbirarily and write $p_0 = p_{\theta_0}$, $p_1 = p_{\theta_1}$. Then

$$\frac{p_1(\{k\})}{p_0(\{k\})} = \left[\frac{\theta_1}{1-\theta_1} \bigg/ \frac{\theta_0}{1-\theta_0}\right] \cdot C$$

with a real number $C = (\frac{1-\theta_1}{1-\theta_0})^n$ independent of k. As $\theta_1 > \theta_0$, the quotient in the square bracket is > 1, hence $\frac{p_1(\{k\})}{p_0(\{k\})}$ increases strictly as k increases. A NP-test, as described in theorem 3.4., is therefore of the form

$$\varphi^*(k) = 1_{\{k_0+1,\dots,n\}} + \gamma 1_{\{k_0\}}$$

for some integer $0 \leq k_0 \in n$, that is, you decide for $p_1 = p_{\theta_1}$ right away if you observe $k > k_0$, and you toss a γ-coin if $k = k_0$. Now consider $E_{p_\theta}(\varphi^*)$ as a function of θ: clearly this is an increasing function of θ. Thus such tests are UMP for any $\theta_0' < \theta_1'$ in $[0,1]$. In particular, they are UMP for $H_0 = [0, \theta_0]$ against $H_1]\theta_0, 1]$.

Let us now look back to those specimens given in §1 which involve a test problem.

2) The problem of the false die is a test problem with $\Omega = \{1,2,3,4,5,6\}$, the one-element null hypothesis

$$H_0 = \{p^{(0)}\} = \left\{\left(\frac{1}{6}, \frac{1}{6}, \frac{1}{6}, \frac{1}{6}, \frac{1}{6}, \frac{1}{6}\right)\right\}$$

and the set H_1 of all DPDs on Ω which are different from $p^{(0)}$ as the counter-hypothesis. It is not worthwhile to write out $H_0 \cup H_1$, as a parametrized family: the set parametrizes itself via the identity mapping. The test, i.e. the critical region which the mumblings of the statistician hint at, is obtained in the following way: the $n(= 60)$ independent throws are modeled by six rows of n IID 0-1-valued RVs

$$\begin{array}{ccc} c_1^{(1)} & \ldots, & c_n^{(1)} \\ \ldots & \ldots & \ldots \\ c_1^{(6)} & \ldots & c_n^{(6)} \end{array}$$

where $c_k^{(6)} = 1$ iff the kth throw results in a six, etc. – all this on a suitable basic set, e.g. Ω^n. The row sums

$$\begin{array}{ccccc} N^{(1)} & = & c_1^{(1)} & +\ldots+ & c_n^{(1)} \\ \ldots & \ldots & \ldots & & \ldots \\ N^{(\sigma)} & = & c_1^{(\sigma)} & +\ldots+ & c_n^{(\sigma)} \end{array}$$

count the number of ones etc. Under the null hypothesis, they are approximately normally distributed, and

$$T = \frac{(N^{(1)} - \frac{n}{\sigma})^2}{\frac{n}{\sigma}} + \ldots + \frac{(N^{(\sigma)} - \frac{1}{\sigma})^2}{\frac{n}{\sigma}}$$

turns out to have a certain distribution derived from $N(0,1)$ and called χ_5^2 = "chi square distribution with 6-1=5 degrees of freedom" – all this approximately only, involving the CLT. To be more specific: if $f_1, \ldots, f_d$ are d independent $N(0,1)$ distributed RVs, then $f_1^2 + \ldots + f_d^2$ is χ_d^2 distributed. The above passage from 6 to $d = 5$ is due to the fact that $N^{(1)}, \ldots, N^{(6)}$ are not independent but fulfil $N^{(1)} + \ldots + N^{(\sigma)} = n$. For details see e.g. Sachs [1984]. The density function of χ_6^2 looks like shown in figure V.3.3.

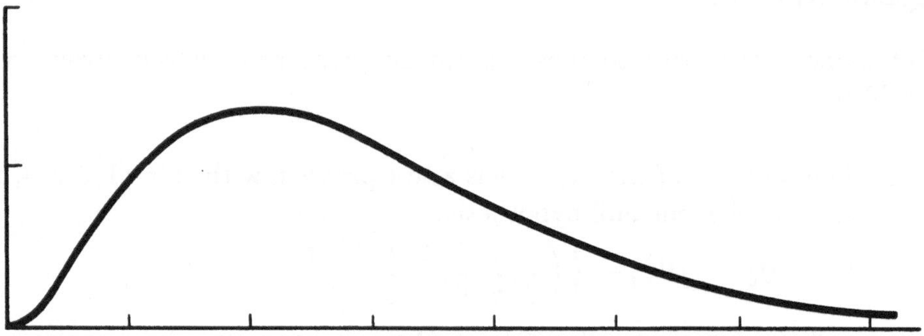

fig. V.3.3

Our statistician tacitly adopted an error probability of first kind $p^{(0)}(K) = 0.05$ and therefore used the critical region $K = \{T \geq 11.07\}$, taking the value $T_0 = 11.07$ from a χ_5^2 table. In short, he used a χ^2-test.

3) For the "tea teasting lady", the null hypothesis reflecting the opinion that the lady has no real tea tasting ability but selects her four cups at random, was given by equidistribution over all $\binom{8}{4}$ possible choices, and the critical region was the set of all choices with at least 3 cups chosen correctly. See Krengel [1988] for a more detailed discussion.
4) The testing of medicament B against medicament A used $\Omega = \{0, 1, \ldots, 20\}$ as a model for the possible numbers of winnings of B over A. The null hypothesis was the symmetric binomial distribution (parameter $\frac{1}{2}$) p, and the critical region could be chosen as $K = \{16, 17, 18, 19, 20\}$ with $p(K) \leq 0.01$.
5) In the test whether a film about juvenile delinquency would have an effect on spectators, a similar test was constructed as in 4), with the number of changes from "more" to "less" as a RV approximately normally distributed under the null hypothesis ("no effect").

Test theory is a vast discipline since decades; the classic on the subject is Lehmann [1959]. There is a classical family of tests based on the normal distribution $N(0, 1)$ and some of its derivates such as χ^2-distributions, Student's T-distributions etc. Their application usually works with a variant of CLT in order to pass to $N(0, 1)$. This device has been criticized ("the normal distribution is a myth"): some test characteristics are too sensitive against small deviations from $N(0,1)$. As a consequence, non-sensitive "robust" methods were established, not only for tests, but also for other statistical procedures, see e.g. Hampel-Rousseauw-Stahel [1985].

In psychology and related empirical disciplines, the non-metric character of many quantities under investigation has motivated the development of statistical methods which largely get along with finite basic sets and thus fit very well into the framework of discrete stochastics. Readers who want to become familiar with such "parameter-free methods" should e.g. consult Siegel [1956], Sachs [1984], Lienert [1973].

4. Outlook

We report here briefly on two more types of statistical decision procedures: *estimation* and *confidence intervals*.

Let a family $(p_\theta)_{\theta\in\Theta}$ of DPDs on a basic set $\Omega \neq \emptyset$ be given. Mappings from the basic parameter sete Θ to other sets are also called *parameters* of the family. The identity mapping $\Theta \to \Theta$ is, of course, the most fundamental parameter, but often other parameters, e.g. real-valued ones, are of particular interest. Take e.g., for a given real-valued RV f on Ω, the expectation

$$a(\theta) = E_{p_\theta}(f),$$

or the variance

$$\sigma^2(\theta) = \sigma^2_{p_\theta}(f).$$

Given a parameter $\alpha : \Theta \to \mathbf{R}$, real-valued RVs designed to guess the true value $\alpha(\theta)$ of that parameter are also called *estimators* of $\alpha(\cdot)$. For an estimator $e : \Omega \to \mathbf{R}$ of $\alpha(\cdot)$, the loss function – in the sense of our general setup (§2) – usually chosen is quadratic deviation, that is

$$L(\theta, \alpha) = (e - \alpha)^2$$

and the resulting risk function thus is

$$R(\theta, e) = E_{p_\theta}((e - \alpha(\theta))^2).$$

A first step to be taken in order to keep the risk low is to achieve

$$E_{p_\theta}(e) = \alpha(\theta) \quad (\theta \in \Theta),$$

that is, to bring the so-called *bias*

$$E_{p_\theta}(e) - \alpha(\theta)$$

down to 0. Such estimators are called *unbiased*. For unbiased estimators, risk boils down to variance:

$$R(\theta, e) = \sigma^2_{p_\theta}(e)$$

The second, and usually more crucial step in achieving a good estimator for $\alpha(\cdot)$ is to make its variance as small as possible, preferably uniformly for all $\theta \in \Theta$. If, as above

$$\alpha(\theta) = E_{p_\theta}(f),$$

and if $f_1, \ldots, f_n$ are, for every $\theta \in \Theta$, IID, distributed like f, then the average

$$e = \frac{1}{n}(f_1 + \ldots + f_n)$$

is an unbiased estimator with risk function = variance, namely,

$$R(\theta, e) = \frac{\sigma^2_{p_\theta}(f)}{n} \quad (\theta \in \Theta).$$

In many cases, this estimator is optimal uniformly for all $\theta \in \Theta$.

Another device of guessing the true value $\alpha(\theta)$ of a parameter $\alpha : \Theta \to \mathbf{R}$ is to chose, depending upon the observed valued $e(\omega)$ or an – say unbiased – estimator e, a so-called *confidence interval*

$$C(\omega) = [e(\omega) - a(\omega),\ e(\omega) + b(\omega)].$$

One guesses $\alpha(\theta)$ to be in that interval if ω was observed. The loss function usually adopted here is

$$L(\theta, \alpha) = \begin{cases} 0 & \text{if} \quad \alpha \in C(\omega) \\ 1 & \text{if} \quad \alpha \notin C(\omega). \end{cases}$$

This leads to the "error probability" risk function

$$\begin{aligned} R(\theta, e) &= p_\theta(\alpha \notin C(\cdot)) \\ &= p_\theta(\{\omega | \alpha < e(\omega) - a(\omega) \text{ or } e(\omega) + b(\omega) < \alpha\}) \end{aligned}$$

One usually choses $a(\omega)$, $b(\omega)$ such that $R(\theta, e) \le 0.01$ or ≤ 0.05, and one is interested in short confidence intervals.

We content ourselves here with these glimpses on paramter estimation and refer the reader to the standard literatur on the subject, recommending Krengel [1988] as a first reading, and Rohatgi [1984] as a comprehensive textbook.

VI. Markov Processes

Classical probability theory investigates independent identically distributed random varialbes: the IID world.

Not everything which comes out of IID random variables is IID again: even in the IID world some dependence phenomena have to be taken into account. If we wish to study depence phenomena in a systematic fashion, we have two methodological options. One is to forget about all the wonderful implications of independence and to plunge headlong into the ocean of dependencies. Another, and certainly not unwise, option is to move out of the realm of IID step by step only and to begin by investigating sort of a first order vicinity of independence. This enterprise goes under the headline

Markov Processes

which is to remind the reader of Andrej Andrejevic Markov (1845 -1922), professor at St.Petersburg/Leningrad, whose seminal publications Markov [1906][1924] mark the early phase of this vast development in stochastic. The most famous of all Markov Processes

Brownian Motion

– to this day an object of intense research, with famous contributions by Brown (1773-1858) [1828], Einstein [1906], Wiener [1923], Kakutani [1944] – is far beyond the scope of this book, but we will do a few steps towards that direction.

The theory of Markov processes investigates, so to say, the microscopic reality below Markov dynamics (ch. II) which, in a way, can be incorporated into the former. This will be done in §2 of the present chapter which begins with a prelude on conditional probabilities and the Bayesian view on statistical inference.

1. Conditional Probabilities

1.1. Examples.

Example 1.1. Consider example no.4 ("inside information") from ch. I §2. The appropriate DPS in that case is

two mines	exactly one mine	no mines
$\frac{6}{1560} = 0.00384\ldots$	$\frac{222}{1560} = 0.14230\ldots$	$\frac{1332}{1560} = 0.8538\ldots$

The inside information excludes the rightmost possibility ("no mine"). By "restricting and renorming", you get the new probabilities

two mines	exactly one mine	no mine
$\frac{6}{228} = 0.0263$	$\frac{222}{228} = 0.9737$	0

Example 1.2. Imagine the following scenario which might have been quite realistic around 1750

> London shipowner A outfits a slow fat vessel for East India. It is to sail there via Cape Town and to return after 8 months. Every third such ship on that route gets lost. But if the ship in question returns safely, it will bring 10^6 £. Shipowner A sells 100 shares of 10^4 £ each, of that profit. Because of the said loss rate, no one will buy such a share for 10^4 £. No, $\frac{2}{3} \cdot 10^4 = 6666.66\ldots$ £ is the price that will do. After 6 months, shareholder B is in urgent need of some cash and wants to sell his 10^4 £ share at the highest possible price. Now, a fast clipper has brought message that the said vessel has made it safely back as far as Cape Town. Four out of five such ships do the passage from Cape Town to London without damage. Thus B can sell his share for $\frac{4}{5} \cdot 10^4 = 8000$ £, making even a modest profit.

Now what is the correct probability model for all that? The vast ocean of possibilities is split into three disjoint events

safe return to London : L

safe return to Cape Town,
but loss during the passage from Cape Town to London : C

no return even to Cape Town : N

Our scenario leads to the following setup for a DPD:

(1)

Event	L	C	N
Probability	$\frac{2}{3}$	$\frac{1}{5}(1-x)$	x

The normalization condition $\frac{2}{3}+\frac{1}{5}(1-x)+x=1$ leads to $\frac{4}{5}x=1-\frac{2}{3}-\frac{1}{5}=\frac{2}{15}$, whence $x=\frac{1}{6}$:

(2)

Event	L	C	N
Probability	$\frac{2}{3}$	$\frac{1}{6}$	$\frac{1}{6}$

In this final setup, the probabilities $\frac{4}{5}, \frac{1}{5}$ are no longer clearly visible, although they play a key rôle in reality. Can we recover them in a transparent fashion, that is, according to a plausible methodological scheme? Well, L and C together represent the case that a ship returns at least to Cape Town. $p(\{L,C\}) = \frac{2}{3}+\frac{1}{6}=\frac{5}{6}$. Of this value, $\frac{2}{3}=\frac{4}{6}$ represents the fraction $\frac{4}{5}$. Mathematically, this calculation is nothing but an application of the scheme "restricting and renorming" displayed in ch. III. §2.

Example 1.3. At an international airport the probability for a passenger to have caught malaria depends on the world region where he comes from. Assume that of the passengers arriving from

S-America	SE Asia	Africa	Near East	others
1,5 %	1 %	2 %	0.5 %	0 %

have caught malaria there and that, on the average, the percentages of arrivals are from

S-America	SE Asia	Africa	Near East	others
5 %	30 %	20 %	10 %	35 %

What is the probability of an arriving passenger with malaria to have caught it in Africa? – Clearly we have, at the outset, to list all possible combinations of proveniences and malaria status. That is, we have to consider the sets $X = \{$S-America, SE-Asia, Africa, Near East, other$\}$ and $X_1 = \{$Malaria, no Malaria$\}$ and to form their cartesian product $\Omega = X_0 \times X_1$. The probability distribution p on Ω has to be computed out of the distribution

$$p^{(0)} = \left(\frac{5}{100}, \frac{30}{100}, \frac{20}{100}, \frac{10}{100}, \frac{35}{100}\right)$$

on X_0, and the transition matrix

$$P = \begin{pmatrix} \frac{1.5}{100} & \frac{98.5}{100} \\ \frac{1}{100} & \frac{99}{100} \\ \frac{2}{100} & \frac{98}{100} \\ \frac{0.5}{100} & \frac{99.5}{100} \\ \frac{0}{100} & \frac{100}{100} \end{pmatrix} \qquad \begin{array}{c}\text{rows: proveniences} \\ \text{columns: malaria status}\end{array}$$

as follows: the probability of a combination j, k of provenience j and Malaria status k is $p_j^{(0)}\ P_{jk}$. Thus p may be written out as

$$\begin{pmatrix} \frac{1.5\cdot 5}{100^2} & \frac{9.5\cdot 5}{100^2} \\ \frac{1\cdot 30}{100^2} & \frac{99\cdot 30}{100^2} \\ \frac{2\cdot 20}{100^2} & \frac{8\cdot 20}{100^2} \\ \frac{0.5\cdot 10}{100^2} & \frac{99.5\cdot 10}{100^2} \\ \frac{0\cdot 35}{100^2} & \frac{100\cdot 35}{100^2} \end{pmatrix} = \begin{pmatrix} 0.00075 & 0.04925 \\ 0.00300 & 0.29700 \\ 0.00400 & 0.19600 \\ 0.00050 & 0.09950 \\ 0 & 0.35000 \end{pmatrix}$$

If we consider "malarians" only, we have to look at the left columns. The "African fraction" there is

$$\frac{400}{75 + 300 + 400 + 50} = \frac{400}{825} = 0.4848485$$

The complete list of such "if" or "conditional" probabilities ("if malaria, then from ...") is

S-America	SE-Asia	Africa	Near East
0.0909091	0.3636364	0.4848485	0.0606060

Clearly we calculated it by "restricting and renorming" again. We could have done it more quickly by realizing that 100^2 cancels out everywhere, and by concentrating on the left columns ("malaria") alone. It is qualitatively plausible that we have to weigh the entries there with the arrival frequencies: if e.g. all arrivals were from Near East, the smaller infection rate of that area would not prevent it from contributing 100 % of the "malarians" at our airport

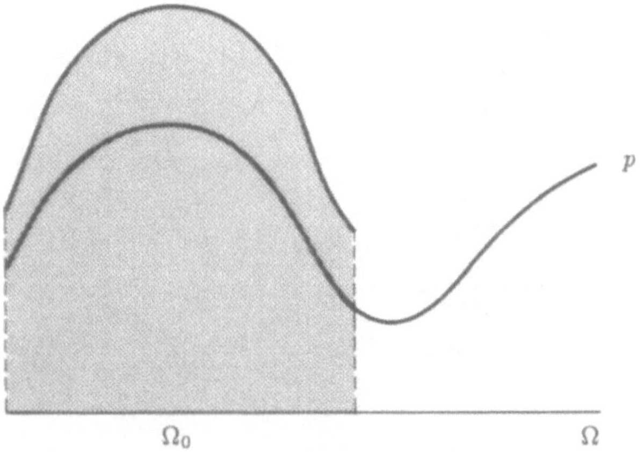

fig. VI.1.1

1.2. Conditional Probabilities.

The above examples motivate the following

Definition 1.4.

Let (Ω, p) be a DPS. Let $\Omega_0 \subseteq \Omega$, $p(\Omega_0) > 0$.

1) For every $E \subseteq \Omega$,
$$p(E|\Omega_0) = \frac{p(E \cap \Omega_0)}{p(\Omega_0)}$$
is called the ***conditional probability*** of E under (the condition) Ω_0.
2) $p(\cdot|\Omega_0)$ defines a DPD on Ω which has no mass at $\Omega \backslash \Omega_0$ any more. It is called the ***conditional probability distribution*** for p under (the condition) Ω_0. It is obviously a DPD again.

If we imagine p as a cake — a mass distribution — sitting on Ω, the passage to $p(\cdot|\Omega_0)$ can be visualized as shown in figure VI.1.1.

cutting off the non-zero chunk sitting over Ω_0
carrying off the rest of the cake
renorming the Ω_0-chunk with the factor $\frac{1}{p(\Omega_0)} > 1$.

We might, of course, as well cut the cake into chunks, take nothing away, renorm each chunk, and see e.g. how to recover p from the result (see figure VI.1.2).

In formulas: starting from a finite disjoint decomposition

$$\Omega = \Omega_1 \cup \Omega_2 \cup \ldots \cup \Omega_{n-1} \cup \Omega_n \qquad \Omega_j \cap \Omega_k = \emptyset \; (j \neq k)$$

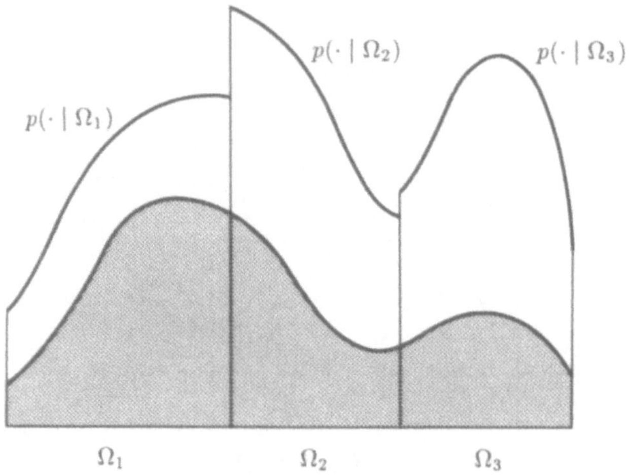

fig. VI.1.2

with

$$p(\Omega_1) > 0, \ldots, p(\Omega_0) > 0$$

we form

$$p(E|\Omega_j) = \frac{p(E \cap \Omega_j)}{p(\Omega_j)} \qquad (E \subseteq \Omega,\ j = 1, \ldots, n)$$

Clearly

$$\begin{aligned} p(E) &= p(E \cap \Omega_1) + \ldots + p(E \cap \Omega_n) \\ &= p(\Omega_1)\tfrac{p(E\cap\Omega_1)}{p(\Omega_1)} + \ldots + p(\Omega_n)\tfrac{p(E\cap\Omega_n)}{p(\Omega_n)} \end{aligned}$$

i.e.

(1) $p(E) = p(\Omega_1)p(E|\Omega_1) + \ldots + p(\Omega_n)p(E|\Omega_n) \qquad (E \subseteq \Omega)$

This result is sometimes called the

complete probability formula

It may also be written as

(2) $p = p(\Omega_1)p(\cdot|\Omega_1) + \ldots + p(\Omega_n)p(\cdot|\Omega_n)$.

In view of $p(\Omega_1) + \ldots + (\Omega_n) = 1$ this is nothing but a representation of p as a convex combination of the conditional probability distributions $p(\cdot|\Omega_1), \ldots, p(\cdot|\Omega_n)$. The hypothesis $p(\Omega_1) > 0, \ldots, p(\Omega_n) > 0$ may be eliminated via an obvious convention like $0 \cdot \frac{a}{0} = 0$ or the permission of choosing

$0 \leq p(E|\Omega_j) \leq 1$ arbitrarily in case $p(\Omega_j) = 0$. It is also obvious how to extend these results to *countable* disjoint decompositions of Ω.

Exercise 1.5. Let D, E be finite non-empty sets. Show that for every probability distribution $m = (m_{jk})_{j \in D,\, k \in E}$ on $D \times E$ there is exactly one probability distribution $p = (p_j)_{j \in D}$ on D and at least one stochastic $D \times E$-matrix $P = (P_{jk})_{j \in D,\, k \in E}$ such that

$$m_{jk} = p_j P_{jk} \qquad (j \in D,\ k \in E).$$

Show that P is unique if $m_{jk} > 0$ $(j \in D,\ k \in E)$. Show that there is a probability distribution $q = (q_k)_{k \in E}$ and a stochastic $E \times D$-matrix $Q = (Q_{kj})_{k \in E,\, j \in D}$ such that

$$m_{jk} = Q_{kj} q_k$$

and hence

$$Q_{kj} = P_{jk} \frac{p_j}{q_k} \quad (j \in D,\ k \in E,\ q_k > 0)$$

represent the terms P_{jk} and Q_{kj} as conditional probabilities for m, and express the p_j and the q_k according to the complete probability formula. Show that $p_i = \sum_{k \in E} m_{jk}$ $(j \in D)$, $q_k = \sum_{j \in D} m_{jk}$ $(k \in E)$. It is suggestive to visualize this situation (taking $D = \{1, \ldots, r\}, E = \{1, \ldots, s\}$) by

	q_1	$\ldots$	q_k	$\ldots$	q_s
p_1	m_{11}	$\ldots$	m_{1k}	$\ldots$	m_{1s}
$\vdots$	$\ldots$	$\ldots$	$\ldots$	$\ldots$	$\ldots$
p_j	m_{j1}	$\ldots$	m_{jk}	$\ldots$	mjs
$\vdots$	$\ldots$	$\ldots$	$\ldots$	$\ldots$	$\ldots$
p	m_{r1}	$\ldots$	m_{rk}	$\ldots$	m_{rs}

writing the row sums p_i and the column sums q_k on the *margin* of the matrix

$$m = (m_{jk})_{j \in E,\, k \in E}$$

p and q are therefore sometimes called the *marginal (distribution)* of the DPD m on $D \times E$. This verbiage is also used in more general product space situations.

Example 1.5. (Hardy-Weinberg). In diploidal genetics, every "locus" in a pair of chromosomes can be occupied by one "allel" of the gene sitting on that locus. Assume that for some specific locus/game) we have two alleles, A and a, so that the three possible "genotypes" for that gene are

AA Aa aa

(genetically, aA and Aa are equivalent). In the process of mating each of the two parents sends one of his alleles, each one with probability $\frac{1}{2}$, into the — say — egg cell. Let us assume that the two parents act independently so that these transition probabilities multiply. The transition matrix from the genotype of the parents to that of the offspring is thus

father/mother's genotype	offspring's genotype		
AA/AA	1	0	0
AA/Aa	1/2	1/2	0
AA/aa	0	1	0
Aa/AA	1/2	1/2	0
Aa/Aa	1/4	1/2	1/4
Aa/aa	0	1/2	1/2
aa/AA	0	1	0
aa/Aa	0	1/2	1/2
aa/aa	0	0	1

Assume now that, in the whole population, the initial frequencies of the three genotypes are

AA	Aa	aa
u	$2v$	w

$$u + 2v + w = 1$$

($2v$ instead of v is suggested by $Aa = aA$, and will simplify some calculations.)
Random mating produces the nine father/mother genotype combinations with the following frequencies:

AA/AA	AA/Aa	AA/aa	Aa/AA	Aa/Aa	Aa/aa	aa/AA	aa/Aa	aa/aa
u^2	$2uv$	uw	$2uv$	$4v^2$	$2vw$	uw	$2vw$	w^2

With (1), (2) and exercise 1.5. in mind, we now weigh the three columns of our transition matrix with these weights (of sum 1!). We calculate the resulting probabilities for the genotypes of the offspring:

for AA: $u^2 \cdot 1 + 2uv \cdot \frac{1}{2} + uw \cdot 0 + 2uv \cdot \frac{1}{2} + 4v^2 \cdot \frac{1}{4} + 0 + 0 + 0 + 0$
$= u^2 + uv + uv + v^2$
$= u^2 + 2uv + v^2 = (u+v)^2 \overset{\text{def}}{=} \bar{u}$

for aa: $(v+w)^2 \overset{\text{def}}{=} \bar{w}$, for symmetry reasons.

for Aa: $1 - \bar{u}\bar{w} = 1 - (u+v)^2 - (v+w)^2$
$= ((u+v)+(v+w))^2 - (u+v)^2 - (v+w)^2 = 2(u+v)(v+w) \overset{\text{def}}{=} 2\bar{v}$

What would be the genotype frequencies $\bar{\bar{u}}, 2\bar{\bar{v}}, \bar{\bar{w}}$ for the overnext generation? Upon replacing $u, 2v, w$ by $\bar{u}, 2\bar{v}, \bar{w}$ in our formulas, we obtain

$$\begin{aligned}
\bar{\bar{u}} &= (\bar{u}+\bar{v})^2 = ((u+v)^2+(u+v)(v+w))^2 \\
&= ((u+v)(u+v+v+w))^2 = (u+v)^2 = \bar{u} \\
\bar{\bar{w}} &= (\bar{v}+\bar{w})^2 = (v+w)^2 = \bar{w} \quad \text{for symmetry reasons} \\
2\bar{\bar{v}} &= 2(\bar{u}+\bar{v})(\bar{v}+\bar{w}) = 2(u+v)(v+w) = 2\bar{v}.
\end{aligned}$$

That is, the frequencies become stable after one generation. This is the famous Hardy-Weinberg Law of genetics (Hardy [1908], Weinberg [1908], independent discoveries).

Exercise 1.6. (Plausible reasoning). Translate the following principles of plausible reasoning into the language of conditional probabilities, and prove them:

1. If $E \Rightarrow F$ and F turns out to be true, then E becomes more plausible (hint: $E \subseteq F \Rightarrow m(E/F) \geq m(E)$)
2. If G is sufficient for E and G turns out to be false, then E becomes less plausible.
3. If C competes incompatibly with E, and if C turns out to be false, the E becomes more plausible.

Do you have more specimens of this kind? See Pólya [1954].

1.3. Conditional Expectation.

A conditional probability distribution $p(\cdot|\Omega_o)$ may, of course, be used in order to form expectations $E_{p(\cdot|\Omega_o)}f$ of real-valued random variables in the same fashion as the DPD p from which it was derived. It would be natural to call the result *conditional expectation*. The customary terminology deviates, however – and for good reasons – a bit from this:
To pass from a real-valued RV f to its expectation $E_p f$ may be considered as a passage from the possibly non-constant function f to a constant function whose constant value is $E_p f$. We might denote this function by $E_p f \cdot 1_\Omega$, where 1_Ω is the indicator function of Ω, namely, the constant 1. In this vein, it seems plausible to focus on the *function* $[E_{p(\cdot|\Omega_o)}f] \cdot 1_{\Omega_o}$, and, in the case of a disjoint decomposition $\Omega = \Omega_1 \cup \ldots \cup \Omega_0$, on the function

$$\left[E_{p(\cdot|\Omega_1)}f\right] \cdot 1_{\Omega_1} + \ldots + \left[E_{p(\cdot|\Omega_n)}f\right] 1_{\Omega_n}$$

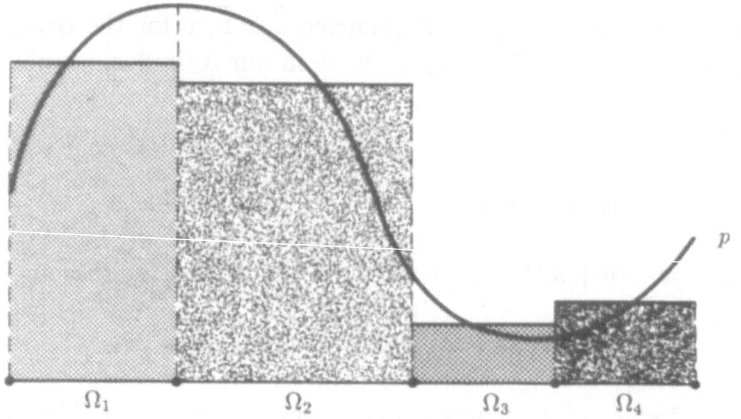

fig. VI.1.3

which replaces, on each Ω_j, the original f by its *average* (for p) over Ω_j: the weights which p has in Ω_j are – after renorming – used in order to average the values of f over Ω_j (figure VI.1.3).

The result is a piecewise (on the Ω_j's) constant function which coincides

with f p-a.e. if each Ω_j contains only one nonzero point mass of the DPD p (in particular, if all Ω_j's are singletons)

with the constant $E_p f$ if $p(\Omega_j) = 0$ for all but one j (in particular, if $n = 1$, $\Omega_1 = \Omega$).

As for notation, it is customary to replace the above decomposition by the corresponding σ-algebra $\mathcal{B}_0$, and thus to define

$$E(f|\mathcal{B}_0) = \sum_j \left[E_{p(\cdot|\Omega_j)f}\right] 1_{\Omega_j}$$

as the conditional expectation (function) of f with respect to $\mathcal{B}_0$. This notion and notation plays, appropriately generalized, a fundamental rôle in measure-theoretical probability theory.

Exercise 1.7. Prove that $E(f|\mathcal{B}_0)$ is, up to modifications on p-nullsets, the only function f_0 with the following two properties:

1) f_0 is constant on every "atom" Ω_j of $\mathcal{B}_0$.
2) $E_p(f_0 1_{F_0} = E_p(f 1_{F_0})$ for every $F_0 \in \mathcal{B}_0$.

1.4. Bayes' Formula.

Let (Ω, p) be a DPS and

$$\begin{aligned}\Omega &= A_1 \cup \ldots \cup A_m \\ \Omega &= B_1 \cup \ldots \cup B_n\end{aligned}$$

two disjoint decompositions of Ω. For simplicity, we shall assume $p(A_1) > 0, \ldots, p(B_n) > 0$. If we express, in

$$p(A_j|B_k) = \frac{p(A_j \cap B_k)}{p(B_k)}, \quad p(B_k|A_j) = \frac{p(A_j \cap B_k)}{p(A_j)}$$

the numerators and denominators by conditional probabilities with respect to the "other" decomposition, we arrive at the so-called *Bayes formulas* (Bayes [1763]):

$$p(A_j|B_k) = \frac{p(B_k|A_j)p(A_j)}{\sum_{i=1}^m p(B_k|A_j)p(A_j)}, \quad p(B_k|A_j) = \frac{p(A_j|B_k)p(B_k)}{\sum_{\ell=1}^n (A_j|B_\ell p(B_\ell)}$$

The reader should, at this stage, carefully reconsider exercise 1.5 in the light of our two Bayes' formulas. I have written them out in "both directions" in order to emphasize the inner symmetry prevailing in the situation envisaged. There is, however, one classical interpretation of Bayes' formula, for which asymmetry is characteristic. We shall explain it in the notation of exercise 1.5 here:

Let D and E be finite nonempty sets. Every $j \in D$ is interpreted as a "hypothesis" about the "true" probability distribution prevailing in E: $P_{j\cdot} = (P_{jk})_{k\in E}$, the j-th row of a certain stochastic $D \times E$-matrix $P = (P_{jk})_{j\in D,\, k\in E}$. We imagine that we are not sure about which hypothesis $j \in D$ to adopt but that we have a certain ("apriori") probability p_j which we ascribe to j. With $A_j = \{j\} \times E \quad (j \in D)$, $B_k = D \times \{k\} \quad (k \in E)$, Bayes' formula yields, for every $k \in E$, a new probability distribution over D: to $j \in D$ ascribe in case k was observed, the "aposteriori" probability

$$\frac{p_j P_{jk}}{\sum_{i\in D} p_i P_{ik}} (= \frac{p_j}{q_k} P_{jk} = Q_{kj} \text{ in the notation of exercise 1.5.})$$

This interpretation represents, in a nutshell, the "Bayesian" view that in "statistical inference" we never should, essentially, do anything else than modifying apriori distributions over parameter sets (here D). This view is, among statisticians, highly controversial (see e.g. Lindley [1978][1988]). The idea presented here were initiated by Bayes [1763], a postumous publication of Rev. Thomas Bayes (1702-1761).

2. Markov Processes

Let $D \neq \emptyset$ be a finite set. A stochastic $D \times D$-matrix $P = (P_{jk})_{j,k \in D}$ may serve as a model for "Markovian" dynamics: if the system under investigation is in state j at time t, we will find it in state k at time $t+1$ – not for sure as a rule, but with a certain probability P_{jk} which will, as a rule, really depend upon j.

The "macroscopic" aspect of this kind of non-deterministic dynamics was thoroughly investigated in ch. II and appendix D. That branch of "Markov Theory" answers questions like

> given an initial probability distribution, how will it evolve during the elapse of (discrete) time?
>
> Answer: asymptotic periodicity, sometimes convergence.

That theory can thus tell us, what the probability is, to find our system in a certain state k at a time t, say, $t = 10^6$. It cannot tell us, what the probability is that state k is reached at time 10 via state j_1 at time 5 and state j_2 at time 8. This is a typical "microscopic" question, and the refined Markovian models to be treated in the remainder of this chapter are precisely of that type.

2.1. Markov Distributions with a finite state space.

Let us begin with a repetition of definition III.§2

Definition 2.1. Let $D \neq \emptyset$ be a finite set, p a probability distribution over D and P a stochastic $D \times D$-matrix. For every natural number the *Markov* (probability) *distribution* m on $\Omega = D^{t+1} = \{\omega = (j_0, j_1, \ldots, j_t) | j_0, \ldots, j_t \in D\}$, with *initial distribution* p and *transition matrix* P is defined by

$$m_\omega = p_{j_0} P_{j_0 j_1} \cdots P_{j_{t-1} j_t} \qquad (\omega = (j_0, \ldots, j_t) \in D^{t+1}).$$

We already know from III. §2.5. that this is in fact a probability distribution. Let us now calculate some probabilities and conditional probabilities for (Ω, m). We will define them conveniently with the help of the *component mappings* (RVs)

$$\begin{aligned} \varphi_u &: \Omega \to D \\ \varphi_u(\omega) &= j_u \qquad (\omega = (j_0, j_1, \ldots, j_t) \in \Omega = D^{t+1}) \end{aligned}$$

We imagine now a system with state space D, which evolves according to p, P, that is, according to m. We will denote the entries of the u-th power P^u of P by $P^{(u)}_{jk}$, $(j, k \in D)$. The probability

1. to find our system in state j at time 0 is
$$\begin{aligned} m(\varphi_0 = j) &= \sum_{\omega=(j,j_1,\ldots,j_t)} m_\omega \\ &= \sum_{j_1,\ldots,j_t \in D} p_j\, P_{jj_1} \ldots P_{j_{t-1}j_t} \\ &= p_j \end{aligned}$$
2. to find our system in state k at time 1 is
$$\begin{aligned} m(\varphi_1 = k) &= \sum_{\omega(j_0,k,j_2,\ldots,j_t)} m_\omega \\ &= \sum_{j_0,j_2,\ldots,j_t \in D} p_{j_0}\, P_{j_0k}\, P_{kj_2} \ldots P_{j_{t-1}j_t} \\ &= \sum_{j_0 \in D} p_{j_0}\, P_{j_0k} \\ &= (pP)_k \end{aligned}$$
3. to find our system in state k at time $1 \leq u \leq t$ is
$$(pP^u)_k$$
(details as an exercise!)
4. to find our system in state j at time 0 and in state k at time 1 is
$$\begin{aligned} m(\varphi_0 = j, \varphi_1 = k) &= \sum_{\omega=(j,k,j_2,\ldots,j_t)} m_\omega \\ &= \sum_{j_2,\ldots,j_t \in D} p_j\, P_{jk}\, P_{kj_2} \ldots P_{j_{t-1}j_t} \\ &= p_j\, P_{jk} \end{aligned}$$
5. to find our system in state j at some time $u < t$ and in state k at time $u+1$ is
$$\begin{aligned} & m(\varphi_u = j,\ \varphi_{u+1} = k) \\ = \ & \textstyle\sum_{j_0,\ldots,j_{u-1},j_{u+2},\ldots,j_t} p_{j_0}\, P_{j_0j_1} \ldots P_{j_{u-1}j}\, P_{jk}\, P_{kj_{u+1}} \ldots P_{j_{t-1}t\ldots j_t} \\ = \ & (pP^u)_j\, P_{jk} \end{aligned}$$
6. to find our system in state i at time 0, in state j at time 1, and in state k at time 2 is (now obviously, isn't it?)
$$\begin{aligned} m(\varphi_0 = i,\ \varphi_1 = j,\ \varphi_2 = k) &= \sum_{\omega=(i,j,k,j_3,\ldots,j_t)} m_\omega \\ &= p_i\, P_{ij}\, P_{jk} \end{aligned}$$
7. to find our system in state j_0 at time 0, j_1 at $1, \ldots, j_u$ at time $u(\leq t)$ is
$$m(\varphi_0 = j_0, \ldots, \varphi_u = k) = p_{j_0}\, P_{j_0j_1} \ldots P_{j_{u-1}j_u}$$

8. to find our system in state j_u at time $0 < u < t$, in state j_{u+1} at time $u+1, \ldots$, in state j_t at time t is
$$m(\varphi_u = j_u, \ldots, \varphi = j_t) = (pP^u)_{j_u} P_{j_u j_{u-1}} \ldots P_{j_t j_t}$$
9. to find our system in state j_{u_1} at time $0 \leq u_1 < t$, in state j_{u_2} at time $u_1 < u_2 < t, \ldots$, in state j_{u_r} at time $u_{r-1} < u_r \leq t$ is
$$m(\varphi_{u_1} = j_{u_1}, \ldots, \varphi_{u_r} = j_{u_r}) = (pP^{u_1})_{j_{u_1}} P^{(u_2-u_1)}_{j_{u_1} j_{u_2}} \ldots P^{(u_r-u_{r-1})}_{j_{u_{r-1}} j_{u_r}}$$

The *conditional* probability

10. to find our system in state k at time 1 *if* it was in state j at time 0 is
$$m(\varphi_1 = k | \varphi_0 = j) = \frac{p_j\, P_{jk}}{p_j} = P_{jk}$$
(we assume $p_j > 0$).
11. to find our system in state k at time $u + 1 (\leq t)$ *if* it was in state j at time u is
$$m(\varphi_{n+1} = k | \varphi_u = j) = \frac{(pP^u)_j\, P_{jk}}{(pP^u)_j} = P_{jk}$$
(we assume $(pP^u)_j > 0$).
12. to find our system in state k at time $v \leq t$ *if* if was in state j at time $0 \leq u < v$, is
$$m(\varphi_v = k | \varphi_u = j) = \frac{(pP^u)_j\, P^{(v-u)}_{jk}}{(pP^u)_j} = P^{(v-u)}_{jk}$$
13. to find our system in state j at time u, and in state k at time $u + 1$ *if* it was in state i at time $u - 1$ (assume $0 < u < t$) is
$$m(\varphi_u = j,\ \varphi_{n+1} = k | \varphi_{u-1} = i) = \frac{(pP^{u-1})_i\, P_{ij}\, P_{jk}}{(pP^{u-1})_i} = P_{ij}\, P_{jk}$$
(assume $(pP^{n-1})_j > 0$).
14. to find our system in state i at time $u - 1$ and in state k at time $u + 1$ *if* it is in state j at time u (assume $0 < u < t$) is
$$\begin{aligned} m(\varphi_{u-1} = i,\ \varphi_{n+1} = k | \varphi_u = j) &= \frac{(pP^{u-1})_i\, P_{ij}\, P_{jk}}{(pP^u)_j} \\ &= \frac{(pP^{u-1})_i\, P_{ij}}{(pP^u)_j} \cdot \frac{(pP^u)_j\, P_{jk}}{(pP^u)_j} \\ &= m(\varphi_{u-1} = i | \varphi_u = j) \\ &\quad \cdot m(\varphi_{u+1} = k | \varphi_u = j) \end{aligned}$$
(assume $(pP^u)_j > 0$).

Exercise 2.2. Prove: if $0 < u < t$ and if $E \subseteq \Omega$ can be defined "in terms of $\varphi_0, \ldots, \varphi_{n-1}$" and $F \subseteq \Omega$ can be defined "in terms of $\varphi_{u+1}, \ldots, \varphi_t$",

then E and F are independent for the conditional probability distribution $m(\cdot|\varphi_u=j)$.
In short: *in a Markov model, past and future are independent, given the present.*

Proposition 2.3. With the above notations, for arbitrary $0 < u \leq t$ and any choice of $j_0, \dots, j_u \in D$ such that $p_{j_0}\, P_{j_0 j_1} \dots P_{j_{u-2} j_{u-1}} > 0$, we have

$$m(\varphi_u = j_u | \varphi_{u-1} = j_{u-1}, \dots, \varphi_0 = j_0) = m(\varphi_u = j_u | \varphi_{u-1} = j_{u-1})$$

PROOF. The left hand member here is

$$= \frac{p_{j_0}\, P_{j_0 j_1} \dots P_{j_{u-2} j_{u-1}}\, P_{j_{u-1} j_u}}{p_{j_0}\, P_{j_0 j_1} \dots P_{j_{u-2} j_{u-1}}} = P_{j_{u-1} j_u},$$

which equals the right hand member. Obviously (yes ?),
$p_{j_0}\, P_{j_0 j_1} \dots P_{j_{u-2} j_{u-1}} > 0$ entails $(pP_{u-1})_{j_{u-1}} > 0$. □

2.2. Markov Processes.

Definition 2.4. Let (Ω, m) be a DPS and t a natural number.

1. A finite sequence $f_0, f_1, \dots, f_t$ of RVs on Ω, all with the same state space X, is called a *stochastic process* with discrete time $\{0, 1, \dots, t\}$ and state space X. The joint distribution of $f_0, f_1, \dots, f_t$ in X^{t+1} (definition in III.2.3) is also called the *distribution of that process.*
2. A stochastic process $f_0, f_1, \dots, f_t$ with state space X is called a *Markov process* in discrete time $\{0, 1, \dots, t\}$ if it has the *Markov property*
$$m(f_u = x_u | f_{u-1} = x_{u-1}, \dots, f_0 = x_0) = m(f_u = x_u | f_{u-1} = x_{u-1})$$
$$(0 < u \leq t, x_0, \dots, x_u \in X,\ m(f_0 = x_0, \dots, f_{u-1} = x_{u-1}) > 0)$$
Note that, as $f_0 = x_0, \dots, f_{u-1} = x_{u-1} \Longrightarrow f_{u-1} = x_{u-1}$,
$m(f_0 = x_0, \dots, f_{u-1} = x_{u-1}) > 0$ entails $m(f_{u-1} = x_{u-1}) > 0$.
3. Markov processes with discrete time and an at most countable state space are also called *Markov Chains.*

Clearly, proposition 2.3 may now be restated as: the special stochastic process $\varphi_0, \varphi_1, \dots, \varphi_t$ considered there is a Markov process with finite state space D. The reader should realize that proposition 2.3. and the considerations preceding it easily carried over to the case of a countable state space X, or even to a "general discrete situation" as envisaged in example III §2.5.– The generalization to a sequence of transition matrices $P^{(1)}, \dots, P^{(t)}$ instead of all the same matrix P for all transitions $0 \to 1, 1 \to 2, \dots, t-1 \to t$ is equally obvious. This enables us to formulate and to prove the following

Theorem 2.5. Let (Ω, m) be a DPS, t a natural number and $f_0, f_1, \ldots, f_t$ a stochastic process with state space X and distribution $\bar{m}$ (in X^{t+1}). Let $\varphi_0, \varphi_1, \ldots, \varphi_t$ be the component mappings $X^{t+1} \to X$, that is $\varphi_u((x_0, x_1, \ldots, x_t)) = x_u$ $(0 \leq u \leq t,\ x_0, x_1, \ldots, x_t \in X)$. Then the following statements hold:

1. The stochastic process $\varphi_0, \varphi_1, \ldots, \varphi_t$ on the DPS $(X^{t+1}, \bar{m})$ has the same distribution as $f_0, f_1, \ldots, f_t$
2. $f_0, f_1, \ldots, f_t$ is a Markov process iff $\bar{m}$ is a Markov probability distribution on X^{t+1}, with initial distribution $p = \varphi_0 m$ and transition matrixes $P^{(u)} = (P^{(u)}_{xy})_{x,y \in X}$:
$$P^{(u)}_{xy} = m(f_u = y | f_{u-1} = x) \quad (u = 1, \ldots, t;\ x, y \in X)$$

The proof is based on the observation

$$\varphi_u \circ (f_0 \times \ldots \times f_t) = f_u \quad (u = 0, 1, \ldots, t)$$

plus some tedious calculations which the reader should, however, carefully write out for himself.

According to this theorem, a Markov process may have a different transition matrix $P^{(u)}$ for every time step $u - 1 \to u$. A Markov process with all transition matrices equal, i.e. with $P^{(1)} = \ldots = P^{(t)} = P$ is called *stationary* with transition matrix P.

Exercise 2.6. Show that, with our previous notations, the Markov property holds for a process iff, for any given $x_u \in X$, $0 < u \leq t$, the conditional probabilities $m(f_t = x_t, \ldots, f_u = x_u | f_{u-1} = x_{u-1},\ f_{u-2} = x_{u-2}, \ldots, f_0 = x_0)$ do not really depend upon $x_{u-2}, \ldots, x_0$.

In the formulation of these definitions and theorems, we employed the conceptual machinery of DPSs in full generality. In the sequel, however, we will restrict ourselves, up to minor exceptions, to Markov processes and distributions with a finite or countable state space which we will denote by $D (= \{i, j, k, \ldots\})$, and not by $X (= \{x, y, z, \ldots\})$.

2.3. Standard Examples.

In this subsection we present some standard examples of stationary Markov processes.

A first class of examples was given in subsection 1: Markov chains given via Markov distributions (with all transition matrices equal) in a product space D^{t+1} of copies of a *finite* state space D, plus the component RVs

$\varphi_0, \varphi_1, \ldots, \varphi_t$. The finite stochastic matrices displayed in ch. II §1 fit into this framework.

If we allow a countable state space D, the same construction – generalized in an obvious fashion – leads to a second bunch of examples. "The drunkard on the cliff" fits in here, with $D = \mathbb{Z}_+ = \{0, 1, \ldots,\}$.

Let us consider the particular case $D = \mathbb{Z}$. If $d_1, \ldots, d_t$ are independent identically distributed (IID) $\mathbb{Z}$-valued RVs, the stochastic process formed by their successive partial sums

$$f_0 = 0,\ f_1 = d_1,\ f_2 = d_1 + d_2,\ f_t = d_1 + \ldots + d_t$$

is called the *random walk with increments* $d_1, \ldots, d_t$. In the special case where the distribution p of d_1 (and likewise of $d_2, \ldots, d_t$) puts probability $p_1 = \alpha$ on $+1$ and $p_{-1} = 1 - \alpha$ on -1 (and 0 else), we speak of *simple random walk*, and of *symmetric simple random walk* if $\alpha = \frac{1}{2}$. In that special case, the passage from f_{u-1} to f_u means that you flip a coin and walk one step upwards if you get head, say, and one step downwards if you get tail. Looking at figure VI.2.1, you see the possibilities for such up-or-down paths and imagine yourself walking in the street lattice of Kyoto or Manhattan, flipping a coin at every corner: random walk. If you rotate the picture to obtain figure VI.2.2., you will feel reminded of Galton's Board and the Pascal Triangle – justly: we treated such situations in ch. IV §4 already.

Thus the IID situation (with state space $\mathbb{Z}$) reappears in our present Markovian context twofold:

as the special case of transition matrices with all rows equal
as random walk, after passage to successive partial sums.

Of course, this constitutes quite an incentive to the probabilist to generalize known facts about random walks to the general Markovian situation.

We will now show that random walks are nothing but Markov chains with an initial distribution "point mass 1 at 0" (because $f_0 \equiv 0$), and transition matrices whose rows are translates of one and the same probability distribution q on $\mathbb{Z}$, namely, the distribution of d_1 (and of $d_2, \ldots, d_t$ likewise):

$$\begin{aligned}
& m(f_0 = 0,\ f_1 = j_1, \ldots, f_t = j_t) \\
= \;& m(d_1 = j_1,\ d_1 + d_2 = j_2, \ldots, d_1 + \ldots + d_t = j_t) \\
= \;& m(d_1 = j_1,\ d_2 = j_2 - j_1, \ldots, d_t = j_t - j_{t-1}) \\
= \;& q_{j_1}\, q_{j_2 - j_1} \cdots q_{j_t - j_{t-1}} \ \text{(independence!)} \\
= \;& P_{0 j_1}\, P_{j_1 j_2} \cdots P_{j_{t-1} j_t}
\end{aligned}$$

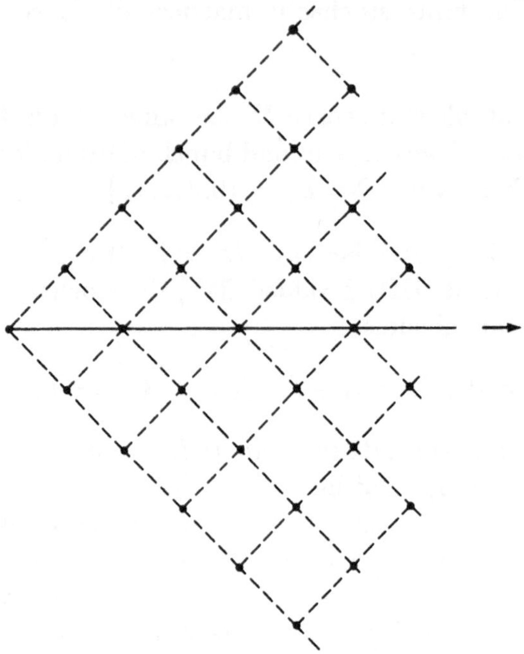

fig. VI.2.1

if we put

$$P_{jk} = q_{k-j} \quad (j, k \in \mathbb{Z}),$$

thus forming the stochastic $\mathbb{Z} \times \mathbb{Z}$-matrix

$$\mathbf{P} = \begin{pmatrix} & & & \vdots & & & \\ \cdots & \cdots & \cdots & \cdots & \cdots & \cdots & \cdots \\ \cdots & q_{-1} & q_0 & q_1 & q_2 & q_3 & \cdots \\ \cdots & q_{-2} & q_{-1} & q_0 & q_1 & q_2 & \cdots \\ \cdots & q_{-3} & q_{-2} & q_{-1} & q_0 & q_1 & \cdots \\ \cdots & \cdots & \cdots & \cdots & \cdots & \cdots & \cdots \\ & & & \vdots & & & \end{pmatrix}$$

Reading the above calculation in the opposite direction, we see that every Markov chain with such a transition matrix is the sequence of partial sums of IID $\mathbb{Z}$ valued RVs.

Further examples of Markov processes with an at most countable state space $D \subseteq \mathbb{Z}$ are obtained as modifications of a *simple random* walk $f_0, f_1, f_2, \ldots$ (say with parameter α) by introduction of *reflecting* resp. *adsorbing barriers*:

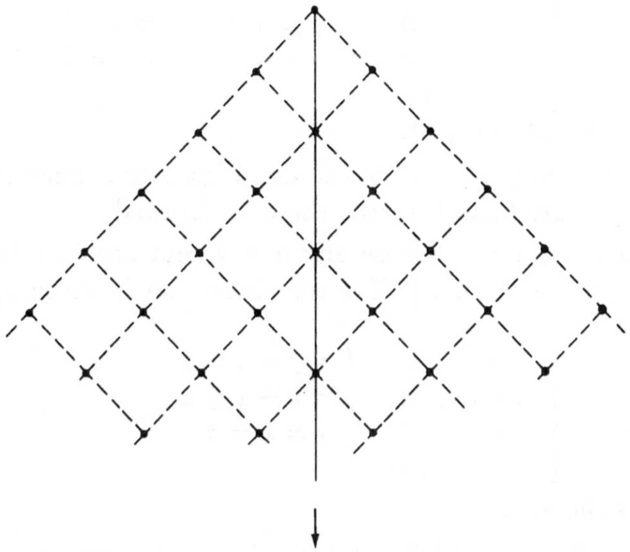

fig. VI.2.2

1. *One adsorbing barrier:* choose any $a \in \mathbb{Z}$ and consider the new state space $D = \{\ldots, a-2, a-1, a\}$. The transition matrix given for any parameter $0 \leq \alpha \leq 1$ by

$$Q_{jk} = \begin{cases} \alpha & if \quad k = j+1 \leq a \\ 1-\alpha & if \quad j = k+1 < a \\ 1 & if \quad j = k = a \\ 0 & else \end{cases}$$

represents the following idea:

> walk randomly – with probability α for "up by 1" and $1-\alpha$ for "down by 1" – as long as you stay $< a$; once you hit a you are bound to stay there – adsorbed at the upper barrier a.

Similarly for new state space $\{a, a+1, \ldots\}$ and adsorption at a again: lower adsorbing barrier. – The reader should recognize "the drunkard on the cliff" (ch. I) as such a random walk with one absorbing barrier, sometimes also called "cemetery" in Markov theory.

2. *Two adsorbing barriers:* choose $a < b$ in $\mathbb{Z}$ and consider the new state space $\{a, a+1, \ldots, b-1, b\}$. The transition matrix given by $0 \leq \alpha \leq 1$

and

$$Q_{jk} = \begin{cases} \alpha & \text{if} \quad a < j,\ k \le b,\ k = j+1 \\ 1-\alpha & \text{if} \quad a \le j,\ k < b,\ k = j-1 \\ 1 & \text{if} \quad j = k = a \text{ or } j = k = b \\ 0 & \text{else} \end{cases}$$

represents the following idea:

> walk randomly as long as you don't hit a or b: once you hit a or b, you are bound to stay there – adsorbed.

3. *One reflecting barrier:* choose any $a \in \mathbf{Z}$ and consider the new state space $\{\ldots, a-2, a-1, a\}$. The transition matrix given by $0 \le \alpha \le 1$ and

$$Q_{jk} = \begin{cases} \alpha & \text{if} \quad k = j+1 \le a \\ 1-\alpha & \text{if} \quad j = k+1 \le a \\ \alpha & \text{if} \quad j = k = a \\ 0 & \text{else} \end{cases}$$

represents the idea:

> walk randomly as long as you stay $< a$; if you are at a, go down by 1 with probability $1-\alpha$ as usual, but stay at a with probability α as if you had walked up but been pushed back by a reflecting (upper) barrier at $a + \frac{1}{2}$.

Exercise 2.7. Modify these examples such as to get a simple random walk with

> one lower reflecting barrier, or
> two reflecting barriers, or
> one reflecting and one adsorbing barrier.

We conclude this subsection with two more examples and exercises.

Example 2.8. (Waiting Line). Let $c_1, \ldots, c_t$ be IID $\mathbf{Z}_+$-valued RVs on a DPS (Ω, m). Define $\mathbf{Z}_+$-valued RVs $f_0, f_1, \ldots, f_t$ recursively by

$$f_0 = 0, \ f_u = max(0, f_{u-1} - 1) + c_u \quad (u = 1, \ldots, t)$$

Interpretation: a single server serves one customer per time unit; c_u new customers arrive between time $u-1$ and time u; f_u is the length of the waiting line at time u. As $f_0, \ldots, f_{u-1}$ are determined by $c_1, \ldots, c_{u-1}$ alone, c_u is independent of their join (exercise III.3.21). Thus we may calculate, for any $j_0, j_1, \ldots, j_u \in \mathbf{Z}_+$, in case $j_{u-1} > 0$

$$\begin{aligned} & m(f_u = j_u,\ f_{u-1} = j_{u-1}, \ldots) \\ = \ & m(c_u = j_u - (j_{u-1} - 1),\ f_{u-1} = j_{u-1}, \ldots) \\ = \ & m(c_u = j_u - j_{u-1} + 1) m(f_{u-1} = j_{u-1}, \ldots) \end{aligned}$$

and hence

$$\begin{aligned} & m(f_u = j_u | f_{u-1} = j_{u-1}, \ldots) \\ = \quad & m(c_u = j_u - j_{u-1} + 1); \end{aligned}$$

in case $j_{u-1} = 0$ we obtain similarly

$$m(f_u = j_u | f_{u-1} = j_{u-1}, \ldots) = m(c_u = j_u)$$

and all this for $u = 1, \ldots, t$. This is the Markov property: $f_0, f_1, \ldots, f_t$ is a stationary Markov chain with state space $\mathbb{Z}_+$.

Exercise 2.9. (Supplementing a store). Let $c_1, \ldots, c_t$ be IID $\mathbb{Z}_+$-valued RVs on a DPS (Ω, m), as in the preceding example. Interpretation: c_u is the number of buyers of – say – a TV in a certain store, on day no. u. Let f_0 be the number of TVs in the store on the evening before opening day no. 1, and assume that the following supply policy is adopted:

> if, on evening no. $u - 1$, less than 4 TVs are left unsold, more 5 will be supplied overnight; if 4 or more TVs are left, no supply takes place.

The number f_u of TVs left unsold on the evening of day no. u can thus be calculated by recursion:

$$f_u = \begin{cases} f_{u-1} + 5 - c_u & if \quad f_{u-1} \leq 3 \\ (f_{u-1} - c_u)_+ & if \quad f_{u-1} > 3 \end{cases}$$

Prove that $f_0, f_1, \ldots, f_t$ is a stationary Markov chain.

Exercise 2.10. Modify exercise 2.9. such that unserved customers return next day, and show that the resulting process is no Markov chain in general.

2.4. Infinite Time.

Let

> $D \neq \emptyset$ be at most countable
> $p = (p_j)_{j \in D}$ a probability vector over D
> $P = (P_{jk})_{j,k \in D}$ a stochastic $D \times D$-matrix

The n-th iterate of P will be denoted by

$$P^n = (P_{jk}^{(n)})_{j,k \in D} \quad (n = 0, 1, \ldots)$$

where, of course, $P^0 = I =$ the unit matrix $(\delta_{jk})_{j,k \in D}$ ($\delta_{jk} = 0$ if $j \neq k$, $=$ 1 if $j = k$). For every natural number $t \geq 1$ we may build the Markov DPS

$(\Omega^{(t)}, m^{(t)})$ as follows:

$$\Omega^{(t)} = D^{t+1} = \{(j_0, \ldots, j_t) | j_0, \ldots, j_t \in D\}$$

$$m^{(t)}_{(j_0,\ldots,j_t)} = p_{j_0} P_{j_0 j_1} \ldots P_{j_{t-1} j_t} \quad (j_0, \ldots, j_t \in D)$$

Every such $(\Omega^{(t)}, m^{(t)})$ models Markov transitions over a fixed finite time interval $\{0, 1, \ldots, t\}$ only. In order to obtain a model for unlimited numbers of transitions, we would have to embed all those $(\Omega^{(t)}, m^{(t)})$ into a probability space with basic set

$$\Omega = D^{\mathbf{Z}_+} = \{\omega = (j_0, j_1, \ldots) | j_0, j_1, \ldots \in D\}.$$

This can be done by means of measure theory, but leads to non-discrete probability distributions as a rule.

Since we want to stay within the framework of discrete stochastics here, we will rather work with the infinite sequence

$$(\Omega^{(1)}, m^{(1)}),\ (\Omega^{(2)}, m^{(2)}), \ldots$$

of DPSs. These DPSs are successive extensions of each other in the following obvious way: if we consider for every $u \in \mathbf{Z}_+$ and every $t \geq u$ the u-th component mapping $\varphi_u : \Omega^{(t)} \to D$ defined by $\varphi_u((j_0, \ldots, j_t)) = j_u$ $(j_0, \ldots, j_t \in D)$, then for every $0 \leq s \leq t$ the join $\varphi_0 \times \ldots \times \varphi_s$ maps $\Omega^{(t)}$ onto $\Omega^{(s)}$ and sends $m^{(t)}$ into $m^{(s)}$; in fact

$$\begin{aligned}
&[(\varphi_0 \times \ldots \times \varphi_s) m^{(t)}]_{(j_0,\ldots,j_s)} \\
&= \textstyle\sum_{j_{s+1},\ldots,j_t} p_{j_0} P_{j_0 j_1} \ldots P_{j_{s-1} j_s} \cdot P_{j_s j_{s+1}} \ldots P_{j_{t-1} j_t} \\
&= p_{j_0} P_{j_0 j_1} \ldots P_{j_{s-1} j_s} \textstyle\sum_{j_{s+1},\ldots,j_t} P_{j_s j_{s+2}} \ldots P_{j_{t-1} j_t} \\
&= p_{j_0} P_{j_0 j_1} \ldots P_{j_{2-1} j_s} \\
&= m^{(s)}_{(j_0,\ldots,j_s)}
\end{aligned}$$

as $\sum_{j_u} P_{j_{u-1} j_u} = 1$ $(u = s, \ldots, t-1)$. That is, for every event $F \subseteq \Omega^{(s)}$ we have

$$\begin{aligned}
m^{(s)}(F) &= m^{(t)}((\varphi_0 \times \ldots \times \varphi_s)^{-1} F) \\
&= m^{(t)}(F \times D^{t-s}) \quad (t \geq s)
\end{aligned}$$

and may thus recover F and $m^{(s)}(F)$ in $(\Omega^{(t)}, m^{(t)})$ again. With this understanding, we will skip the upper indices and write m instead of $m^{(t)}$. If the initial distribution p is ϵ_j = point mass 1 at j, we will write P_j instead of m, e.g.

$$P_j(\varphi_1 = j_1, \ldots, \varphi_t = j_t) = P_{jj_1} \ldots P_{j_{t-1}j_t}.$$

This technique allows us to consistently deal with probabilities of events involving transitions over time intervals only, but with no limitations as to the length of such time intervals. We will, however, find it necessary to form even probabilities of some events involving in an infinity of transition steps. Fortunately, only a very few of such events will be of real interest to us here, and these are of such a simple structure that we can introduce their probabilities as limiting values of probabilities of events involving only a finite time interval each.
Take e.g. the probability to hit k at time $t \geq 1$, starting from j:

$$\begin{aligned} P_j(\varphi_t = k) &= \sum_{j_1, \ldots, j_{t-1} \in D} P_{jj_1} \ldots P_{j_{t-1}k} \\ &= P_{jk}^{(t)}. \end{aligned}$$

It involves $\varphi_1, \ldots, \varphi_t$ only. The same is true for the probability to hit k at time t *but not earlier*. Let us denote it by $F_{jk}^{(t)}$ and thus obtain

$$\begin{aligned} F_{jk}^{(t)} &= P_j(\varphi_1 \neq k, \ldots, \varphi_{t-1} \neq k, \ \varphi_t = k) \\ &= \sum_{j_1, \ldots, j_{t-1} \neq k} P_{jj_1} \ldots P_{j_{t-1}k} \end{aligned}$$

Within $(\Omega^{(t)}, m^{(t)})$, the $F_{jk}^{(1)}, \ldots, F_{jk}^{(t)}$ are probabilities of pairwise disjoint events; thus we obtain

$$F_{jk}^{(t)} + \ldots + F_{jk}^{(t)} \leq 1.$$

We conclude the same for the infinite sum

$$F_{jk}^{(1)} + F_{jk}^{(2)} + \ldots \leq 1$$

We will now *interpret* this infinite sum as

the probability to *ever* hit k, starting from j

This is not the probability of an event in one of our DPSs $(\Omega^{(1)}, m^{(1)})$, $(\Omega^{(2)}, m^{(2)}), \ldots$; only measure theory could provide us with such an event ($\subseteq D^{\mathbf{Z}_+}$). The reader will, however, see that such "extended" probabilities will not lead to any real difficulties, and I assure him that things could be perfectly put in order by measure theory.

2.5. Recurrent and Transient States.

Within the framwork estabilished so far, we obtain, denoting expectation under P_j by E_j,

$$\begin{aligned} P_{jk}^{(t)} &= P_j(\varphi_t = k) \\ &= E_j 1_{\{\varphi_t = k\}} \\ \sum_{u=1}^{t} P_{jk}^{(u)} &= E_j(\sum_{u=1}^{t} 1_{\{\varphi_u = k\}}) \\ &= E_j(\text{number of visits to } k, \text{ until time } t) \end{aligned}$$

and thus find the following definition and interpretation justified

$$\begin{aligned} P_{jk}^{(*)} &= \sum_{t=1}^{\infty} P_{jk}^{(t)} \\ &= \text{expected number of visits to } k, \text{ after starting from } j. \end{aligned}$$

We also recall and define

$$\begin{aligned} F_{jk}^{(t)} &= P_j(\varphi_1 \neq k, \ldots, \varphi_{t-1} \neq k,\ \varphi_t = k) \\ &= P_j \text{ (first visit to } k \text{ at time } t) \\ F_{jk}^{(*)} &= \sum_{t=1}^{\infty} F_{jk}(t) \\ &= P_j \text{ (visit to } k \text{ at some time } t \geq 1) \end{aligned}$$

The reader should be aware of the fact that these probabilities and expectations depend upon the stochastic matrix $P = (P_{jk})_{j,k \in D}$ only, and upon nothing else.

Proposition 2.11.

$$P_j(\text{at least } r \text{ returns to } j) = (F_{jj}^{(*)})^r \quad (r = 1, 2, \ldots)$$

PROOF. For $r = 1$ the assertion is nothing but the definition of $F_{jj}^{(*)}$. Assume now that we are through for r. Then

$$\begin{aligned} &P_j \text{ (at least } r+1 \text{ returns to } j) \\ &= \textstyle\sum_{t=1}^{\infty} P_j \text{ (}r\text{th return to } j \text{ at time } t, \text{ at least one return more thereafter)} \\ &= \textstyle\sum_{t=1}^{\infty} \sum_{u=1}^{\infty} P_j \text{ (}r\text{th return to } j \text{ at time } t, (r+1)\text{th return at time } t+u) \end{aligned}$$

Now the event $\{r\text{th return to } j \text{ at time } j\}$ can be read from the values of

$\varphi_0, \varphi_1, \ldots, \varphi_t$, and {next return to j at time $t+u$} $= \{\varphi_{t+1} \neq j, \ldots, \varphi_{t+u-1} \neq j,\ \varphi_{t+u} = j\}$. The Markov property – or a tedious direct calculation – now implies the above probability to be

$$\begin{aligned}
&= \sum_{t=1}^{\infty}\sum_{u=1}^{\infty} P_j \text{ (}r\text{th return to } j \text{ at time } t) \\
&\quad \cdot P_j \text{ first return to } j \text{ at time } u) \\
&= (\sum_{t=1}^{\infty} P_j \text{ (}r\text{th return to } j \text{ at time } t)) \\
&\quad \cdot (\sum_{u=1}^{\infty} P_j \text{ (first return to } j \text{ at time } u)) \\
&= (F_{jj})^{(*)})^r\, F_{jj}^{(*)} = (F_{jj}^{(*)})^{r+1}.
\end{aligned}$$

□

Definition 2.12. Let $P = (P_{jk})_{j,k \in D}$ be a stochastic $D \times D$-matrix. A state $j \in D$ is called

1. *recurrent*, if
   P_j (return to j infinitely often) $= 1$
2. *transient*, if
   P_j (return to j infinitely often) < 1.

Recall that this definition involves the stochastic matrix $P = (P_{jk})_{j,k \in D}$, and nothing more, although we have used some intermediary constructions of DPSs in order to justify the verbiage "return" etc.

Proposition 2.13. One returns to j infinitely often with probability 1 iff one returns to j at all, and with probability 1.

PROOF. By proposition 2.11 we have

$$\begin{aligned}
&P_j \text{ (return to } j \text{ infinitely often)} \\
&= \lim_{r\to\infty} P_j \text{ (return to } j \text{ at least } r \text{ times)} \\
&= \lim_{r\to\infty} (F_{jj}^{(*)})^r
\end{aligned}$$

and this is $= 1$ iff $F_{jj}^{(*)} = 1$. □

Theorem 2.14. (Recurrence Theorem).

$$F_{jj}^{(*)} = 1 \iff P_{jj}^{(*)} = \infty$$

PROOF. By proposition 2.12

$$\begin{aligned} F_{jj}^{(*)} = 1 &\iff P_j \text{ (return to } j \text{ infinitely often)} = 1 \\ &\iff P_j \text{ (return to } j \text{ at least } r \text{ times)} = 1 \quad (r = 1, 2, \ldots) \\ &\iff E_j \text{ (number of returns to } j) \geq r \quad (r = 1, 2, \ldots) \\ &\iff P_{jj}^{(*)} \geq r \quad (r = 1, 2, \ldots) \\ &\iff P_{jj}^{(*)} = \infty. \end{aligned}$$

The reader should be aware that these probabilities and expectations involve a few limiting procedures which pose no real problems, however, since all terms are ≥ 0. □

Let us investigate recurrence and transience in a few special situations.

Example 2.15. Let D be finite. Markovian dynamics (ch. II) tells us that, for a given stochastic $D \times D$-matrix P, D splits into "cycles" and a "remainder" $N \subseteq D$ which is "emptied" with exponential speed:

$$P_{jk}^{(t)} \leq Ae^{-\alpha t} \quad (j, k \in N,\ t = 1, 2, \ldots)$$

for some $A, \alpha > 0$. Clearly this involves

$$P_{jj}^{(*)} = \sum_{t=1}^{\infty} P_{jj}^{(t)} \leq A \sum_{t=1}^{\infty} e^{-\alpha t} < \infty \quad (j \in N),$$

that is, all $j \in N$ are transient states. In order to find out what happens in the cycles, we may w.l.o.g. assume that D itself is one such cycle, i.e. that

$$D = D_0 \cup \ldots \cup D_{d-1}$$

with pairwise disjoint nonempty setes $D_0, \ldots D_{d-1}$ and some $d \geq 1$. From ch. II we know

$$\lim_{n \to \infty} P_{jk}^{(nd)} > 0$$

if j, k belong to the same D_ν. Clearly this involves

$$P_{jj}^{(*)} \geq \sum_{n=1}^{\infty} P_{jj}^{(nd)} = \infty \quad (j \in D),$$

that is, if we go back to our initial general situation, every $j \in D \backslash N$ is recurrent.

Example 2.16. Let P be simple random walk on $\mathbb{Z}$, i.e.

$$P_{jk} = \begin{cases} p & \text{for} \quad k = j+1 \\ 1-p & \text{for} \quad k = j-1 \\ 0 & \text{else} \end{cases}$$

with some $0 < p < 1$. Clearly

$$P_{00}^{(t)} = \begin{cases} 0 & \text{if} \quad t \text{ is odd} \\ \binom{2n}{n} p^n (1-p)^n & \text{if} \quad t = 2 \text{ is even} \end{cases}$$

Write $q = 1 - p$ and apply Stirling's formula (p. □□□) in order to obtain

$$P_{00}^{(2n)} \sim \frac{1}{\sqrt{\pi n}} (4pq)^n$$

In case $p \neq q$ we obtain $4pq < 1$ and thus

$$P_{00}^{(*)} = \sum_{t=1}^{\infty} P_{00}^{(t)} = \sum_{n=1}^{\infty} < \infty,$$

that is, 0 is a trasient state. By translation invariance of P the transience of all $j \in \mathbb{Z}$ follows. If $p = q = \frac{1}{2}$, the divergence of $\sum \frac{1}{\sqrt{n}}$ yields $P_{00}^{(*)} = \infty$ and thus 0 (and likewise every $j \in \mathbb{Z}$) turns out to be recurrent.

Example 2.17. Define simple random walk on $\mathbb{Z}^d = \mathbb{Z} \times \ldots \times \mathbb{Z}$ as the independent combination of d one-dimensional simple random walks all with the same parameter $0 < p < 1$. As return from $(0, \ldots 0) \in \mathbb{Z}^d$ to $(0, \ldots 0)$ is tantamount to simultaneous return of all the d component random walks, we obviously get transience if $p \neq q$, i.e. $p \neq \frac{1}{2}$. If $p = \frac{1}{2}$, we see

$$P_{(0,\ldots,0),(0,\ldots,0)}^{(2n)} \sim \left(\frac{1}{\sqrt{\pi n}} \right)^d .$$

As $\sum \frac{1}{n^{\frac{d}{2}}}$ diverges iff $d \leq 2$, we get recurrence for $d \leq 2$, and transience for $d \geq 3$. – This phenomenon of a passage from recurrence to transience upon passage from dimension ≤ 2 to dimension ≥ 3 can be found in many much more general but still random-walk-like situations, such as with Brownian Motion; it reflects a close relation of such processes to potential theory where a similar passage phenomenon is well-known.

The next proposition is sort of a prelude to our next subsection.

Proposition 2.18. For any $t \geq 1$

(1) $P_{jk}^{(t)} = \sum_{u=1}^{t} F_{jk}^{(u)} P_{kk}^{(t-u)} \quad (j, k \in D).$

PROOF. This formula is quite plausible: in order to go from j to k in t steps, you have to make your first arrival at k at some time $u \leq t$ and then to return to k after $t-u$ steps. The calculation runs as follows:

$$\begin{aligned} P_{jk}^{(t)} &= \sum_{i_1,\ldots,i_{t-1} \in D} P_{ji_1} \cdots P_{i_{t-1}k} \\ &= \sum_{u=1}^{t} \sum_{i_1,\ldots,i_{u-1} \neq k} P_{ji_1} \cdots P_{i_{u-1}k} \sum_{i_{u+1},\ldots,i_{t-1} \in D} P_{ki_{n+1}} \cdots P_{i_{t-1} \geq k} \\ &= \sum_{u=1}^{t} F_{jk}^{(u)} P_{kk}^{(t-u)}. \end{aligned}$$

□

Corollary 2.19. Starting from anywhere the expected number of visits to a transient state is finite. That is: if k is transient, then

$$P_{jk}^{(*)} < \infty$$

and in particular

$$\lim_{t \to \infty} P_{jk}^{(t)} = 0$$

for all $j \in D$.

PROOF.

$$\begin{aligned} P_{jk}^{(*)} &= \sum_{t=1}^{\infty} \sum_{u=1}^{t} F_{jk}^{(u)} P_{kk}^{(t-u)} \\ &= \sum_{u=1}^{\infty} \sum_{s=0}^{\infty} F_{jk}^{(u)} P_{kk}^{(s)} \\ &= F_{jk}^{(*)}(1 + P_{kk}^{(*)}), \end{aligned}$$

and this is finite, because k transient $\Longrightarrow P_{kk}^{(*)} < \infty$. □

2.6. The Renewal Theorem.

Definition 2.20. Let $f = (f_1, f_2, \ldots)$ be a probability vector over $\mathbb{N}$, and let $u = (u_1, u_2, \ldots) \in \mathbb{R}^{\mathbb{N}}$. We say that u shows *renewal* under f if the *renewal equation*

(2) $u = f * u,$

more precisely if, with the "artificial" value $u_0 = 1$,

(3) $u_t = \sum_{s=1}^{t} f_s u_{t-s} \quad (t = 1, 2, \ldots)$

holds.

Thus proposition 2.18 tells us that

$$u = (P_{jk}^{(t)})_{t \in \mathbb{N}}$$

shows renewal under

$$f = (F_{jk}^{(t)})_{n \in \mathbb{N}}$$

if k is recurrent (which implies that f is indeed a probability vector), and that (2) even holds if k is not recurrent (and thus $f_1 + f_2 + \ldots < 1$).

Now, the verbiage "renewal" is motivated by phenomena which take place in a context which does not apriori belong to Markov theory:

Example 2.21. (Renewal of light bulbs). A bulb in an electric lamp will burn out after a life span T which we assume to be a $\mathbb{N}$-valued random variable with distribution $f = (f_1, f_2, \ldots)$. As soon as a bulb burns out, we replace it by a new one. Let us assume that the life spans $T_1, T_2, \ldots$ of the successive bulbs are IID, all with distribution f. Let now

$u_t =$ the probability that some bulb burns out at time t.

We can't define the infinity of RVs $T_1, T_2, \ldots$ on one and the same DPS (Ω, m) (see the last sections of III.2.4 and III.3.3), but we can work with an infinite sequence of DPSs in such a fashion as to make

$u_t = m($ there is some $n \in \mathbb{N}$ such that $T_1 + \ldots + T_n = t)$

meaningful (see subsection 2.4 above). Now we can calculate

$$u_t = \sum_{s=1}^{t} m(T_1 = s \text{ and there is some } n \text{ such that } T_2 + \ldots T_n = t - s)$$

$$= \sum_{s=1}^{t} m(T_1 = s)m(T_2 + \ldots + T_n = t - s \text{ for some } n)$$

$$= \sum_{s=1}^{t} f_s u_{t-s}$$

(the second = is due to the independence of $T_1, T_2, \ldots$) and see that $u = (u_1, u_2, \ldots)$ shows renewal under f.

We now investigate the asymptotic behavior of the components $u_1, u_2, \ldots$ of a vector u showing renewal under the probability vector $f = (f_1, f_2, \ldots)$. For this, it will be important to see whether f really "occupies" (in a sense still to be defined) all of $\mathbb{N}$ or only a fraction of it. To be precise, we will, for any vector

$$x = (x_1, x_2, \ldots) \in \mathbb{R}^{\mathbb{N}}$$

define

$d_x =$ the greatest common divisor of all $t \in \mathbb{N}$ with $x_t \neq 0$.

If $d_x = 1$, we will say that x "really occupies" $\mathbb{N}$. In case $d_x = 2$ we will consider only half of $\mathbb{N}$ as really occupied by x etc.

In our renewal context, we will be interested in d_f and d_u. Now, as we had introduced the artificial value $u_0 = 1$, the right side of

$$(4) \quad u_t = \sum_{s=1}^{t} f_s u_{t-s}$$

always contains a last term $f_t u_{t-t} = f_t$. Thus $f_t > 0 \Longrightarrow u_t > 0$, that is, u occupies at least as much as f, and thus $d_u \leq d_f$ follows. But u cannot really occupy more than f: if $f_t > 0$ holds only for $t = nd$, then we see from (4) that the same is true for u_t, and thus

$$d_u = d_f$$

follows. This is quite plausible: if the life span of a bulb is always a multiple of d, then burn-outs happen only at multiples of d.

We now arrive at the goal of this subsection,

Theorem 2.22. (Renewal Theorem). If $u = (u_1, u_2, \ldots) \in \mathbb{R}^{\mathbb{N}}$ shows renewal under the probability vector $f = (f_1, f_2, \ldots)$, and if $d_f = d_u = 1$, that is, f occupies all of $\mathbb{N}$, then

$$\lim_{t \to \infty} u_t = \frac{1}{\mu}$$

where

$$\mu = \sum_{s=1}^{\infty} s f_s$$

denotes the expectation of f.

PROOF.

1. The renewal equations plus $u_0 = 1$ clearly imply
   $$0 \leq u_t \leq 1 \quad (t = 1, 2, \ldots)$$
2. Let $\bar{u} = \limsup_t u_t$ and $1 \leq t_1 \leq t_2 < \ldots$ be such that
   $$(5) \quad \lim_{\nu \to \infty} u_{t_\nu} = \bar{u}$$
   We shall show
   $$\lim_{t \to \infty} u_t = \bar{u},$$
   and we shall do this by showing that by (4) the convergence (5) entails more convergences
   $$\lim_\nu u_{t_\nu - r} \quad (r = 1, 2, \ldots)$$
   In fact, we easily derive from (4), for any given $r > 0$,
   $$\begin{aligned} \bar{u} &= \lim u_{t_\nu} = \lim \left[f_r u_{t_\nu - r} + \sum_{\substack{1 \leq s \leq t_\nu \\ s \neq r}} f_s u_{t_\nu - s} \right] \\ &= \liminf_\nu f_r u_{t_\nu - r} + \limsup_\nu \left[\sum_{\substack{1 \leq s \leq t_\nu \\ s \neq r}} f_s u_{t_\nu - s} \right] \\ &\leq f_r \liminf_\nu u_{t_\nu - r} + \sum_{\substack{1 \leq s \leq n_\nu \\ s \neq r}} f_s \limsup_\nu u_{t_\nu - s} \\ &\leq f_r \liminf_\nu u_{t_\nu - r} + \sum_{\substack{1 \leq s \leq n_\nu \\ s \neq r}} f_s \bar{u}. \end{aligned}$$
   If $f_r > 0$, this implies
   $$\liminf_\nu u_{t_\nu - r} \geq \bar{u}$$
   and hence
   $$\lim_\nu u_{t_\nu - r} = \bar{u}$$
   Iterating this procedure, we get
   $$\lim_\nu u_{t_\nu - (r_1 + \ldots + r_n)} = \bar{u}$$

for all $n > 0$ and all $r_1, \ldots, r_n$ with $f_{r_1}, \ldots, f_{r_n} > 0$. By lemma II.4.12 $d_f = 1$ implies the possibility of representing all sufficiently large natural numbers r in the form $r_1 + \ldots + r_n$ with $f_{r_1}, \ldots, f_{r_n} > 0$, hence we can find $R \in \mathbb{N}$ such that

$$\lim_{\nu \to \infty} u_{t_\nu - r} = \bar{u}$$

for alle $r \geq R$.

3. Put

$$g_n = f_{n+1} + f_{n+2} + \ldots \quad (n \in \mathbb{N})$$

thus obtaining $g_0 = 1$,

$$f_n = g_{n-1} - g_n \quad (n \in \mathbb{N})$$

and furthermore

$$\sum_{n \in \mathbb{N}} n f_n = \sum_{n \in \mathbb{N}} g_n.$$

Let us now apply (4):

$$\begin{aligned} g_0 u_t = u_t &= \sum_{s=1}^{t} f_s u_{t-s} \\ &= -\sum_{s=1}^{t} (g_s - g_{s-1}) u_{t-s} \end{aligned}$$

yields

$$(6) \quad \sum_{s=0}^{t} g_s u_{t-s} = \sum_{s=1}^{t} g_{s-1} u_{t-s} = \sum_{s=1}^{t-0} g_s u_{t-1-s}$$

If we put

$$G_t = \sum_{s=0}^{t} g_s u_{t-s} \quad (t = 0, 1, \ldots)$$

we have

$$G_t = G_{t-1} \qquad (t = 1, 2, \ldots)$$

from (5) and thus

$$G_t = G_0 = g_0 u_0 = 1 \quad (t = 1, 2, \ldots).$$

For $t = t_\nu - M$ this yields

$$(7) \quad \sum_{s=0}^{t_\nu - M} g_s u_{t_\nu - M - s} = 1.$$

4. Let us now prove

$$\bar{u} = \frac{1}{\mu}.$$

Well, if

$$(8) \quad \mu = \sum_{s=1}^{\infty} g_s < \infty,$$

then for every $\epsilon > 0$ we may find some S with

$$g_1 + \ldots + g_s > \mu - \epsilon$$

If $t_\nu - M \geq S$, we obtain from (7)

$$(9) \quad 1 \geq \sum_{s=0}^{S} g_s u_{t_\nu - (M+s)}$$

As

$$\lim_\nu u_{t_\nu - (M+s)} = \bar{u},$$

we conclude, passing to $\nu \to \infty$ in (9),

$$1 \geq \bar{u} \sum_{s=0}^{S} g_s \geq \bar{u}(\mu - \epsilon).$$

As $\epsilon > 0$ is arbitrary,

$$\bar{u} \leq \frac{1}{\mu}$$

follows. But we also have $0 \leq u_0, u_1, \ldots \leq 1$ and $g_{S+1} + g_{S+2} + \ldots < \epsilon$, and thus from (7)

$$\begin{aligned} 1 &= \sum_{s=0}^{t_\nu - M} g_s u_{t_\nu - (M+s)} \\ &\leq \sum_{s=0}^{S} g_s u_{t_\nu - (M+s)} + \epsilon \end{aligned}$$

which yields, upon $\nu \to \infty$,

$$1 \leq \epsilon + \sum_{s=0}^{S} g_s \bar{u} \leq \epsilon + \mu \bar{u}.$$

As $\epsilon > 0$ was arbitrary,

$$\bar{u} \geq \frac{1}{\mu}$$

follows.

We can now carry through, for $\liminf_t u_t$, an analogous program as for $\limsup_t u_t$, leading to

$$\liminf_t u_t = \frac{1}{\mu},$$

which proves the theorem in case $\mu < \infty$. – In the still remaining case

$$\mu = \sum_{s=1}^{\infty} g_s = \infty$$

we get arbitrarily large values of

$$\sum_{s=1}^{S} g_s$$

by choosing S sufficiently large. From (7) we then obtain, upon $\nu \to \infty$,

$$G\bar{u} \le 1$$

for arbitrarily large G, and thus

$$\bar{u} = 0,$$

which entails

$$\lim_{t\to\infty} u_t \le \limsup_t u_t = \bar{u} = 0.$$

That is, we are through in this case too.

□

Exercise 2.23. Show that in case $d_u = d_f > 1$, the result of theorem 2.22 generalizes to

$$\lim_{n\to\infty} u_{nd_u} = \frac{d_u}{\mu}$$

where $\mu = \sum_{t\ge 1} t f_t$ as before.

With proposition 2.18 in mind, the reader should be able to apply the renewal theorem to recurrent states in Markov chains, and e.g. to solve

Exercise 2.24. Let $D \in \emptyset$ be finite and P a stochastic $D \times D$-matrix. Let

$$D = N \cup \bigcup_{\nu=1}^{n} \left[\bigcup_{\varrho=0}^{r_\nu - 1} C_{\nu\varrho} \right]$$

be the disjoint decomposition of D into n cycles of lengths $r_1, \ldots, r_n$, and the (possibly empty) remainder set N of all transient states. Employing our previous notations, we define for every recurrent j $(\in D \backslash N)$

$$\begin{aligned} \mu_j &= \sum_{t=1}^{\infty} t F_{jj}^{(t)} \\ &= \text{expected return time to } j \end{aligned}$$

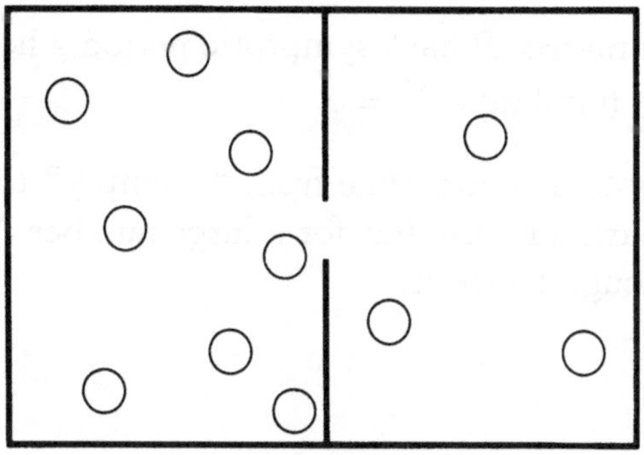

fig. VI.2.3

a) Show that for $j \in \bigcup_{\varrho=1}^{r_\nu} C_{\nu\varrho}$

$$\lim_{n\to\infty} P_{jj}^{(nr_\nu)} = \frac{r_\nu}{\mu_j}$$

b) Show that for $j \in C_{\nu o}$, $0 \le \ell < r_\nu$, $k \in C_{\nu\ell}$

$$\lim_{n\to\infty} P_{jk}^{(nr_\nu+\ell)} = \frac{r_\nu}{\mu_j}$$

Exercise 2.25. (Ehrenfest-Ehrenfest [1911] Model). Consider a container split into two equal chambers A and B by a diaphragm with a hole (figure VI.2.3).

Assume the container to contain N molecules, define $D = \{0, 1, \ldots N\}$, where $j \in D$ represents the state "j molecules in A, $N - j$ in B". Define transition probabilities

$$P_{jk} = \begin{cases} \frac{j}{N} & \text{for} \quad k = j-1 \\ \frac{N-j}{N} & \text{for} \quad k = j+1 \\ 0 & \text{else} \end{cases}$$

Interpretation: state j may change into

state $j - 1$: one molecule goes from A to B
state $j + 1$: one molecule goes from B to A

and these are the only two possibilities. The probability of a transition is proportional to the number of candidate molecules for that transition: $\frac{j}{N}$ for $A \to B$, and $\frac{N-j}{N}$ for $B \to A$. Prove that

$$\mu_j = \frac{2^N}{\binom{N}{j}} \quad (j \in D)$$

Hint: Our transition matrix P has asymptotic period 2 here, and $\lim P_{jj}^{(2n)}$ is a vector $q = (q_j)_{j \in D}$ fulfilling $qP^2 = q$.

Corollary: μ_0 = expected return time from "A empty" to "A empty" $= 2^N$. That is: you will return for sure, but for a large number N of molecules, you will not live long enough to see it.

VII. Elements of Information Theory

Information theory deals with the quantitative aspects of storage and transmission of messages. Some of these aspects involve the inner structure of messages: the grammar of their language, the acoustic spectrum of their sound etc. We will not enter into questions of this kind. For us here, messages are elements of a set, and when we handle them, our sole concern is to keep distinct messages distinct, i.e. to make the mappings resulting from our manipulations *injective* (= one-to-one). Aiming at this, we face two obstacles:

lack of capacity of storage or transmission procedures
noise, i.e. distortion of messages during transmission.

We all know how annoying noise in a telephone connection can be, and every one of us has used , many times in his life, one remedy: spelling the message, and using code words for the letters. This device certainly increases the chance for one-one-ness, but at quite some cost on the side of transmission capacity. At any rate, we see that we will have to calculate, to estimate and to impose bounds on certain – e.g. error – *probabilities*, and that we will have to measure storage and transmission *capacities*.

Let us try to get some first ideas about these two tasks.

1. The binary symmetric channel.

Assume that our messages are given as 0-1-sequences of length n, and that we transmit them symbol by symbol, independently, with a probability $0 \leq p \leq \frac{1}{2}$ that during transmission, the symbol will be distorted (0 into 1, 1 into 0): transmission through a *binary symmetric channel* (BSC). The probability that the first symbol will be distorted, and the remaining $n-1$ ones will be transmitted correctly, is $p(1-p)\dots(1-p) = p(1-p)^{n-1}$, and the same formula applies to distortion of symbol no. 2 and no other, bringing the probability of getting exactly one symbol distorted to $np(1-p)^{n-1}$. Clearly, $\binom{n}{k}p^k(1-p)^{n-k}$ is the probability of getting exactly k symbols distorted and the remaining $n-k$ symbols undistorted during transmission. We may consider our messages as vectors x in the linear space $GF(2)^n$ and the distortion of precisely k symbols as the result of the addition, to x, of a *distortion pattern* $e \in GF(2)^n$ with exactly k components 1, the probability of e being $p^k(1-p)^{n-k}$. The probability of getting at most k distortions is, of course, $\sum_{0\leq j\leq k}\binom{n}{j}p^j(1-p)^{n-j}$. The set of all the distortion patterns involved in this formula may be viewed as the ball of radius k around $0 \in GF(2)^n$ if we introduce the ***Hamming distance*** $d(x,y)$ of two messages $x, y \in GF(2)^n$ as the number of components in which x and y differ:

$$d(x,y) = \sum_{i=1}^{n} | x_i - y_i | \quad (x = (x_1, \ldots, x_n), y = (y_1, \ldots, y_n))$$

where $| 0 |$ is the real number 0, and $| 1 |$ is the real number 1, for the two elements 0,1 of $GF(2)$. Clearly, $GF(2)^n$ becomes a metric space in this way. If two messages x, y have $d(x,y) \geq 2k+1$, then the Hamming balls of radius k around them are disjoint, and if no more than k distortions happen, x and y remain distinct after transmission.

2. Capacities.

Our spelling procedure is only one out of several types of procedures evolving in time. We will be particularly interested in one quantitative aspect of such procedures: their

$$\text{rate} = \frac{\text{number of performances in time } t}{t}$$

and here, again, in the maximal rate which the requirements allow to which the procedure is subjected. If such a maximal rate can be characterized, maybe after passage to one or more limits, by a certain constant $C \geq 0$, i.e. if, $\sim$ denoting a suitable type of approximation,

$$\text{maximal number of performances in time } t \sim Ct,$$

then C is called the *capacity* of the given type of procedure.

Thus, if we spell one letter per time unit (second), say, from the alphabet $\{0,1\}$, the capacity is clearly 1. This very elementary procedure is, so to say, the *clock* for all other procedures to be considered in this chapter, and we all are accustomed to denote its capacity by

$$1 \text{ bit per second.}$$

We may express the capacities of other procedures in bits/second as well if we agree to *encode* them into our binary clockwork. Take e.g. the procedure of spelling one letter per second, but from an alphabet of length $a \geq 2$. If $a = 2$, then the encoding into our 0-1-clock is obvious and yields a capacity of 1 bit/second again. If $a > 2$, we may encode our symbols into as many distinct 0-1-words say, of a given length r. The smallest r for which this is possible is given by

$$2^{r-1} < a \le 2^r$$

or, equivalently

$$r - 1 < \log_2 a \le r$$

$$\text{i.e.} \quad r = \lceil \log_2 a \rceil$$

Here $\lceil x \rceil$ stands generally for the least integer $\geq x$, and $\log_2$ denotes the logarithm to the base 2.

Such an encoding thus results in a new procedure which achieves in r seconds what the original procedure did in 1 second: we have to set our binary clockwork to the pace r in order to let it do equivalent work per second. Thus the rate of our original procedure can be measured by $r = \lceil \log_2 a \rceil$ bits/second – under this particular encoding. We might consider other encodings, such as the following: chose some s and encode the a^s words of length s from our general alphabet into 0-1-words of length r, this time with

$$r = \lceil \log_2 a^2 \rceil = \lceil s \log_2 a \rceil$$

of course. Now the binary procedure achives in r seconds what the original one did in s; thus we now calculate a rate of

$$\frac{r}{s} = \frac{1}{s} \lceil s \log_2 a \rceil \text{bits/second},$$

and if we allow s to tend to infinity, a capacity of $\log_2 a$ bits/second results in the limit.

The present chapter is devoted to a somewhat systematic introduction into the basic ideas of information theory, at which we have just thrown a few first glances. Modern information splits into two subdisciplines

algebraic (and combinatorial) coding theory
probabilistic information theory.

The first of these is by far more important for down-to-the-earth applications than the latter, because algebraic procedures can well be implemented on computers; a classic on this subject is Gallager [1968], a brilliant recent account is Lint [1982]; Jacobs [1983] contains a brief introduction; for the history of algebraic coding theory, one should consult the beautiful little book Thompson [1983]. Modern information theory begins with the fundamental paper Shannon [1948]. Further classics on probabilistic information theory are

Feinstein [1958], Chintschin [1956], Wolfowitz [1964]; a more recent heavyweight monograph is Csiszár-Körner [1981].

We will shortly deal with combinatoric/algebraic information theory in §1 and devote the rest of the chapter to probabilistic information theory. In §2 we show that the binary encoding capacity of a "source" with independent symbols is given by entropy: the source coding theorem. §3 and §4 are devoted to code lengths for noisy channels with independent signals. It should be emphasized that the independence assumptions mentioned here are not always realistic: we do not choose the next letter independently of what we have spoken before, and also channels may have a memory. Already in Shannon [1948] models involving dependence were introduced, and there is quite some literature which makes use of so-called ergodic theory in order to cope with dependence problems; information theory in turn had a strong impact on the evolution of ergodic theory; details are beyond the scope of this book; see e.g. Chintschin [1956]. Although the special case of Markovian dependence would largely be accessible to the methods used in this book, we refrain from going into details.

1. Combinatorial and Algebraic Coding Theory

Algebraic resp. combinatorial coding theory sides her probabilistic sister in tackling the problems arising from the presence of random noise, but with overall emphasis on algorithmic implementability, which leads to a nearly total absence of probabilistic results. I believe, nevertheless, that the reader should not see the probabilistic branch of coding theory without having thrown a few glances on the other branch which at present dominates the applications. This section is to provide him with such a minimal sightseeing tour.

We shall deal here with the alphabet {0,1} exclusively, although many of the results can easily be generalized to arbitrary finite alphabets. We shall constantly work with finite 0-1-words like

$$0,\ 1,\ 00,\ 01,\ 10,\ 11,\ 10110$$

and denote the empty word by $\Box$. Words are written comma-free. The set $\{0,1\}^*$ of all finite 0-1-words is conveniently displayed in an infinite binary tree (figure VII.1.1.).

If we move to the right in the tree, starting from one word $w = w_1 \dots w_n$, we encounter the *successors* of w, i.e. the words arising from w by adjunction of more symbols 0 and 1, i.e. words $v = w_1 \dots w_n w_{n+1}, \dots w_{n+k}$ where every

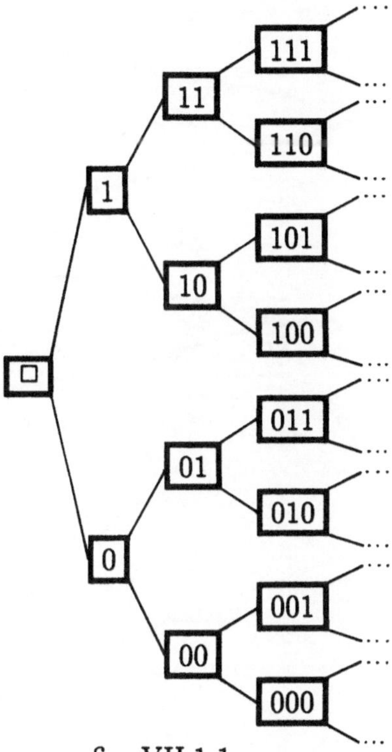

fig. VII.1.1.

$w_{n+k} = 1$ means "up" and 0 means "down". Column no. n of our tree displays all 2^n 0-1-words of length n in lexicographic order.

1.1. Prefix Codes and Kraft's Inequality.

Any finite set of 0-1-words is also called a *code*, and the number of its words is called the *length* of this code. The length of the words in a code may vary, but if none of these words reappears as the inital section of another word in the code, the code is called *prefix code*. If we look up our code in the binary tree, this means that we encounter no other code word if we walk from a given code word to the right (along branches of the tree, of course). Clearly, every code with constant word length is a prefix code. 01, 110, 1001, 1011 is a simple example of a prefix code with variable word lengths. If the length of a given prefix code $C \leq \{0,1\}$ is N, and if $\ell_1 \leq \ell_2 \leq \ldots \leq \ell_N$ is a listing of word lengths in the code, we may draw the picture in figure VII.1.2.

We see that, according to the definition of a prefix code, the column no. ℓ_N contains the following N *disjoint* sets of words of length ℓ_N,

the $2^{\ell_N - \ell_1}$ successors of the code word of length ℓ_1

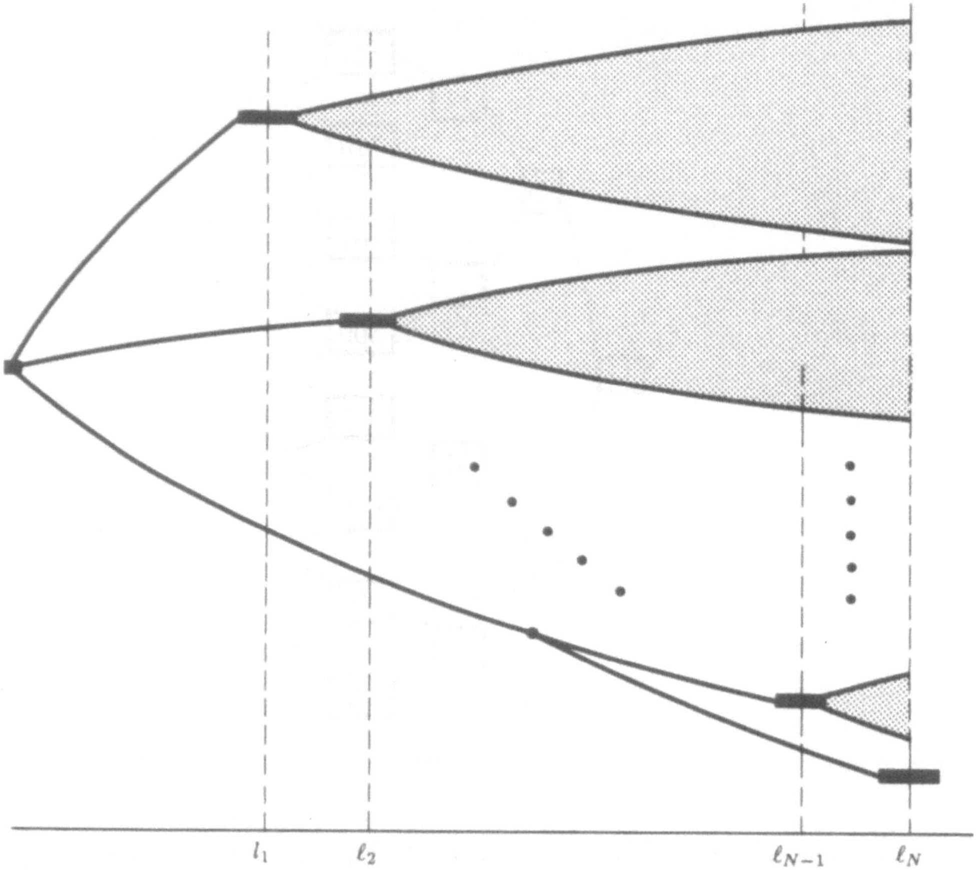

fig. VII.1.2

the $2^{\ell_N-\ell_2}$ successors of the code word of length ℓ_2
etc.

Thus

(1) $2^{\ell_N-\ell_1}+\ldots+2^{\ell_N-\ell_N}\leq 2^{\ell_N}$,

and division by 2^{ℓ_N} yields

Kraft's inequality for prefix codes (Kraft [1949]):

(2) $$\sum_{k=1}^{N}\frac{1}{2^{\ell_N}}\leq 1$$

Proposition 1.1. If the natural numbers $\ell_1\leq\ldots\leq\ell_N$ satisfy Kraft's inequality (2), there exist a *prefix* code with $\ell_1,\ldots,\ell_N$ as its list of word lengths.

Proof. Choose any word of length ℓ_1 and "block" all its successors, in particular all its $2^{\ell_N-\ell_1}$ successors of length ℓ_N. By (1), which is equivalent to (2), we may find a word of length ℓ_2 which hasn't been blocked yet. Block all its $2^{\ell_N-\ell_2}$ successors of length ℓ_N etc. We may continue in this fashion, and a prefix code with the given list of word lengths results. □

What is the use of a prefix code? Well, if we take the words of a code as spell-words for the spelling of messages formulated in an alphabet of length $\leq N$, the code words are concatenated into long 0-1-sequences. The problem is how to decipher such a lengthy sequence, that is, how to recognize from which sequence of code words it was built. If the code was a prefix code, the solution is unambiguous. For instance, if we know that the sequence

$$1011110101111001$$

was built with the prefix code 01.110, 1001, 1011, it could only have been built as follows

$$1011 \quad 110 \quad 1011 \quad 110 \quad 01.$$

If the code words all have the same length, the solution is, of course, extremely simple.

1.2. Error Detecting and Error Correcting Codes.

Let us consider codes (with constant word length n, i.e. $C \leq \{0,1\}^n$). Let $d(v,w)$ denote the Hamming distance of any two 0-1-words $v = v_1 \dots v_n, w = w_1 \dots w_n$ of length n:

$$d(v,w) = | \{k \mid 1 \leq k \leq n, \ v_k \neq w_k\} |$$

and let

$$B_e(w) = \{v \mid d(v,w) \leq e\}$$

denote the *e-ball* around the word w.

Definition 1.2. Let $C \leq \{0,1\}^n$ be any code. We shall say that

1. C *detects* up to e errors if $v, w \in C, v \neq w \Longrightarrow d(v,w) \geq e+1$
2. C *corrects* up to e errors, if $v, w \in C, v \neq w \Longrightarrow d(v,w) \geq 2e+1$

In fact, if a code word $w \in C$ is distorted in at most e components, a 0-1word $v \in \{0,1\}^n$ with $d(v,w) \leq e$ results.

If C *detects* up to e errors, v is either w or not in C, and the latter case indicates ("detects") the fact of distortion; if there is a code word $w' \in C$

with $d(w, w') = e + 1$, a distortion in $e + 1$ components may result in $v = w'$, in which case the fact of distortion remains undetected.

If C *corrects* up to e errors, the e-balls $B_e(w)$ $w \in \ell)$ around the code words are pairwise disjoint, and a distortion of w in at most e components doesn't lead out of $B_e(w)$, hence not into another $B_e(w')$, and we may correctly conclude that v arose by distortion of w and hence correct it back into w. If more than e components are distorted, the correction may well lead to a wrong w.

Richard W. Hamming, back in 1947, had access to a computer only on weekends. He remembers (Thompson [1983]):

> Two weekends in a row I came in and found that all my stuff had been dumped and nothing was done .... And so I said "Damn it, if the machine can detect an error, why can't it locate the position of the error and correct it?"

Since every $B_e(w)$ contains $\sum_{k \leq e} \binom{n}{k}$ 0-1-words, a code of length N, and correcting up to e errors, is possible only if $N \sum_{k \leq e} \binom{n}{k} \leq 2^n$, i.e.

$$N \leq \frac{2^n}{\sum_{k \leq e} \binom{n}{k}}$$

(Hamming [1950]).

1.3. Linear Codes.

It was certainly one of the most important ideas in coding theory to identify the binary alphabet $\{0,1\}^n$ with the binary field $GF(2)$, and thus $\{0,1\}$ with the vector space $GF(2)^n$: every 0-1-word became a vector, and the ivory-tower-toy $GF(2)$ became a powerful tool of applied mathematics.

Definition 1.3. Linear subspaces of $GF(2)^n$ are called *linear codes.*

The advantage of restricting attention to linear codes is obvious: in order to decide whether a 0-1-word is a code word, we need not go through a whole list but only check whether this vector satisfies the system of linear equations defining that subspace. And with $GF(2)$, this is extremely simple: We only have to check parities: 1 odd, 0 even.

In fact, for every linear code $C \leq GF(2)^n$ there is a $GF(2)$-matrix M with n columns and m rows, such that

$$C = \{w \mid Mw^T = 0\}.$$

Here w^T denotes the transpose of w, a column vector of n components . This *parity check matrix* M of C – and we know from our previous remark why we call it that way – is, of course, not unique, but we shall always assume that its m rows are linearly independent and hence $n - m$ is the dimension of the linear subspace C of $GF(2)^n$.

Let us now investigate the possibility of constructing ***linear*** codes which correct up to e errors. Since the Hamming distance $d(v, w)$ of vectors in $GF(2)^n$ is obviously translation invariant, the requirement that any two different words v, w in C differ in at least $2e + 1$ symbols is tantamount to the requirement that any $0 \neq w \in C$ differs from $0 \in GF(2)^n$ in at least $2e + 1$ places, i.e. contains at least $2e + 1$ components 1. For the parity check matrix M of a linear code C this means that $Mw^T \neq 0$ unless w contains at least $2e + 1$ symbols 1. In other words, we have

Lemma 1.4. A linear code $C \leq GF(2)^n$ with parity check matrix M corrects up to e errors if and only if any $2e$ columns of M are linearly independent.

Under which conditions do such parity check matrices exist, and how may we construct them? An answer is given by the following theorem and its proof.

Theorem 1.5. (Varshamov [1957], Gilbert [1952], Sacks [1958]).

1. Let $C \leq GF(2)^n$ be a d-dimensional linear code correcting up to e errors. Then
$$\sum_{k \leq e} \binom{n}{k} \leq 2^{n-d}.$$
2. Let
$$\sum_{1 \leq k \leq 2e-1} \binom{n-1}{k} < 2^{n-d}$$
Then there exists a linear code of dimension d which corrects up to e errors.

PROOF. 1) is obvious from Hamming's abovementioned inequality because C has length 2^d.
2) We construct a parity check matrix with $n - d$ rows such that any $2e$ of it columns are linearly independent. We start by choosing any non-zero column $c^{(1)} \in GF(2)^{n-d}$. Assume that we have already constructed r columns $c^{(1)}, \ldots, c^{(r)}$ such that any $2e$ among them are linearly independent. Form all sums of less than $2e$ among the columns $c^{(1)}, \ldots, c^{(r)}$. We may select them in $\sum_{k=1}^{2e-1} \binom{r}{k}$ fashions, and this number is $\leq \sum_{k=1}^{2e-1} \binom{n-1}{k}$ as long as we haven't reached the full number of n columns. According to (3) there is at least one

column in $GF(2)^{n-d}$ which is not representable as one of the above sums: choose it as $c^{(r+1)}$. Clearly, any $2e$ among the $c^{(1)}, \ldots, c^{(r+1)}$ are independent. Continuing in this fashion we arrive at the desired result. □

Assume now that we have a linear code $C \leq GF(2)^n$ which corrects up to e errors. Let v be a 0-1-word arising form some $w \in C$ by distortion of at most e symbols. How to determine w?

Clearly we have to find the closest neighbor of v within C. This can be done as follows: form $v + x$ for every $x \in C$; the number of ones in $v + x$ tells us what the Hamming distance between v and x is; w is the x with the least numbers of components 1 in $v + x$.

2. Source Coding

We turn to stochastics again and consider finite random experiments whose outcomes are messages: information *sources*. Their outputs are to be encoded, that is, transformed into binary code words in a one-to-one fashion. We want to know something about the average length of the resulting code words as well as about the question at which pace (bits/second) the encoding mechanism has to be set in order to cope with the production of messages if the source repeats its work over and over again, say, independently. Both questions are related to each other. The key concept here is *entropy*.

2.1. Entropy.

Definition 2.1. Let $p = (p_1, \ldots, p_a)$ be any probability vector: $p_1, \ldots, p_a \geq 0$, $p_1 + \ldots + p_a = 1$. Then

$$H(p) = -\sum_{k=1}^{a} p_k \log_2 p_k$$

is called the (binary) *entropy* of p (we set $-0 \log_2 0 = 0$ by continuous extrapolation).

In order to establish the fundamental properties of entropy, we first consider the function

$$f(x) = \begin{cases} -x \log x & \text{for } x > 0 \\ 0 & \text{for } x = 0 \end{cases}$$

involved here, and observe:

$f(x)$ is defined and continuous for $x \geq 0$

$f(x) = 0$ for $x = 0$ and for $x = 1$
$f(x) > 0$ for $0 < x < 1$
$f(x)$ is strictly concave:
$\sum_{j=1}^{r} \alpha_j f(x_j) \leq f(\sum_{j=1}^{r} \alpha_j x_j)$
$(r > 0, \alpha_1, \dots, \alpha_r \geq 0, \alpha_1 + \dots + \alpha_r = 1)$
with equality iff all x_j with $\alpha_j > 0$ are equal
$f(x)$ attains its unique maximum at $x = \frac{1}{2}$, and $f(\frac{1}{2}) = \frac{1}{2}$.

The simple proofs of these facts are left to the reader as an exercise. Now we can prove

Proposition 2.2. Let a be any natural number and

$$P_1 = \{p \mid = (p_1 \dots, p_a), p_1, \dots, p_a \geq 0, p_1 + \dots + p_a = 1\}$$

the simplex of all probability vectors with a components. The function

$$H \ : \ P_a \longrightarrow \mathbf{R}$$
$$p \longrightarrow H(p)$$

has the following properties:

1. H is continuous.
2. $H(p) \geq 0$ with $H(p) = 0$ iff p is Dirac, that is iff of the a components of the probability vector p exactly one is 1, and the rest 0.
3. H is strictly concave:
$$\sum_{j=1}^{r} \alpha_j H(p^{(j)} \leq H(\sum_{j=1}^{r} \alpha_j p^{(j)}$$
$$(r > 0, \alpha_1, \dots, \alpha_r \geq 0, a_1 + \dots, +\alpha_r = 1)$$
with equality iff all $p^{(j)}$ with $\alpha_j > 0$ are equal.
4. H attains its unique maximum at the barycenter $(\frac{1}{a}, \dots, \frac{1}{a})$ of P_a, and its value there is $\log_2 a$.

The proof is again left as an exercise for the reader.

If a random variable X has a finite state space, and its distribution is p, we also write $H(X)$ instead of $H(p)$. The addition formula for $\log_2$ then entails

$$H(X \times Y) = H(X) + H(Y)$$

if X and Y are *independent* finite-state random variables and $X \times Y$ is their cartesian join (ch.III.2.3). Entropy is characterizable by its properties mentioned (see e.g. Faddejew [1956], Kannappan [1972][1972a], Aczél-Daroczy [1963]).

In the sequel, we will often carry out calculations related to the notion of entropy. This would necessitate, in principle, to treat the case $p_k = 0$ separately at every stage. We will refrain from this in order to avoid clumsiness, and thus rely on the reader's flexibility of mind. The same will apply to similar calculations in §3.

2.2. Expected Length of Code Words.

Theorem 2.3. Let $p = (p_1, \ldots, p_a)$ be any probability vector. If $w^{(1)}, \ldots$ $\ldots, w^{(a)}$ is any numbering of the words of any prefix code of length a, and if ℓ_k denotes the length of the 0-1-word $w^{(k)}$, then the expected length of code words

$$E_p(\ell) = \sum_{k=1}^{a} \ell_k p_k$$

fulfils

$$H(p) \leq E_p(\ell)$$

There is a prefix code C of length a such that, under suitable numbering $w^{(1)}, \ldots, w^{(a)}$ of its words, and hence of their lengths $\ell_1, \ldots, \ell_a$, the expected length fulfils

$$(1) \quad E_p(\ell) \leq H(p) + 1$$

Before we proceed to the proof of this theorem, we point out what it means and implies. It means that we can find a prefix code and encode the outcomes $k = 1, \ldots, a$ of the random experiment governed by p into the words of C in such a fashion that the resulting random variable "code word length" has an expectation fulfilling (1). If we repeat our random experiment, say n times, independently, then the sum of the code word lengths shining up in this process is, by the law of large numbers, approximately

$$\leq n(H_a(p) + 1),$$

that is, $H_a(p)+1$ bounds, up to random deviations, the pace at which a paper strip has to be set in order to store all those code words.

Our proof of theorem 2.3. begins with

Lemma 2.4. Let $p = (p_1, \ldots, p_a), q = (q_1, \ldots, q_a)$ be two probability vectors with the same number a of components. Then

$$(2) \quad H(p) \leq - \sum_{k=1}^{a} p_k \log_2 q_k$$

with equality iff $q = p$ (we set $0 \log_2 0 = 0$).

PROOF.

$$
\begin{aligned}
H(p) + \sum_{k=1}^{a} p_k \log_2 q_k &= \sum_{p_k>0} p_k \log_2 \frac{q_k}{p_k} = \sum_{p_k>0} p_k \log_2(1 + \frac{q_k - p_k}{p_k}) \\
&\leq \sum_{p_k>0} (q_k - p_k) = \sum_{p_k>0} q_k - \sum_{p_k>0} p_k \qquad (3)\\
&= \sum_{p_k>0} q_k - 1
\end{aligned}
$$

because $\log(1+x) \leq x$. If $\sum_{p_k>0} q_k < 1$, then the last expression is < 0. In all other cases $q_k > 0 \Rightarrow p_k > 0$ and the last expression is $= 0$. This proves (2). In order to prove the last statement of our lemma, we observe that equality in (2) entails $q_k > 0 \Rightarrow p_k > 0$. If $p_k > 0, q_k \neq p_k$, then $<$ holds in (3) as $\log(1+x) < x$ if $x \neq 0$. This does it. □

PROOF. of theorem 2.3.

1) Let C be any prefix code of length a, $w^1, \ldots, w^{(a)}$ be any numbering of its words, and ℓ_k the length of word $w^{(a)}$. Kraft's inequality states

$$\sum_{k=1}^{a} \frac{1}{2^{\ell_k}} \leq 1$$

Let A denote the value of the left side here. Then $0 < A \leq 1$, and

$$q_k = \frac{1}{A 2^{\ell_k}} \quad (k = 1, \ldots, a)$$

defines a probability vector. We apply lemma 2.4., and obtain

$$
\begin{aligned}
H(p) &\leq -\textstyle\sum_{k=1}^{a} p_k \log_2 q_k = \sum_{k=1}^{a} p_k [\log_2 A + \log_2 2^{\ell_k}] \\
&= \log_2 A + (\log_2 2) \cdot \textstyle\sum_{k=1}^{a} \ell_k p_k
\end{aligned}
$$

As $A \leq 1 \Rightarrow \log A \leq 0$, and $\log 2 = 1$, we obtain

$$H_a(p) \leq \sum_{k=1}^{a} \ell_k p_k,$$

with equality iff $p = q$. If p has the property that $\log_2 p_k$ is an integer $-\ell_k$ for every $k = 1, \ldots, a$, then the integers $\ell_1, \ldots, \ell_a$ fulfil Kraft's inequality

$$1 = p_1 + \ldots + p_a = \frac{1}{2^{\ell_1}} + \ldots + \frac{1}{2^{\ell_a}}$$

and thus there is a prefix code with word lengths $\ell_1, \dots, \ell_a$ and $E_p(\ell) = H(p)$ (proposition 1.1). – If p is an arbitrary probability vector, let

$$\ell_k = \lceil -\log_2 p_k \rceil \quad (k = 1, \dots, a)$$

where generally $\lceil x \rceil$ denotes the least integer $\geq x$. Clearly Kraft's inequality

$$\sum_{k=1}^{a} 2^{\ell_k} \leq \sum_{k=1}^{a} p_k = 1$$

holds and we may thus find (proposition 1.1) a prefix code with word lengths $\ell_1, \dots, \ell_a$. We have

$$\begin{aligned} E_p(\ell) = p_1\ell_1 + \dots + p_a\ell_a &\leq p_1(-\log_2 p_1 + 1) \\ &\qquad + \dots + p_a(-\log_2 p_a + 1) \\ &= H_p(p) + 1 \end{aligned}$$

□

The reader is invited to carry these results over to alphabets of any finite length ≥ 2.

2.3. Coding Probable Messages Only.

Let $A = \{1, \dots, a\}$ be a finite alphabet and $p = (p_1, \dots, p_a)$ a probability vector over a. We consider (A, p) as the model for the random generation of one letter from A. For a given integer $t > 0$, let (Ω, m) be the probability space modeling t independent repetions of the experiment modeled by (A, p): random generation of messages of length t, over the alphabet A, with independent letters. Thus

$$\begin{aligned} \Omega &= A^t = \{\omega \mid \omega = \omega_1 \dots \omega_t;\ \omega_1, \dots, \omega_t \in A\} \\ m &= p \times \dots \times p \quad (t \text{ factors }) \\ m_\omega &= p_{\omega_1} \cdots p_{\omega_t} \quad (\omega = \omega_1 \dots \omega_t \in \Omega) \end{aligned}$$

Let us now look at the entropies $H(p)$ and $H(m)$. Clearly (see the remark in subsection 1)

$$\begin{aligned} H(m) &= -\textstyle\sum_{\omega_1, \dots, \omega_t \in A} p_{\omega_1} \cdots p_{\omega_t} \log_2(p_{\omega_1} \cdots p_{\omega_t}) \\ &= \textstyle\sum_{u=1}^{t} [-\sum_{\omega_1, \dots, \omega_{u-1}, \omega_{u+1}, \dots, \omega_t} p_{\omega_1} \cdots p_{\omega_{u-1}} p_{\omega_{u+1}} \cdots p_{\omega_t} \\ &\qquad \textstyle\sum_{\omega_u} p_{\omega_u} \log p_{\omega_u}] \\ &= tH(p) \end{aligned}$$

The defining formula

$$H(p) = \sum_{k\in A}(-\log_2 p_k)p_k$$

suggests an interpretation of $H(p)$ as the expectation of the random variable

$$h_p(k) = \begin{cases} -\log_2 p_k & \text{if } p_k > 0 \\ 0 & \text{if } p_k = 0 \end{cases} \quad s$$

and the same idea applies to $H(m)$. Clearly

$$\begin{aligned} h_m(\omega) &= -\log_2(p_{\omega_1}\cdots p_{\omega_n}) \\ &= \textstyle\sum_{u=1}^t(-\log_2 p_{\omega_u}) \\ &= \textstyle\sum_{u=1}^t h_p(\omega_u) \end{aligned}$$

represents h_m as a sum of IID random variables, suggesting an application of the law of large numbers (LLN), i.e. of Tchebyshev's inequality: let

$$\sigma^2 = \sum_{k\in A}(-\log_2 p_k - H(p))^2 p_k$$

be the variance of h_p. Then Tchebyshev's inequality sounds, for any $\epsilon > o$

$$m(|\sum_{u=1}^{t} h_p(\omega_k) - tH(p)| \geq \epsilon) \leq t\frac{\sigma^2}{\epsilon^2}$$

or, equivalently,

$$(4) \quad (m \mid \sum_{u=1}^{t} h_p(\omega_k) - tH(p) \mid < \epsilon) \geq 1 - t\frac{\sigma^2}{\epsilon^2}$$

The set M figuring implicitly on the left side may be represented as

$$M = M_+ \cap M_-$$

where

$$\begin{aligned} M_+ &= \{\omega \mid h_m(\omega) < tH(p) + \epsilon\} \\ M_- &= \{\omega \mid h_m(\omega) > tH(p) - \epsilon\}. \end{aligned}$$

For $\omega \in M_+$ we have the following chain of equivalent inequalities:

$$\begin{aligned} h_m(\omega) &< tH(p) + \epsilon \\ -\log_2 m_\omega &< tH(p) + \epsilon \\ m_\omega &> 2^{-(tH(p)+\epsilon)} \end{aligned}$$

Summing over all $| M_+ |$ elements $\omega \in M_+$, we obtain

$$1 >| M_+ | 2^{-(tH(p)+\epsilon)}$$

i.e.

$$(5) \quad | M | \leq | M_+ | < 2^{tH(p)+\epsilon}$$

Similarly, for $\omega \in M$, hence $\omega \in M_-$, we get

$$m_\omega < 2^{-(tH(p)-\epsilon)}$$

and hence, for any $M' \subseteq \Omega$,

$$m(M \cap M') <| M \cap M' | 2^{-(tH(p)-\epsilon)} \leq | M' | 2^{-(tH(p)-\epsilon)}$$

$$(6) \quad | M' |> \frac{1}{m(M \cap M')} 2^{tH(p)-\epsilon}.$$

(4) and (5) lead to the following conclusion: for large t, the probability m is almost entirely concentrated on a set M with $| M |< 2^{tH(p)+\epsilon}$ messages: $H(p)$ is the pace to which we have to set our binary clock in order to encode binarily, not all messages, but probabilistically nearly all of them.
Replacing ϵ by $K\sqrt{t}$, with $K > 0, K^2 = \frac{\sigma^2}{\epsilon^2}$, we may concentrate our estimates into the following

Theorem 2.5. Let $(\Omega, m) = (A, p)^t$. Then, for every $\epsilon > 0$, there is a set $M \subseteq \Omega$ such that

$$m(M) \geq 1 - \epsilon$$

$$| M |< 2^{t(H(p)+\frac{K}{\sqrt{t}})}$$

Every set $M' \subseteq \Omega$ fulfilling

$$m(M') \geq 1 - \epsilon$$

contains

$$| M' |> \frac{1}{1-2\epsilon} 2^{t(H(p)-\frac{K}{\sqrt{t}})}$$

messages.

Our way of disposing of ϵ is not the only possible one; for other substitutions we may obtain other interesting estimates. In any case, $H(p)$ is the basic quantitiy measuring the productivity of our random source of messages in bits/second. The methods displayed in the proof have been widely generalized. Any generalization of LLN that applies to $h_m = h_{p^t}$ for $t \to \infty$, yields an analogous theorem. The sharpest estimates ever obtained for the case of independence are in Strassen [1964a]

$$2^{tH(p)+K_1\sqrt{t}-\frac{1}{2}\log t+K_2(t)+0(1)}$$

with a constant K_1 and a bounded function $K_2(t)$ of t.

Since our customary natural and computer languages produce their symbol sequences with a certain coherence among neighboring symbols, generalizations which go beyond independence are of particular interest. The next simplest case is Markovian dependence, and in fact several of our above results carry over to it; we will refrain from going into any details. So-called ergodic theory provides a framework wide enough to model all sorts of dependence, and Chintschin [1956] is a testimony of the efforts in this direction. Results of practical importance have, however, nearly always used independence assumptions.

3. Noisy Channels

In this section we deal with quantitative aspects of the transmission of messages from a sender through a receiver through a *channel.* Their situation may roughly be plotted as

$$\text{sender} \longrightarrow \text{channel} \longrightarrow \text{receiver}$$

Channels use to be *noisy*: the messages are subject to random distortions during transmission. That is, instead of an ideal one-to-one transmission, we face a probabilistic situation shown in figure VII.3.1.

Here A denotes the sender's, and B the receiver's alphabet, and the channel is represented by a mapping which associates with a letter $j \in A$ not a unique letter $k \in B$, but a probability distribution $P(j, \cdot)$ on B, depending upon j. That is, a noisy channel is, from our probabilistic viewpoint here, nothing but a *stochastic* $A \times B$*-matrix*

$$\begin{aligned} P = \quad & (P(j,k))_{j\in A, k\in B} \\ & P(j,k) \geq 0 && \text{for all } j,k \\ & \textstyle\sum_{k\in B} P(j,k) = 1 && \text{for all } j \in A \end{aligned}$$

The theory of noisy channels which we shall present in this section, is nothing but an investigation of such stochastic matrices, guided by intuitions stimulated by the task of fighting message distortion. As for notation, we will

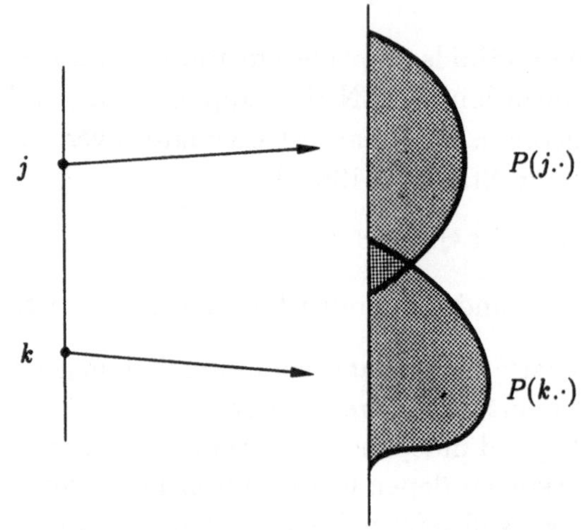

fig. VII.3.1

have to make use of many upper and lower indices in various combinations. In order to reduce clumsiness, we will adopt the following

Notational Convention (this section only): Probability vectors will be denoted $p = (p(j))_{j\in A}$ instead of $(p_j)_{j\in A}$ etc. Likewise, stochastic matrices will be de noted $P = (P(j,k))_{j\in A, k\in B}$ instead of $(P_{jk})_{j\in A,k\in B}$ etc. We will also use the notations

$$p(E) = \sum_{j\in E} p(j) \quad (E \subseteq A)$$

and the like.

$$P(j,F) = \sum_{k\in F} P(j,k) \quad (j \in A, F \subseteq B)$$

Now, every probabilist will, certainly, be inclined to act according to the following two routines:

1. Embedding $P = (P(j,k))_{j\in A,k\in B}$ into a probability space $(A \times B, \tilde{p})$: choose any probability vector $p = (p(j))_{j\in A}$ and define $\tilde{p} = p \times P$, i.e.

   $$\tilde{p}(j,k) = p(j)P(j,k) \qquad (j \in A, k \in B)$$

   Then $\tilde{p} = (\tilde{p}(j,k))_{j\in A,k\in B}$ is a probability distribution (vector) over $A \times B$, and the entries of P appear as conditional probabilities

   $$P(j,k) = \tilde{p}(k \mid j) \qquad\qquad (p(j) > 0).$$

   p is, of course, the first *marginal* of $\tilde{p}$:

   $$p(j) = \sum_{k\in B} \tilde{p}(j,k)$$

and the second marginal q can be calculated by

$$q(k) = \sum_{j\in A} \tilde{p}(j,k) = \sum_{j\in A} p(j)P(j,k)$$

i.e. by the vector-matrix equation

$$pP = q$$

Clearly every probability $\tilde{p}$ on $A \times B$ can be obtained by suitable choice of p and P (see exercise VI.1.5).

2. *Time expansion*, that is independent repetition of the transmission procedure:
for an arbitrary natural number t, let

$$\begin{aligned} A^t &= \{\eta = \eta_1 \ldots \eta_t | \eta_1, \ldots, \eta_t \in A\} \\ B^t &= \{\omega = \omega_1 \ldots \omega_t | \omega_1, \ldots, \omega_t \in B\} \\ m^{(t)} &= p \times \ldots \times p \quad (t \text{ factors, probability on } A^t) \\ P^{(t)} &= (P^{(t)}(\eta,\omega))_{\eta\in A^t, \omega\in B^t}, \text{ given by} \end{aligned}$$

$$P^{(t)}(\eta,\omega) = P(\eta_1,\omega_1)\ldots P(\eta_t,\omega_t)$$

$$(\eta = \eta_1 \ldots \eta_1 \in A^t, \omega = \omega_1 \ldots \omega_t \in B^t;$$

stochastic matrix from A^t to B^t).

And clearly one will apply routine 1) in this new situation again.

Our final aim will be the proof of a so-called *coding theorem*. Coding theorems deal with so-called *ϵ-codes* (for $\epsilon \geq 0$, and mostly for $\epsilon > 0$) and tell us which lengths of such ϵ-codes may be achieved. In the time-expanded situation sketched in 2) above, a *code of length N* is a string

$$\begin{array}{ccc} (\eta^{(1)}, E^{(1)}), & \ldots, & (\eta^{(N)}, E^{(N)}) \\ \eta^{(1)}, & \ldots, & \eta^{(N)} \in A^t \\ E^{(1)}, & \ldots, & E^{(N)} \subseteq B^t \end{array}$$

such that the $E^{(i)}$ are mutually disjoint. If the inequalities

$$P^{(t)}(\eta^{(i)}, E^{(i)}) \geq 1 - \epsilon \qquad (i = 1, \ldots, N)$$

(recall that

$$P^{(t)}(\eta, F) = \sum_{\omega\in F} P^{(t)}(\eta,\omega) \quad (\eta \in A^t, F \subseteq B^t))$$

hold, the code is called an *ϵ-code*. It is not required apriori that the $\eta^{(i)}$ be pairwise distinct, but if we have an ϵ-code with $\epsilon < \frac{1}{2}$, they obviously are. Interpretation: if we use the $\eta^{(i)}$ as spell-words, their distorted versions after transmission still belong to disjoint sets, enabling us to recognize which $\eta^{(i)}$

was sent – up to an *error probability* $\epsilon \geq 0$. Before we enter into proofs of a coding theorem proper, we define a notion closely related to entropy and investigate its basic properties.

3.1. Information.

Definition 3.1. Let A, B be nonempty finite sets and $\tilde{p}$ a probability vector over $A \times B$, with p marginals p (on A) and q (on B). Then the function $\tilde{h}$ defined on $A \times B$ by

$$\tilde{h}(j,k) = \begin{cases} \log_2 \frac{\tilde{p}(j,k)}{p(j)q(k)} & \text{if } \tilde{p}(j,k) > 0 \\ 0 & \text{if } \tilde{p}(j,k) = 0 \qquad (j \in A, k \in B) \end{cases}$$

is called the *information function* on $A \times B$, and its expectation

$$I(\tilde{p}) = \sum_{\tilde{p}(j,k)>0} \tilde{p}(j,k) \log \frac{\tilde{p}(j,k)}{p(j)q(k)}$$

is called the *information* of $\tilde{p}$.

Calculations involving information should, in principle, treat the case $\tilde{p}(j,k) = 0$ separately at each stage. We will, as in §2, refrain from doing this, and thus rely on the reader's flexibility of mind.

Information is closely related to entropy, as the following calculation shows:

$$\begin{aligned} I(\tilde{p}) &= \textstyle\sum_{j\in A,k\in B} \tilde{p}(j,k)\log_2 \tilde{p}(j,k) - \sum_{j,k} \tilde{p}(j,k) \log_2 p(j) \\ &\qquad - \textstyle\sum_{j,k} \tilde{p}(j,k) \log q(k) \\ &= H(p) + H(q) - H(\tilde{p}) \\ &= H(p \times q) - H(\tilde{p}) \end{aligned}$$

suggesting to consider $I(\tilde{p})$ as a measure of deviation from independence. In fact, in the case $\tilde{p} = p \times q$ of independence we get $I(\tilde{p}) = 0$.

What happens if $\tilde{p} \neq p \times q$? We calculate a bit differently, representing

$$\begin{aligned} \tilde{p} &= p \times P \\ \tilde{p}(j,k) &= p(j)P(j,k) \quad (j \in A, k \in B) \end{aligned}$$

with some stochastic matrix P (after exercise VI.1.5, this is always possible):

$$\begin{aligned} I(\tilde{p}) &= \textstyle\sum_{j,k} p(j)P(j,k) \log_2 \frac{p(j)P(j,k)}{p(j)q(k)} \\ &= H(q) + \textstyle\sum_j p_j \sum_k P(j,k) \log_2 P(j,k) \\ &= H(q) - \textstyle\sum_j p_j H(P(j,\cdot)). \end{aligned}$$

Now the strict concavity of entropy entails

$$\begin{aligned} \textstyle\sum_j p_j H(P(j,\cdot)) &\leq H(\textstyle\sum_j p_j P(j,\cdot)) \\ &= H(q) \end{aligned}$$

with equality iff all rows $P(j,\cdot)$ endowed with $p(j) > 0$ are equal, i.e. iff $\tilde{p} = p \times q$. That is

$$\begin{aligned} I(\tilde{p}) &\geq 0 \\ I(\tilde{p}) &= 0 \text{ iff } \tilde{p} = p \times q, \end{aligned}$$

confirming the previous suggestion.

As p and q depend continuously upon $\tilde{p}$, the formula $I(\tilde{p}) = H(p) + H(q) - H(\tilde{p})$ proves that $I(\tilde{p})$ is a continuous function of $\tilde{p}$.

The quantity $\sum_j p_j H(P(j,\cdot)$ is usually called *conditional entropy*, and, somewhat sloppily, denoted by $H(q \mid p)$:

$$H(q \mid p) = \sum_{j \in A} p_j H(P(j,\cdot))$$

Theorem 3.2. Let A, B be nonempty finite sets and $\tilde{p}$ be a probability on $A \times B$ with marginals p and q. Then $I(\tilde{p})$ is a continuous function of $\tilde{p}$. We have

$$0 \leq I(\tilde{p}) \leq H(p)$$

with

1) $I(\tilde{p}) = 0$ iff $\tilde{p} = p \times q$
2) $I(\tilde{p}) = H(p)$ iff there is 0-code of length $\mid \{j \mid j \in A, p(j) > 0\} \mid$.

PROOF. 1) was already proved above. For the proof of 2) we choose a representation $\tilde{p} = Q \times q$, i.e.

$$\tilde{p}(j,k) = q(k)Q(k,j) \quad (j \in A, k \in B)$$

with a stochastic $B \times A$-Matrix $Q = (Q(k,j))_{k \in B, j \in A}$:

$$\begin{aligned} Q(k,j) &\geq 0 \\ \textstyle\sum_j Q(k,j) &= 1. \end{aligned}$$

An analogous calculation as before yields

$$I(\tilde{p}) = H(p) - H(p \mid q)$$

where

$$H(p \mid q) = \sum_k q(k) H(Q(k,\cdot)).$$

Now we conclude

$$I((\tilde{p}) \leq H(p)$$

with equality iff

$$H(p \mid q) = 0,$$

i.e. iff

$$\sum_k q(k) H(Q(k,\cdot)) = 0$$

i.e. iff

$$q(k) > 0 \Rightarrow H(Q(k,\cdot)) = 0$$

Now $H(Q(k,\cdot)) = 0$ iff $Q(k,\cdot)$ is Dirac, i.e. concentrated on one simple point $j = \varphi(k) \in A$. Clearly the points j occurring in this form make up precisely the support $\{j \mid j \in A, p(j) > 0\}$ of p, as $p = \sum_{q(k)>0} q(k) Q(k,\cdot)$. Now we subdivide $\{k \mid q(k) > 0\}$ into regions of constancy of φ:

$$E^{(j)} = \{k \mid k \in B, \varphi(k) = j\} \quad (j \in A, p(k) > 0)$$

We calculate, for $p(j) > 0$,

$$\begin{aligned} p(j) &= \textstyle\sum_k q(k) Q(k,j) = \sum_{q(k)>0} q(k) Q(k,j) \\ &= \textstyle\sum_{\varphi(k)=j} q(k) \cdot 1 + \sum_{\varphi(k)\neq j} q(k) \cdot 0 \\ &= \textstyle\sum_{k \in E^{(j)}} q(k) = q(E^{(j)}) \end{aligned}$$

but also

$$= \sum_{\varphi(k)=j} p(j) P(j,k) = p(j) P(j, E^{(j)})$$

and hence finally $\qquad P(j, E^{(j)}) = 1 \qquad (p(j) > 0),$

which essentially finishes the proof of 2). The reader is invited to fill in the details which are still missing. □

Theorem 3.2. deals with the extreme cases $I(\tilde{p}) = 0$ and $I(\tilde{p}) = H(p)$. In the remaining subsections we shall prove coding theorems tackling in particular the non-extreme cases $0 < I(\tilde{p}) < H(p)$. We shall employ two different approaches – maximal codes, random codes – leading to the same final conclusion.

3.2. Maximal Codes.

Let P be a stochastic matrix from A to B. If we choose any probability vector p over A and form $\tilde{p} = p \times P$; $\tilde{p}$ depends continuously upon p, and hence so does $I(\tilde{p})$. Let us choose some p for which the value of $I(\tilde{p})$ attains its maximum C:

$$C = I(\tilde{p}) = I(p \times P) = \max\{I(p' \times P) | p' \text{ DPD on } A\}.$$

We shall now pass over to the time-expanded situation and consider, for some integer $t > 0$, the product spaces A^t, B^t and the t-factor products

$$\begin{array}{lcll} P^{(t)} & = & P \times \ldots \times P & \text{(stochastic matrix from } A^t \text{ to } B^t) \\ p^{(t)} & = & p \times \ldots \times p & \text{(probability vector over } A^t) \\ \tilde{p}^{(t)} & = & \tilde{p} \times \ldots \times \tilde{p} & \text{(probability vector over } A^t \times B^t) \end{array}$$

Obviously

$$I(\tilde{p}^{(t)}) = tI(\tilde{p})$$

and similarly, for the information function $\tilde{h}^{(t)}$ of $\tilde{p}^{(t)}$

$$(1) \quad \tilde{h}^{(t)}(\eta, \omega) = \tilde{h}(\eta_1, \omega_1) + \ldots + \tilde{h}(\eta_t, \omega_t)$$

$$(\eta = \eta_1 \ldots \eta_t \in A^t, \omega = \omega_1 \ldots \omega_t \in B^t)$$

$\tilde{p}^{(t)}$-almost everywhere.

Theorem 3.3. Let $c > 0$ and $\tilde{M} \subseteq A^t \times B^t$ a set such that

$$(\eta, \omega) \in \tilde{M} \Rightarrow \tilde{p}^{(t)}(\eta, \omega) > 0 \text{ and } \tilde{h}^{(t)}(\eta, \omega) \geq c$$

Let $0 < \epsilon < \frac{1}{2}$ and N the maximal length of an ϵ-code $(\eta^{(1)}, E^{(1)}), \ldots, (\eta^{(N)}, E^{(N)})$ for $P^{(t)}$, which has the additional property that every $E^{(i)}$ is contained in the $\eta^{(i)}$-section of $\tilde{M}$, i.e. $\{\eta^{(i)}\} \times E^{(i)} \subseteq \tilde{M} (i = 1, \ldots, N)$. Then *Feinstein's inequality*

$$n \geq 2^c(\tilde{p}^{(t)}(\tilde{M}) - (1 - \epsilon))$$

(Feinstein [1954]) holds.

PROOF. Let us assume that $(\eta^{(1)}, E^{(1)}), \ldots, (\eta^{(N)}, E^{(N)})$ is an ϵ-code with the property mentioned, and of maximal length among all such codes. For any $\eta \in A^t$, let $\tilde{M}_\eta$ denote the η-section of $\tilde{M}$:

$$\tilde{M}_\eta = \{\omega \mid (\eta, \omega) \in \tilde{M}\}.$$

Let furthermore $E = E^{(1)} \cup \ldots \cup E^{(N)}$.
Then clearly

(2) $P^{(t)}(\eta, \tilde{M}_\eta \backslash E) < 1 - \epsilon$

for all $\eta \notin \{\eta^{(1)}, \ldots, \eta^{(N)}\}$ because otherwise we could choose at least one such η as $\eta^{(N+1)}$ and the corresponding $\tilde{M}_\eta \backslash E$ as $E^{(N+1)}$ and thus get a longer ϵ-code with the required properties. (2) holds for $\eta = \eta^{(i)}(i = 1, \ldots, N)$ as well since $P^{(t)}(\eta^{(i)}, E) \geq P^{(t)}(\eta^{(i)}, E^{(i)}) \geq 1 - \epsilon > \epsilon$ as $\epsilon < \frac{1}{2}$. By combination with the probabilities $p^{(t)}(\eta)$ we obtain from (1)

$$\begin{aligned} 1 - \epsilon &\geq \textstyle\sum_\eta p^{(t)}(\eta) P^{(t)}(\eta, \tilde{M}_\eta \backslash E) \\ &\geq \textstyle\sum_\eta p^{(t)}(\eta) P^{(t)}(\eta, \tilde{M}_\eta) - \sum_\eta p^{(t)}(\eta) P^{(t)}(\eta, E) \\ &= \tilde{p}^{(t)}(M) - q^{(t)}(E) \end{aligned}$$

and thus

(3) $q^{(t)}(E) \geq \tilde{p}^{(t)}(M) - (1 - \epsilon)$

Let us now deal with the left side of this inequality:
because of the particular property of our code, we have the following sequence of equivalent inequalities for every $\eta^{(i)}$ and $\omega \in E^{(i)}$

$$\tilde{h}^{(t)}(\eta^{(i)}, \omega) \geq c$$

$$\log_2 \frac{p^{(t)}(\eta^{(i)}) P^{(t)}(\eta^{(i)}, \omega)}{p^{(t)}(\eta^{(i)}) q^{(t)}(\omega)} \geq c$$

$$P^{(t)}(\eta^{(i)}, \omega) \geq 2^c q^{(t)}(\omega),$$

By summation over $\omega \in E^{(i)}$ we get

$$1 \geq P^{(t)}(\eta^{(i)}, E^{(i)}) \geq 2^c q^{(t)}(E^{(i)}).$$

Summing finally over $i = 1, \ldots, N$, we find

$$N \geq 2^c q^{(t)}(E).$$

By combination with (3), our theorem follows. □

Neither in the formulation nor in the proof of this theorem we have made explicit use of the time (= product) structure built into the model. Now we focus on precisely this structure, representing $\tilde{h}^{(t)}$ as a sum of t IID random variables and applying Tchebyschev's inequality. Let σ^2 denote the variance of $h(\eta_u, \omega_u)$ $(u = 1, \ldots, t)$, and remember $C = I(p \times P) = E_p(h)$. Then, for every $\delta > 0$,

$$\begin{aligned} & \tilde{p}^{(t)}(\{(\eta,\omega) \mid \tilde{h}^{(t)}(\eta,\omega) \geq tC - \delta\}) \\ = \;& \tilde{p}^{(t)}(\{(\eta,\omega) \mid \textstyle\sum_{u=1}^{t} h(\eta_u,\omega_u) \geq tC - \delta\}) \\ \geq \;& 1 - \frac{\sigma^2}{t\delta^2}. \end{aligned}$$

We may therefore apply Theorem 3.3. (Feinstein's inequality) with the specifications

$$\begin{aligned} c &= tC - \delta \\ \tilde{M} &= \{(\eta,\omega) \mid \tilde{h}^{(t)}(\eta,\omega) \geq tC - \delta\} \end{aligned}$$

and obtain ϵ-codes with a length

$$\begin{aligned} N &\geq e^{tC-\delta}(1 - \tfrac{\delta^2}{t\delta^2} - (1-\epsilon)) \\ &= e^{tC-\delta}(\epsilon - \tfrac{\sigma^2}{t\delta^2}) \end{aligned}$$

Setting $\delta^2 = \frac{2\sigma^2}{t\epsilon}$, we obtain $\epsilon - \frac{\sigma^2}{t\delta^2} = \frac{\epsilon}{2}, tC - \delta = t(C - \frac{K'}{\sqrt{t}})$ with $K' = \sqrt{2\sigma^2/t^2\epsilon}$, and thus

Theorem 3.4. (Coding Theorem). Let A, B be finite nonempty alphabets and P a stochastic matrix, i.e. a noisy channel, from A to B. Let

$$C = \max I(p \times P)$$

Then for every t, there is an ϵ-code from A^t to B^t with length

$$N \geq 2^{t(C - \frac{K}{\sqrt{t}})}$$

where $K > 0$ is suitable constant.

For the proof we have only to observe that it suffices to choose $K > K'$ sufficiently large in order to get $2^{(K-K')} \geq \frac{2}{\epsilon}$. If we are content to work with $t \geq t_0$ only, for some specified $t_0 > 0$, it suffices to ensure $2^{(K-K')\sqrt{t_0}} \geq \frac{2}{\epsilon}$.

This theorem shows that $C = \max_p I(p \times P)$ is a lower bound for the *transmission rate* of our channel if we define it via maximal lengths of ϵ-codes. There is a so-called *converse of a coding theorem* showing that C *is* actually the transmission rate: the maximal length N is an ϵ-code for $P^{(t)}$ is $\leq 2^{(t + \frac{M}{\sqrt{t}})}$ for some constant $M > 0$. The proof of this result is intricate and will not be given here; the standard reference is Wolfowitz [1964].

3.3. Random Codes.

We shall now present another deduction of theorem **3.3** (coding theorem) which makes use of an entirely different method: the ***method of random codes.***

Let us consider the same situation as before: A, B, P and *a* p on A with $I(p\times P)$ maximal; again we fix some integer $t > 0$ and form the time expansion $A^t, B^t, P^{(t)}, p^{(t)}$ of our model.

Contrary to our previous approach we now fix, for the time being, the length N of the codes $(\eta^{(1)}, E^{(1)}), \ldots, (\eta^{(N)}, E^{(N)})$ from A^t to B^t, which we will consider, and decide to form them at random:

a) choose $\eta^{(1)}, \ldots, \eta^{(N)}$ in N independent repetitions of the random experiment $(A^t, p^{(t)})$
b) once $\eta^{(1)}, \ldots, \eta^{(N)}$ has been determined, choose the sets $E^{(1)}, \ldots, E^{(N)} \subseteq B^t$ in a purely mechanical way: $E^{(i)}$ is a subset of the $\eta^{(i)}$-section of
$$\tilde{M} = \{(\eta, \omega) \mid \tilde{h}^{(t)}(\eta, \omega) > tI(p \times P) - \delta\},$$
namely,
$$E^{(i)} = \tilde{M}_{\eta^{(i)}} \setminus \bigcup_{\nu \neq i} \tilde{M}_{\eta^{(\nu)}}$$

Clearly, this way we get a code, i.e. pairwise disjoint sets $E^{(i)}$. It may happen that $E^{(i)} = E^{(\nu)} = \emptyset$. We will later on eliminate such phenomena by passage to suitable sub-codes, i.e. by elimination of some $(\eta^{(i)}, E^{(i)})$. More precisely, we will proceed as follows: the average error probability

$$\bar{\epsilon}(\eta^{(1)}, \ldots, \eta^{(N)}) = \frac{1}{N} \sum_{i=1}^{N} P^{(t)}(\eta^{(I)}, B^t \setminus E^{(i)})$$

of our random code $(\eta^{(1)}, D^{(1)}), \ldots, (\eta^{(N)}, E^{(N)})$ is a random variable $\geq 0, \geq 1$ on the N-fold cartesian product $(A^t)^N$ endowed with the N-fold product probability vector $(p^{(t)})^N$, i.e. within the natural model of random choice of $\eta^{(1)}, \ldots, \eta^{(N)}$. If for some $0 < \epsilon < \frac{1}{2}$ we can achieve a sharp estimate of the expectation of that random variable, namely

$$E(\bar{\epsilon}) < \epsilon^2,$$

we may conclude – and this is a typical specimen of a ***probabilistic existence proof*** – that there is at least one $\eta^{(1)}, \ldots, \eta^{(N)}$ such that

$$\bar{\epsilon}(\eta^{(1)}, \ldots, \eta^{(N)} = \frac{1}{N} \sum_{i=1}^{N} P(\eta^{(i)}, B^t \setminus E^{(i)}) < \epsilon^2$$

and this in turn allows the obvious conclusion that at most ϵN of the i yield $P(\eta^{(i)}, B^t \backslash E^{(i)}) \geq \epsilon$. If we eliminate precisely these, we obtain an ϵ-code of length $\geq (1-\epsilon)N$, with now all $\eta^{(i)}$ pairwise distinct (and all of its $E^{(i)}$ now nonempty, of course.)

Our realization of this program will be a bit tricky. The central point is an efficient upper estimate of the expectation of the probability $P^{(t)}(\eta^{(i)}, \bigcup_{\nu\neq i} \tilde{M}_{\eta^{(\nu)}})$, seen as a random variable over $((A^t)^N, (p^{(t)})^N)$. We will denote expectations over this space by E simply. We will write I instead of $I(p^{(t)} \times P^{(t)}) = tI(p \times P)$ and recall

$$\tilde{M} = \{(\eta, \omega) \mid \eta \in A^t, \omega \in B^t, \tilde{h}^{(t)}(\eta, \omega) > I - \delta\}.$$

We will also use such simple facts as: x^N is isotone over $\mathbf{R}$ if N is odd, and 2^x is ≥ 1 for $x \geq 0$ and ≥ 0 for $x < 0$. Now let the machinery go into action:

$$E(P^{(t)}(\eta^{(i)}, \bigcup_{\nu\neq i} \tilde{M}_{\eta^{(\nu)}}))$$

$$= \sum_{\eta^{(1)},\ldots,\eta^{(i-1)},\eta^{(i+1)},\ldots,\eta^{(N)}}$$

$$\prod_{\nu\neq i} p^{(t)}(\eta^{(\nu)})[\sum_{\eta^{(i)}} p^{(t)}(\eta^{(i)}) P^{(t)}(\eta^{(i)}, \bigcup_{\nu\neq i} \tilde{M}_{\eta^{(\nu)}})]$$

$$= \sum_{\eta^{(1)},\ldots,\eta^{(i-1)},\eta^{(i+1)},\ldots,\eta^{(N)}} \prod_{\nu\neq i} p^{(t)}(\eta^{(\nu)})[q^{(t)}(\bigcup_{\nu\neq i} \tilde{M}_{\eta^{(\nu)}})]$$

$$\leq E(q^{(t)}(\bigcup_{\nu=1}^{N} \tilde{M}_{\eta^{(\nu)}}))$$

$$= E(q^{(t)}(sup_\nu \tilde{h}^{(t)}(\eta^{(\nu)}, \omega) > I - \delta)).$$

Now, this is the product probability $(p^{(t)})^N \times q^{(t)}$ of a set $\subseteq (A^t)^N \times B^t$, calculated via sections of that set. Now we may invert the order of $(p^{(t)})^N$ and $q^{(t)}$ in this calculation ("Fubini-type argument")

$$\begin{aligned} &= E_{q^{(t)}}[(p^{(t)})^N(\{(\eta^{(1)}, \ldots, \eta^{(N)}) \mid \sup_\nu \tilde{h}^{(t)}(\eta^{(\nu)}, \omega) > I - \delta\})] \\ &= 1 - E_{q^{(t)}}[(p^{(t)})^N(\{(\eta^{(1)}, \ldots, \eta^{(N)}) \mid \tilde{h}^{(t)}(\eta^{(\nu)}, \omega) \leq I - \delta \\ &\qquad (\nu = 1, \ldots, N)\})] \end{aligned}$$

Now we make use of the fact that $\eta^{(1)}, \ldots, \eta^{(N)}$ were chosen in N *independent* experiments, that is, we exploit the product structure of $(p^{(t)})^N$:

$$= \quad 1 - E_{q^{(t)}}[p^{(t)}(\{\eta \mid \tilde{h}^{(t)}(\eta,\omega) \leq I - \delta\})^N]$$

$$= \quad 1 - E_{q^{(t)}}[1 - p^{(t)}(\{\eta \mid \tilde{h}^{(t)}(\eta,\omega) > I - \delta\}))^N]$$

If N is odd, we may continue – tricky, isn't it? –

$$\leq \quad 1 - E_{q^{(t)}}[(1 - E_{p^{(t)}}(2^{\tilde{h}^{(t)}(\eta,\omega)-(I-\delta)})^N]$$

Bernoulli's inequality yields

$$\leq \quad 1 - E_{q^{(t)}}(1 - N2^{-(I-\delta)}E_{p^{(t)}}(2^{\tilde{h}^{(t)}(\eta,\omega)}))$$

$$= \quad N2^{-(I-\delta)}$$

because $\tilde{h}^{(t)}(\eta,\omega) = \log_2 \frac{P^{(t)}(\eta,\omega)}{q^{(t)}(\omega)}$ entails

$$E_{p^{(t)}}(2^{\tilde{h}^{(t)}}) = \frac{1}{q^{(t)}(\omega)} E_{p^{(t)}}(P^{(t)}(\eta,\omega)) = \frac{q^{(t)}(\omega)}{q^{(t)}(\omega)} = 1.$$

Now we conclude

$$\begin{aligned} &E(P^{(t)}(\eta^{(i)}, E^{(i)})) \\ &\quad = E(P^{(t)}(\eta^{(i)}, \tilde{M}_{\eta^{(i)}} \setminus \textstyle\bigcup_{\nu \neq i} \tilde{M}_{\eta^{(\nu)}}) \\ &\quad \geq (P^{(t)}(\eta^{(i)}, \tilde{M}_{\eta^{(i)}})) - E(P^{(t)}(\eta^{(i)}, \textstyle\bigcup_{\nu \neq i} \tilde{M}_{\eta^{(\nu)}})) \\ &\quad \geq \tilde{p}^{(t)}(\tilde{M}) - N2^{-(I-\delta)} \end{aligned}$$

Let us now use $I = tI(p \times P)$ and specify

$$N = \langle 2^{tI(p\times P)-2\delta} \rangle$$

where $\langle x \rangle$ generally denotes the largest odd integer $\leq x$. Then

$$(4) \quad E(P^{(t)}(\eta^{(i)}, E^{(i)})) \geq \tilde{p}^{(t)}(\tilde{M}) - 2^{-\delta}.$$

From this we now easily deduce the coding theorem 3.4 via Tchebyshev's inequality and a suitable specification of δ. Recall that

$$\tilde{M} = \{(\eta,\omega) \mid \tilde{h}^{(t)}(\eta,\omega) > tI(p \times P) - \delta\}$$

and that $\tilde{h}^{(t)}$ is a sum of t IID random variables with expectation $I(p \times P)$ each. Tchebyshev's inequality yields

$$\tilde{p}^{(t)}(\tilde{M}) \geq 1 - \frac{Kt}{\delta^2}$$

where $K > 0$ is any bound for the variance of h. If we set, for a given $\epsilon > 0$

$$\delta = \sqrt{\frac{2Kt}{\epsilon}}$$

we get

$$\tilde{p}^{(t)}(M) \geq 1 - \frac{\epsilon^2}{2}$$

and thus, using (5) and recalling the definition of the average error probability

$$\overline{\epsilon} = \frac{1}{N} \sum_{i=1}^{N} (1 - P^{(i)}(\eta^{(i)}, E^{(i)}))$$

the estimate

$$E(\overline{\epsilon}) \leq \frac{\epsilon^2}{2} + 2^{-\sqrt{\frac{2Kt}{\epsilon}}} < \epsilon^2$$

for t sufficiently large (or for all t right away if we enlarge K suitably). Discarding ϵN of the $\eta^{(i)}$, we obtain an ϵ-code of length

$$\begin{aligned} &\geq (1 - \epsilon)N \\ &\geq (1 - \epsilon)(2^{tI(p \times P) - 2\delta} - 1) \\ &\geq (1 - \epsilon)(2^{tI(p \times P) - L'\sqrt{t}} - 1) \end{aligned}$$

for some constant L'. Enlarging L' to a suitable $L > L'$ we may continue with

$$\geq 2^{t(I(p \times P) - \frac{L}{\sqrt{t}})}.$$

This proves theorem 3.4 once more.

VIII. Fluctuation Theory

Fluctuation theory deals with the sequence of partial sums of a sequence of real-valued RVs and transforms it into new RVs such as

> the position of the maximum of the first n partial sums
> the number of strictly positive ones among the first n partial sums.

Such transforms are non-linear, not even algebraic; on the other hand they are of utmost interest, in particular for certain applications. As a consequence, fluctuation theory abounds in techniques both intricate and beautiful, with a strongly combinatorial touch. Historically, flucatuation theory as a coherent discipline originated with the papers Erdös–Kac [1947] and Andersen [1950][1953][1953/54], incorporating a number of older isolated results into its new edifice.

This chapter gives an introduction to fluctuation theory, with a heavy accent on combinatorial methods invented by Erik Sparre Andersen (*1919). We present results largely in a purely combinatorial form first, and get their stochastic counterparts by "filling in" probability structures. In §1 we prove Andersen's combinatorial equivalence principle. §2 presents the finite arcsin distributions and their asymptotic behavior; we also prove the combinatorial arcsin law of Andersen. In §3 we "fill in" probability theory and thus obtain the stochastic counterparts of the previous combinatorial results. §4 contains a side-step into the domain of random walks, where certain crucial assumptions of Andersen's theory are, as a rule, not fulfilled; nevertheless some similar results can be obtained: finite arcsin formulas and the "ballot theorem". In §5 we treat some very deep and far-reaching results of Bohnenblust, E.S. Andersen and Frank Spitzer.

1. The Combinatorial Arcsin Law of Erik Sparre Andersen

1.1. Paths with given increments.

Let t be a natural number and

$$c_1, \ldots, c_t$$

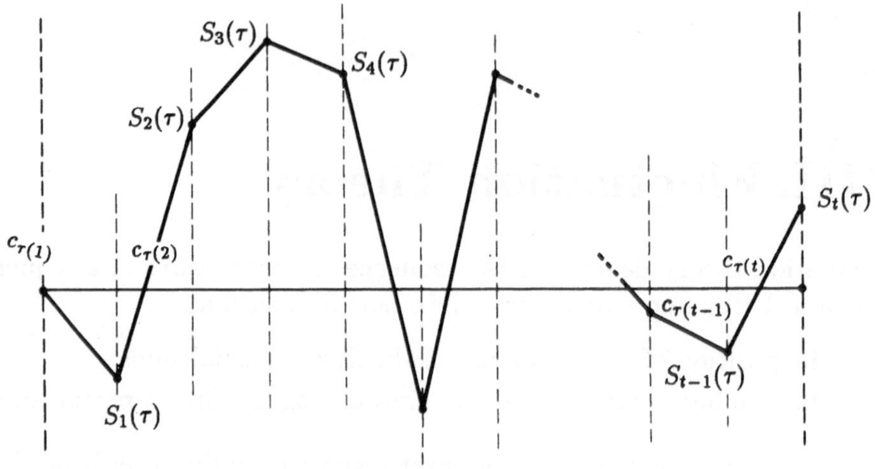

fig. VIII.1.1

real numbers. For every permutation τ of $\{1,...,t\}$, we define

$$S_o(\tau) = 0, \quad S_u(\tau) = c_{\tau(1)} + \ldots + c_{\tau(u)} \qquad (1 \le u \le t).$$

The proper way of visualizing this new sequence is a ***path*** with vertices

$$(0,0), (1, S_1(\tau), \ldots, (t, S_t(\tau)),$$

i.e. with the increments $c_{\tau(1)}, \ldots, c_{\tau(t)}$ (figure VIII.1.1).

We will investigate the following quantities which reflect some geometrical features of such a path shown in figure VIII.1.2.

$$\begin{aligned}
max(\tau) &= \text{position of the first maximum} \\
&= min\{u \mid 0 \le u \le t, \quad S_u(\tau) \ge S_v(\tau) \quad (v = 0, \ldots, t)\} \\
Max(\tau) &= \text{position of the last maximum} \\
&= max\{u \mid 0 \le u \le t, \quad S_u(\tau) = \max_{0 \le v \le t} S_v(\tau)\} \\
pos(\tau) &= \text{number of strictly positive vertices} \\
&= \mid \{u \mid 0 \le u \le t, S_u(\tau) > 0\} \mid \\
Pos(\tau) &= \text{number of nonnegative vertices} \\
&= \mid \{u \mid 0 \le u \le t, \; S_u(\tau) \ge 0\} \mid
\end{aligned}$$

All these quantities depend upon the choice of $c = (c_1, \ldots, c_t)$ and might thus more correctly be denoted by $max^c(\tau)$ etc. We refrain from doing so for the time being, but will come back to such a notation in later sections.

In fig. VIII.1.2 I have plotted, for $t = 3, c_1 = -1, c_2 = -2, c_3 = 4$, the paths for all τ and listed the values of $pos(\tau)$ etc.

permutation	path	pos	max
−1, −2, 4	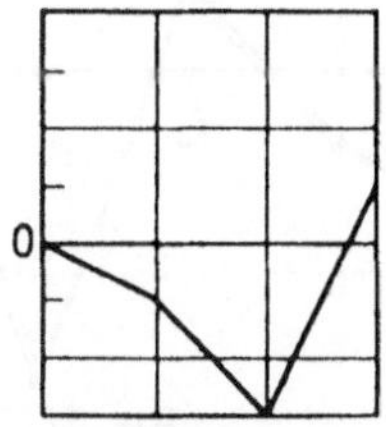	1	3
4, −1, −2	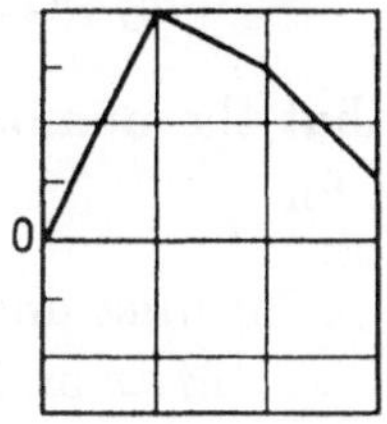	3	1
−2, 4, −1	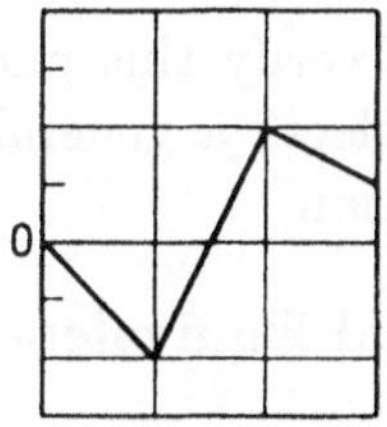	2	2
−2, −1, 4	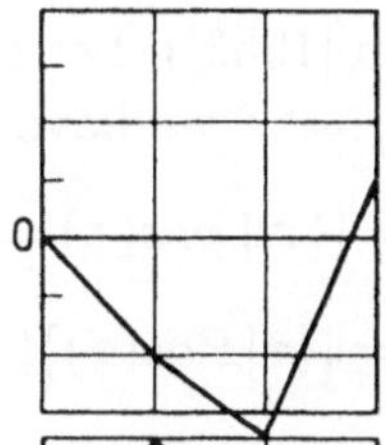	1	3
4, −2, −1	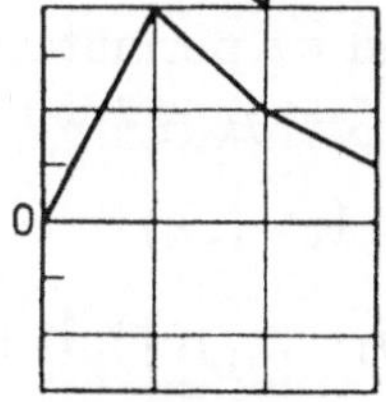	3	1
−1, 4, −2	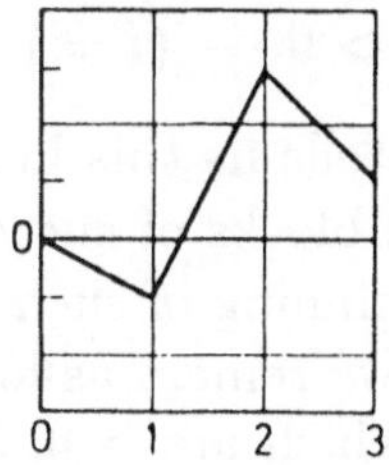	2	2

fig. VIII.1.2
(From: K. Jacobs, Selecta Mathematica I; Springer-Verlag, Heidelberg)

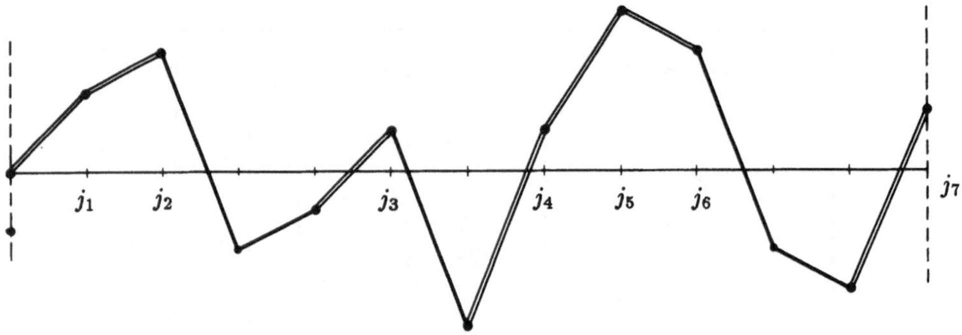

fig. VIII.1.3

The reader will observe that the number of $\tau's$ for which $max(\tau)$ attains a certain value, is the same for

max and pos
Max and Pos.

The reader is invited to verify this phenomenon in some more cases. This phenomenon actually reflects a general law which we shall formulate and prove in the next subsection.

1.2. The Combinatorial Equivalence Principle.

Theorem 1.1. (Andersen [1953]). Let t be a natural number and $c_1, \ldots, c_t \in \mathbf{R}$. Then, for every $k = 0, \ldots, t$ we have

$$|\{\tau | max(\tau) = k\}| = |\{\tau \mid pos(\tau) = k\}|$$

$$|\{\tau | Max(\tau) = k\}| = |\{\tau | Pos(\tau)\}|$$

PROOF. Let $0 \leq k \leq t$ and τ a permutation of $\{1, \ldots, t\}$ such that $pos(\tau) = k$. Thus we have $0 < j_1 < \ldots < j_k \leq t$ with

$$S_{j_\nu}(r) > 0 \qquad (\nu = 1, \ldots, k)$$

(and $S_j(\tau) \leq 0$ for $j \notin \{j_1, \ldots, j_k\}$). It is easily seen that

$$(1) \quad c_{\tau(j_1)} + \ldots + c_{\tau(j_i)} > 0 \qquad (i = 1, \ldots, k).$$

In fact, picture VIII.1.3 tells us this immediately: the sum appearing in (1) runs over some complete blocks of successive $c_{\tau(j_\nu)}$ plus an initial section of such a block; but the beginning of such a block leads out of the non-positive into the positive, where we remain as long as the block lasts, hence our sum is a sum of some sums which are > 0. Similarly, if $0 < \ell_1 < \ldots < \ell_{t-k} \leq t$ are such that $S_{\ell_\nu}(\tau) \leq 0 \quad (\nu = 1, \ldots, t-k)$, then

$$(2) \quad c_{\tau(\ell_1)} + \ldots + c_{\tau(\ell_i)} \leq 0 \quad (i = 1, \ldots, t-k).$$

Let us define a new permutation σ_τ of $\{1,\ldots,t\}$ by

$$\sigma_\tau(1) = \tau(j_k), \ldots, \sigma_\tau(k) = \tau(j_1)$$

$$\sigma_\tau(k+1) = \tau(\ell_1), \ldots, \sigma_\tau(t) = \tau(\ell_{t-k}),$$

that is, first the $\tau(j_\nu)$ are placed in reverse order, and then the $\tau(\ell_\nu)$ follow in their original order. From (1) and (2) we infer: if, in the path based on σ_τ, we go from position k to the left, we go strictly down, and when going to the right, we never get higher than we were at k. That is, the path based on σ_τ has its first maximum at k:

$$max(\sigma_\tau) = k.$$

Now the mapping $\tau \to \sigma_\tau$ is injective: let $\tau \neq \tau'$;, if $pos(\tau) \neq pos(\tau') \Rightarrow max(\sigma_\tau) \neq max(\sigma_{\tau'}) \Rightarrow \sigma_\tau \neq \sigma'_{\tau'}$; if $pos(\tau) = pos(\tau') = k$, but $\{j_1, \ldots, j_k\}$ is a different set for τ' than for τ, then clearly $\sigma_\tau \neq \sigma_{\tau'}$, follows again, if, finally, $pos(\tau) = pos(\tau')$ and $\{j_1, \ldots, j_k\}, \{\ell_1, \ldots, \ell_{t-k}\}$ are the same pairs of set for τ and for τ' although $\tau \neq \tau'$, then either $\tau(j_\nu) \neq \tau'(j_\nu)$ for some ν and $\sigma_\tau(k-\nu+1) \neq \sigma_{\nu'}(k-\nu+1)$, and hence $\sigma_\tau \neq \sigma_{\tau'}$ follows, or else a similar argument works for the ℓ_ν.

Thus the mapping $\tau \to \sigma_\tau$ permutes the set of all permutations of $\{1,\ldots,t\}$, sending every τ with $pos(\tau) = k$ into a σ_τ with $max(\tau) = k$. Obviously a similar proof works for Pos and Max, and our theorem is proved. □

2. Arcsin

2.1. The Finite Arcsin Distribution.

Proposition 2.1. For every $t = 1, 2, \ldots$

$$a_k^{(t)} = \binom{2k}{k}\binom{2(t-k)}{t-k}\frac{1}{2^{2t}} \quad (k = 0, 1, \ldots, t)$$

defines a probability distribution $a^{(t)}$ on $\{0, 1, \ldots, t\}$. It is called the *finite arcsin distribution of order t*.

PROOF. We first show

$$a_k^{(t)} = \binom{-\frac{1}{2}}{k}\binom{-\frac{1}{2}}{t-k}(-1)^t.$$

In fact the right member here can be written out

$$= \frac{(-\frac{1}{2})(-\frac{1}{2}-1)\dots(-\frac{1}{2}-k+1)(-\frac{1}{2}-1)\dots(-\frac{1}{2}-(t-k)+1)}{k! \quad (t-k)!}(-1)^t$$

$$= \frac{1\cdot 3\cdots(2k-1)\cdot 1\cdot 3\cdots(2(t-k)-1)}{k!2^k \quad (t-k)!2^{t-k}}$$

$$= \frac{1\cdot 2\cdot 3\cdots(2k-1)\cdot 2k\cdot 1\cdot 2\cdot 3\cdots(2(t-k)-1)(2(t-k))}{k!k!2^{2k} \quad (t-k)!(t-k)!2^{2(t-k)}}$$

$$= \binom{2k}{k}\binom{2(t-k)}{t-k}\frac{1}{2^{2t}}$$

$$= a_k^{(t)}.$$

Now

$$\frac{1}{\sqrt{1-x}} = \sum_{k=0}^{\infty}\binom{-\frac{1}{2}}{l}(-1)^k x^k$$

yields

$$\sum_{t-o}^{\infty} x^t = \frac{1}{1-x} = \left(\frac{1}{\sqrt{1-x}}\right)^2$$

$$= \sum_{t=0}^{\infty}\sum_{k=0}^{t}\left[\binom{-\frac{1}{2}}{k}\binom{-\frac{1}{2}}{t-k}(-1)^k(-1)^{t-k}\right]x^t$$

$$= \sum_{t=0}^{\infty}\left(\sum_{t=0}^{t} a_k^{(t)}\right)x^t.$$

By comparison of coefficients,

$$\sum_{k=0}^{t} a_k^{(t)} = 1$$

follows for all t. As all $a_k^{(t)}$ are ≥ 0, our proposition is proved. □

For $t = 2, 3, 4, 5$ the $a^{(t)}$ may be plotted as shown in figure VIII.2.1

Curiously enough, the middle values — around $\frac{t}{2}$ — are much smaller than the extremal ones.

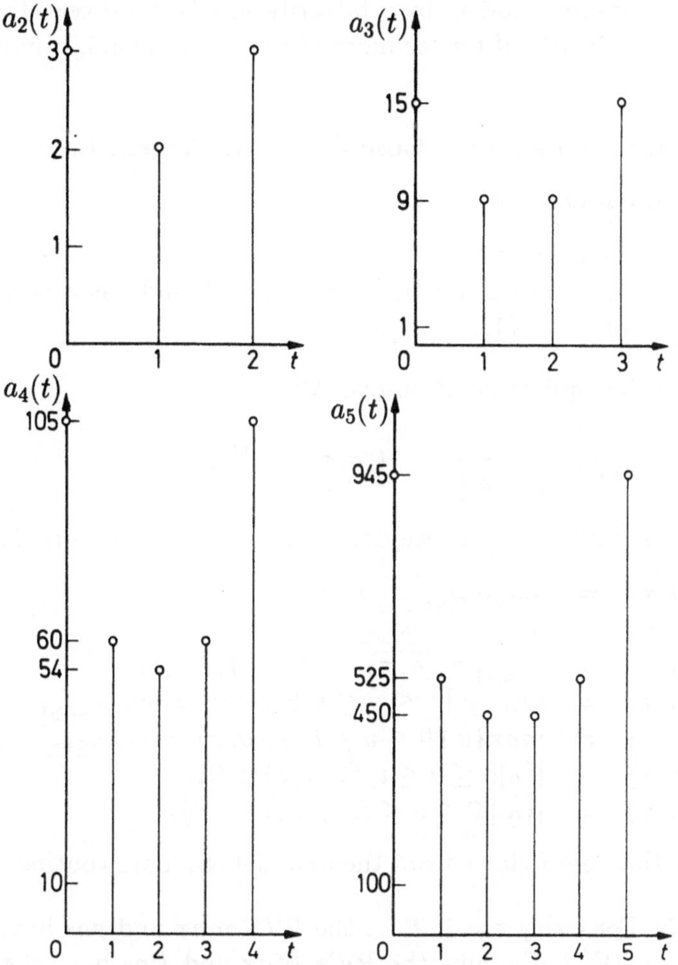

fig. VIII.2.1

(From: K. Jacobs, Selecta Mathematica I; Springer-Verlag, Heidelberg)

2.2. σ-Independent Increments.

Definition 2.2. Let t be a natural number. t real numbers $c_1, \ldots, c_t$ are called *σ-independent* if for any choice of $\sigma_1, \ldots, \sigma_t \in \{-1, 0, 1\}$, not all $\sigma_u = 0$,

$$\sum_{u=1}^{t} \sigma_u c_u \neq 0$$

follows.

Equality in (1) defines a hyperplane in $\mathbf{R}^t$. Hence all $(c_1, \ldots, c_t)$ not in the union of all those $2^t - 1$ hyperplanes are σ-independent. Any $(c_1, \ldots, c_t) \in \mathbf{R}^t$

can thus be made σ-independent by arbitrarily small alterations of some of the components c_u. – Clearly, if two or more of the c_u are equal, σ-independence fails.

2.3. Erik Sparre Andersen's Combinatorial Arcsin Law.

For any natural number t let

$$\begin{aligned} \Omega^{(t)} &= \{-1,1\}^t \times S_t \\ &= \{(\sigma,\tau) \mid \sigma \text{ is a } t\text{-tuple of signs} \pm 1, \text{and } \tau \text{ is a permutation} \\ &\quad \text{of} \quad \{1,\dots,t\}\}. \end{aligned}$$

Let $p^{(t)}$ denote the equidistribution over $\Omega^{(t)}$.

$$p^{(t)}(\sigma,\tau) = \frac{1}{\mid \Omega^{(t)} \mid} = \frac{1}{2^t t!} \qquad ((\sigma,\tau) \in \Omega^{(t)}).$$

For any real numbers $c_1,\dots,c_t$, we define the following RV's on $\Omega^{(t)}$:

$$\begin{aligned} x_u(\sigma,\tau) &= \sigma_{\tau(u)} c_{\tau(u)} \qquad (u = 1,\dots,t) \\ S_0(\sigma,\tau) &= 0 \\ S_u(\sigma,\tau) &= \textstyle\sum_{v=1}^{u} x_u(\sigma,\tau) \qquad (u = 1,\dots,t) \\ max(\sigma,\tau) &= min\{u \mid 0 \leq u \leq t, S_u(\sigma,\tau) = \max_{0\leq v\leq t} S_v(\sigma,\tau)\} \\ Max(\sigma,\tau) &= max\{u \mid 0 \leq u \leq t, S_u(\sigma,\tau) = \max_{0\leq v\leq t} S_v(\sigma,\tau)\} \\ pos(\sigma,\tau) &= |\{u | 0 \leq u \leq t, S_u(\sigma,\tau) > 0\}| \\ Pos(\sigma,\tau) &= |\{u \mid 0 \leq u \leq t, S_u(\sigma,\tau) \geq 0\}| \end{aligned}$$

The following theorem follows from theorem 1.1. by bare routine.

Theorem 2.3. For every $t = 1,2,\dots$ the RV's *max* and *pos* have the same distribution over $(\Omega^{(t)}$. Equally, the RV's *Max* and *Pos* have the same distribution.

PROOF.

$$\begin{aligned} p^{(t)}(max = k) &= \frac{1}{2^t t!} |(\sigma,\tau) \mid max(\sigma,\tau) = k\}| \\ &= \frac{1}{2^t t!} \sum_{\sigma} \sum_{\tau} |\{(\sigma,\tau) \mid max(\sigma,\tau) = k\}| \end{aligned}$$

Applying theorem 1.1. to the t real numbers $\sigma_1 c_1,\dots,\sigma_t c_t$ for every $\sigma = (\sigma_1,\dots,\sigma_t) \in \{-1,1\}^t$, we continue with

$$\begin{aligned} &= \frac{1}{2^t t!} \sum_{\sigma} \sum_{\tau} |\{(\sigma,\tau) \mid pos(\sigma,\tau) = k\}| \\ &= p^{(t)}(pos = k). \end{aligned}$$

The proof for *Max* and *Pos* follows in the same way. □

Having shown the equality of certain distributions, the obvious question arises what these distributions really look like. A universal answer can be given in the case of σ-independence:

Theorem 2.4. (E.S. Andersen's finite arcsin law)

Let t be a natural number and let $c_1, \ldots, c_t \in \mathbf{R}$ be σ-independent. Then the four RV's

$$max, pos, Max, Pos$$

on $(\Omega^{(t)}, p^{(t)})$ all have the same distribution, namely, the finite arcsin distribution of order t.

$$p^{(t)}(max = k) = p^{(t)}(pos = k) = \binom{2k}{k}\binom{2(t-k)}{t-k}\frac{1}{2^{2t}}$$
$$(k = 0, 1, \ldots, t)$$

and the same for Max and Pos.

PROOF. The σ-independence of $c_1, \ldots, c_t$ obviously implies $S_u(\sigma,\tau) - S_v(\sigma,\tau) \neq 0$ $(0 \leq v < u \leq t)$; thus the numbers

$$S_o(\sigma,\tau), S_1(\sigma,\tau), \ldots, S_t(\sigma,\tau)$$

are pairwise different, hence the RV's max and Max are the same, and so are pos and Pos. Theorem 2.3. now implies that all these four RV's have the same distribution. σ-independence also implies that all c_u are nonzero, and since we couple them with arbitrary signs σ_u anyhow, we may assume $c_1 > 0, \ldots, c_t > 0$. We now proceed by induction over $t = 1, 2, \ldots$. For $t = 1$ we have $\mid \Omega^{(1)} \mid = 2^1 \cdot 1! = 2$

$$a_o^{(1)} = \tbinom{0}{0}\tbinom{2}{1}\tfrac{1}{2^{2\cdot 1}} = \tfrac{1}{2} = \tbinom{1}{1}\tbinom{0}{0}\tfrac{1}{2^{2\cdot 1}} = a_1^{(1)}$$

$$p^{(1)}(max = 0) = p^{(1)}(\sigma_1 = -1) = \tfrac{1}{2} = a_o^{(1)}$$

$$p^{(2)}(max = 1) = p^{(1)}(\sigma_1 = 1) = \tfrac{1}{2} = a_1^{(1)}$$

The little pictures show the two possible situations. Having thus settled the case $t = 1$, we now assume $t > 1$ and may make free use of our statement in all cases $t' < t$. Now, for any $0 \leq k \leq t, max(\sigma,\tau) = k$ means the following: the path with successive increments $\sigma_{\tau(1)}c_{\tau(1)}, \ldots, \sigma_{\tau(k)}c_{\tau(k)}$ attains its maximum at its last vertex no. k, and the path with successive increments $\sigma_{\tau(k+1)}c_{\tau(k+1)}, \ldots, \sigma_{\tau(t)}c_{\tau(t)}$ has its maximum at its very beginning. Now, if we assume $0 < k < t$, and hence $0 < t - k < t$, we may argue as follows: there are $\binom{t}{k}$ ways of splitting the $c_1, \ldots, c_t$ into two portions of k resp. $t - k$ members, and for every splitting of this kind, the induction hypothesis tells us, in how many arrangements the first portion produces a max in the last position, and in how many cases the second portion produces a max in position zero:

$$\begin{aligned}
|\{(\sigma,\tau)|\max(\sigma,\tau)=k\}| &= \binom{t}{k}\left[\binom{2k}{k}\binom{0}{0}\frac{1}{2^{2k}}\cdot 2^k k!\right] \\
&\cdot \left[\binom{0}{0}\binom{2(t-k)}{t-k}\frac{1}{2^{2(t-k)}}\cdot 2^{t-k}(t-k)!\right] \\
&= \binom{2k}{k}\binom{2(t-k)}{t-k}\frac{1}{2^{2t}}\cdot 2^t\cdot t!
\end{aligned}$$

Division by $2^t t!$ yields the desired result

$$p^{(t)}(max=k)=a_k^{(t)} \qquad (0<k<t).$$

In order to settle the remaining extreme cases $k=0, k=t$, we first observe the obvious symmetry

$$a_k^{(t)}=a_{t-k}^{(t)} \qquad (k=0,1,\ldots,t)$$

and the a little bit less obvious symmetry

$$p^{(t)}(max=k)=p^{(t)}(max=t-k) \qquad (k=0,1,\ldots,t)$$

which follows from

$$p^{(t)}(pos=k)=p^{(t)}(pos=t-k) \qquad (k=0,1,\ldots,t)$$

via the equivalence principle. The latter equality, however, follows by passage to opposite signs: $(\sigma_1,\ldots,\sigma_t)\to(-\sigma_1,\ldots,-\sigma_t)$. Since both $p^{(t)}(max=0)+p^{(t)}(max=t)=2p^{(t)}(max=0)$ and $a_o^{(t)}+a_t^{(t)}=2a_o^{(t)}$ make up the same remainder

$$1-\sum_{0<k<t}p^{(t)}(max=k)=1-\sum_{0<k<t}a_k^{(t)},$$

they are equal, and our induction proof is complete. □

The following example shows that the conclusion of theorem 2.4. may become false if the hypothesis of σ-independence fails.

Example 2.5. Let $(c_1,c_2,c_3)=(1,2,3)$. As e.g. $-c_1-c_2+c_3=0$, these three reals are not σ-independent

We get

k	0	1	2	3
$\mid\{Pos=k\}\mid$	17	9	9	13

which is not conformal with *arcsin*.

Exercise 2.6. Discuss the cases $(c_1, c_2, c_3) =$

(1, 1, 1)
(1, 1, 2)
(1, 1, 3)
(1, 2, 4)

in the same way as (1,2,3) above.

2.4. The Asymptotic Behavior of the Arcsin Distributions.

We will now prove that the sequence $a^{(1)}, a^{(2)}, \ldots$ of finite *arcsin* distributions, when suitably put on the unit interval, converges to a certain continuous probability distribution which in fact involves the classical arcsin function, thus justifying , in a way, this name for the discrete distributions $a^{(t)}$.

Lemma 2.7. The function $\rho(x)$ defined on the unit interval $[0, 1]$ by

$$\rho(x) = \begin{cases} \frac{1}{\pi} \cdot \frac{1}{\sqrt{x(1-x)}} & (0 < x < 1) \\ 0 & (x = 0, x = 1) \end{cases}$$

is Riemann integrable in the generalized sense and fulfils

$$\int_0^1 \rho(x)dx = 1 \qquad \int_0^x \rho(y)dy = \frac{2}{\pi} arcsin\sqrt{x} \qquad (0 \le x \le 1).$$

PROOF. Everybody knows that

$$\frac{d}{dx}\ arcsin \quad x = \frac{1}{\sqrt{1-x^2}}.$$

Hence

$$\frac{d}{dx}[\frac{2}{\pi}\ arcsin\ \sqrt{x}] = \frac{2}{\pi}\frac{1}{\sqrt{1-x}} \cdot \frac{1}{2\sqrt{x}} = \rho(x).$$

The lemma now follows easily. □

Theorem 2.8. For the above function $\rho(x)$ and the sequence $a^{(1)}, a^{(2)}, \ldots$ of finite *arcsin* distribution, the following holds:

$$(1) \quad \lim_{t \to \infty} \sum_{\alpha \le \frac{k}{t} \le \beta} a_k^{(t)} = \int_\alpha^\beta \rho(x)dx$$

uniformly for all α, β with $0 \le \alpha \le \beta \le 1$.

PROOF. We use Stirling's formula

$$n! = \sqrt{2\pi n}(\frac{n}{e})^n F(n) \qquad (k = 0, 1, \ldots)$$

$$\lim_{n\to\infty} F(n) = 1$$

for $k, 2k, t-k, 2(t-k)$ in the place of n and get, after an easy calculation,

$$\begin{aligned} a_k^{(t)} &= \binom{2k}{k}\binom{2(t-k)}{t-k}\frac{1}{2^{2t}} = \frac{(2k)![2(t-k)]!}{k!k!(t-k)!(t-k)!2^{2t}} \\ &= \frac{1}{\pi}\cdot\frac{1}{\sqrt{\frac{k}{t}(1-\frac{k}{t})}}\cdot\frac{1}{t}\cdot\frac{F(2k)\cdot F(2(t-k))}{F(k)\cdot F(k)\cdot F(t-k)\cdot F(t-k)}. \end{aligned}$$

Let us now choose some δ with $0 < \delta < \frac{1}{2}$ and consider, for every t, only those k which fulfil

$$\delta \le \frac{k}{t} \le 1 - \delta.$$

Then $t\delta \le k, t\delta \le t-k$, and hence all terms $F(k)$ etc. tend to 1 uniformly for all such k, as $t \to \infty$. Therefore there is a sequence $G_{\alpha\beta}\ (t)(t = 1, 2, \ldots)$ such that for any δ given as above

$$\lim_{t\to\infty} G_{\alpha\beta}(t) = 1 \quad \text{uniformly for} \quad \delta \le \alpha \le \beta \le 1 - \delta$$

and

$$\sum_{\alpha \le \frac{k}{t} \le \beta} a_k^{(t)} = \left[\frac{1}{t}\sum_{\alpha\le\frac{k}{t}\le\beta}\frac{1}{\pi}\cdot\frac{1}{\sqrt{\frac{k}{t}\cdot(1-\frac{k}{t})}}\right]\cdot G_{\alpha\beta}(t)$$

hold. The term in square brackets obviously tends to $\int_\alpha^\beta \rho(x)dx$ (Riemann sums!). Thus we obtain the desired result (2) uniformly for $\delta \le \alpha \le \beta \le 1-\delta$, however $0 < \delta < \frac{1}{2}$ has been chosen.

Uniformity of convergence for *all* $0 \le \alpha \le \beta \le 1$ is now easily established via estimates of the contributions at the ends of the interval $[0, 1]$: for any given $\epsilon > 0$ choose $0 < \delta < \frac{1}{2}$ such that

$$\int_\delta^{1-\delta} \rho(x)dx > 1 - \frac{\epsilon}{5}.$$

Find $t_0 > 0$ such that $t \geq t_0$ implies

$$\left| \sum_{\alpha' \leq \frac{k}{t} \leq \beta'} a_k^{(t)} - \int_{\alpha'}^{\beta'} \rho(x)dx \right| < \frac{\epsilon}{5}$$

uniformly for $\delta \leq \alpha' \leq \beta' \leq 1 - \delta$. Choose now any α, β with $0 \leq \alpha \leq \beta \leq 1$ and define

$$\alpha' = max\{\alpha, \delta\}, \quad \beta' = min\{\beta, 1-\delta\}.$$

We now obtain for $t \geq t_0$

$$\begin{aligned}
& \left| \sum_{\alpha \leq \frac{t}{t} \leq \beta} a_k^{(t)} - \int_{\alpha}^{\beta} \rho(x)dx \right| \\
\leq & \left| \sum_{\alpha' \leq \frac{k}{t} \leq \beta'} a_k^{(t)} - \int_{\alpha'}^{\beta'} \rho(x)dx \right| \\
+ & \sum_{\alpha \leq \frac{k}{t} < \alpha'} a_k^{(t)} + \sum_{\beta' < \frac{k}{t} \leq \beta} a_k^{(t)} + \int_{\alpha}^{\alpha'} \rho(x)dx + \int_{\beta'}^{\beta} \rho(x)dx \\
< & \frac{\epsilon}{5} + 2 \cdot \frac{\epsilon}{5} + \left(1 - \sum_{\delta \leq \frac{k}{t} \leq 1-\delta} a_k^{(t)} \right) \\
\leq & \frac{3}{5}\epsilon + \left(1 - \int_{\delta}^{1-\delta} \rho(x)dx \right) + \left| \int_{\delta}^{1-\delta} \rho(x)dx - \sum_{\delta \leq \frac{k}{t} \leq 1-\delta} a_k^{(t)} \right| \\
< & \frac{3}{5}\epsilon + \frac{\epsilon}{5} + \frac{\epsilon}{5} = \epsilon
\end{aligned}$$

independently of the choice of α, β. □

The functions $\rho(x)$ and $\frac{2}{\pi}$ arcsin $\sqrt{x}$ look as shown in figures VIII.2.2 and 2.3.

The shape of $\rho(x)$ leads, for large t, to the following surprising observation:

> The middle values – around $\frac{t}{2}$ – of pos and max are much less probable than the extremal ones (near 0 or t).

This phenomenon became visible already for small values of t (see fig.2.1).

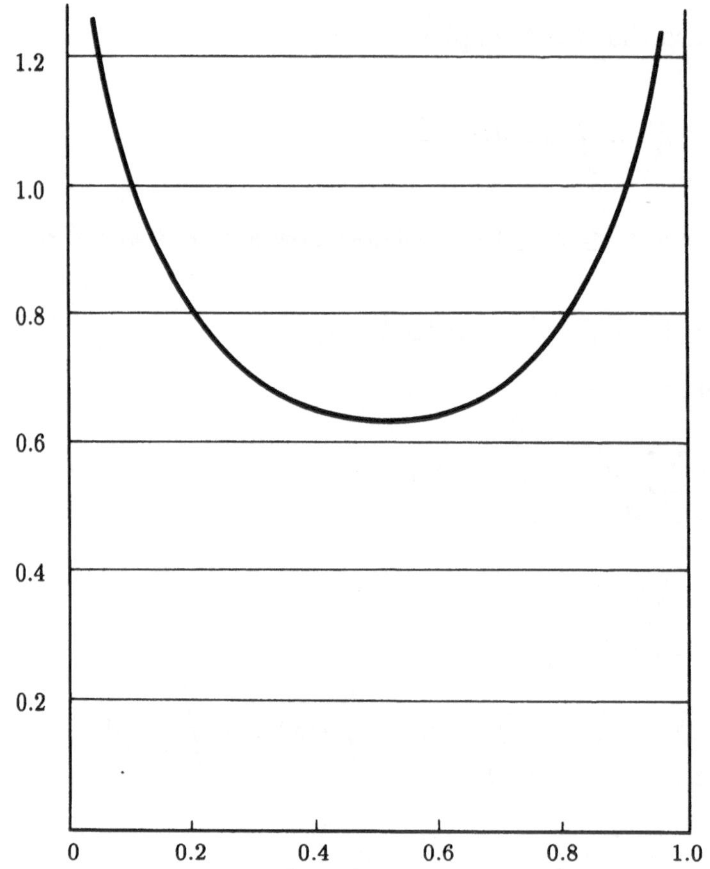

fig. VIII.2.2

3. Symmetrically Distributed Random Variables

Our setup in the previous sections was largely combinatorial. Now we carry over our results to the most general probabilistic situation for which they can be established: σ-independent symmetrically distributed RV's. Our restriction to discrete probability distributions causes quite some inconvenience here: the standard case of IID cannot be subsumed under our present theory. With continuous probability distributions this inconvenience would not arise. In the next section, we will modify our combinatorial setup somewhat and prove certain arcsin type results for IID random variables which attain values ± 1 only.

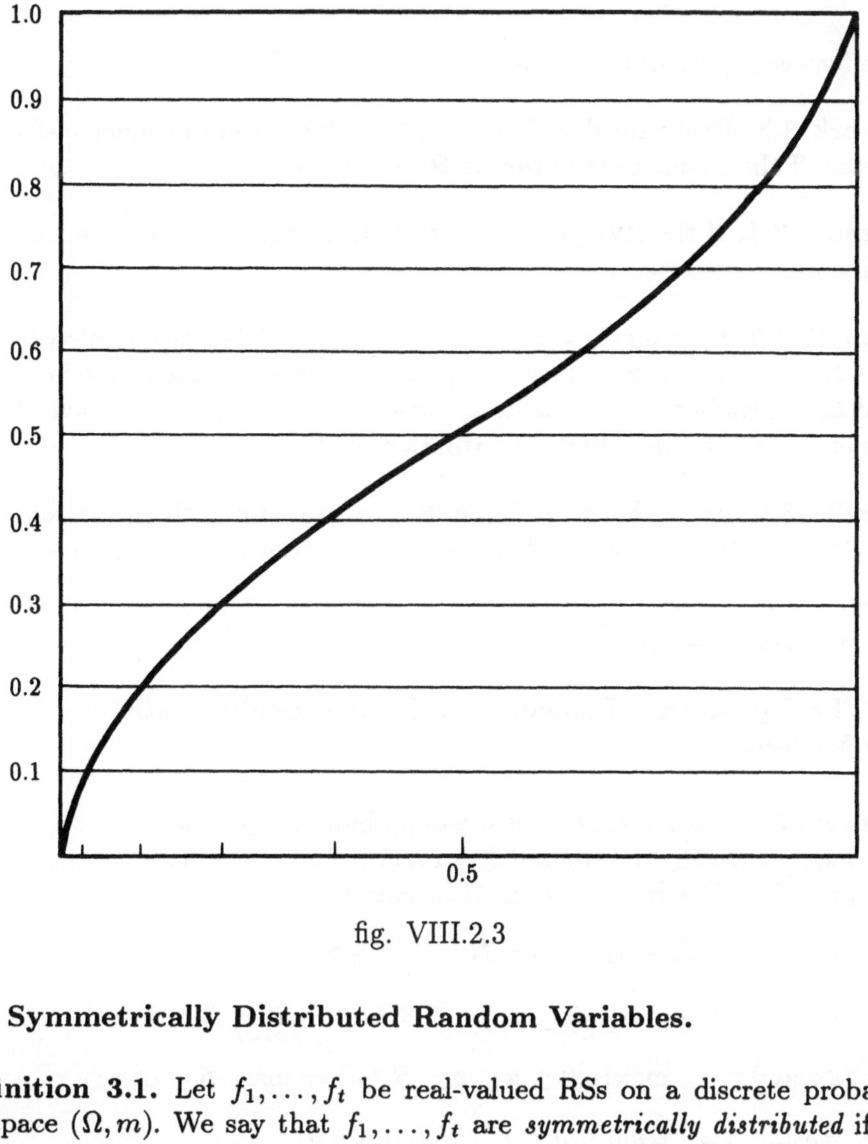

fig. VIII.2.3

3.1. Symmetrically Distributed Random Variables.

Definition 3.1. Let $f_1, \ldots, f_t$ be real-valued RSs on a discrete probability space (Ω, m). We say that $f_1, \ldots, f_t$ are *symmetrically distributed* if for any permutation τ of $\{1, \ldots, t\}$ the RV's $f_{\tau(1)}, \ldots, f_{\tau(t)}$ have the same joint distribution as $f_1, \ldots, f_t$.

We may reformulate this as follows:

Definition 3.2. For any permutation τ of $\{1, \ldots, t\}$ let $T_\tau : \mathbf{R}^t \to \mathbf{R}^t$ be defined by

$$T_\tau(x_1, \ldots, x_t) = (x_{\tau(1)}, \ldots, x_{\tau(t)}).$$

A discrete probability distribution p in $\mathbf{R}^t$ is called *symmetric* if

$$T_\tau p = p$$

holds for every permutation τ of $\{1,\dots,t\}$.

Remark 3.3. Real-valued RVs $f_1,\dots,f_t$ on (Ω,m) are symmetrically distributed iff their joint distribution in $\mathbf{R}^t$ is symmetric.

Example 3.4. If the RVs $f_1,\dots,f_t$ are IID, they are clearly symmetrically distributed.

Example 3.5. For every $x=(x_1,\dots,x_t)\in\mathbf{R}^t$ let q^x be the equidistribution over $\{T_\tau x \mid \tau \text{ permutes } \{1,\dots,t\}\}$; the power of the latter set is a divisor of $t!$, depending upon equalities between certain x_u. Every such q^x is a symmetric discrete probability distribution in $\mathbf{R}^t$.

Exercise 3.6. Prove that a DPD p in $\mathbf{R}^t$ is symmetric iff there is a sequence $x^{(1)},x^{(2)},\dots\in\mathbf{R}^t$ and real numbers $\alpha_1,\alpha_2,\dots\geq 0$ such that $\alpha_1+\alpha_2+\dots=1$ and

$$p=\alpha_1 q^{x^{(1)}}+\alpha_2 q^{x^{(2)}}+\dots.$$

3.2. The Equivalence Principle for Symmetrically Distributed Random Variables.

Theorem 3.7. Let (Ω,m) be a discrete probability space and $f_1,\dots,f_t$ real-valued symmetrically distributed RVs on (Ω,m). Let the RVs $S_0,S_1,\dots,S_t$, max, pos, Max, Pos be defined on Ω as follows:

$$S_0\equiv 0,\qquad S_u=f_1+\dots+f_u\qquad(1\leq u\leq t)$$

$$\begin{aligned}
max(\omega) &= \min\{u|0\leq u\leq t,\quad S_u(\omega)=\max_{0\leq v\leq t} S_v(\omega)\}\\
Max(\omega) &= \max\{u|0\leq u\leq t,\quad S_u(\omega)=\max_{0\leq v\leq t} S_v(\omega)\}\\
pos(\omega) &= |\{u|0\leq u\leq t,\quad S_u(\omega)>0\}|\\
Pos(\omega) &= |\{u|0\leq u\leq t,\quad S_u(\omega)\geq 0\}|
\end{aligned}$$

Then

max and *pos* have the same distribution
Max and *Pos* have the same distribution

PROOF. Define, for every $c=(c_1,\dots,c_t)\in\mathbf{R}^t$ and every permutation τ of $\{1,\dots,t\}$, the real numbers $max^c(\tau)$ etc. in the same fashion as $max(\tau)$ etc. in §1 (we didn't write out c there). Define further

$$max^\tau(\omega)=max^{(f_1(\omega),\dots,f_t(\omega))}(\tau)\quad\text{etc.}$$

As

$$max^{(f_1(\omega),\ldots,f_t(\omega))}(\tau) = max^{(f_{\tau(1)}(\omega),\ldots,f_{\tau(t)}(\omega))} \quad (id)$$

and $f_1, \ldots, f_t$ are symmetrically distributed, all $t!$ random variables max^τ have the same distribution; the same result holds for Max^τ etc. Now for any $k = 0, \ldots, t$ we get

$$\begin{aligned} m(max = k) &= \frac{1}{t!}\sum_\tau m(max^\tau = k) && \text{(symmetry)} \\ &= \frac{1}{t!}\sum_\tau E_m(1_{\{max^\tau = k\}}) \\ &= \frac{1}{t!}E_m(\sum_\tau 1_{\{max^\tau = k\}}) \end{aligned}$$

Now for every ω the value

$$\sum_\tau 1_{\{max^\tau = k\}}\,(\omega)$$

counts how often $max^{(f_1(\omega),\ldots,f_t(\omega))}(\tau) = k$; by theorem 1.1. (combinatorial equivalence principle) this happens equally often as $pos^{(f_1(\omega),\ldots,f_t(\omega))} = k$, i.e.

$$\sum_\tau 1_{\{pos^\tau = k\}}(\omega)$$

times. Thus

$$m(max = k) = \frac{1}{t!}E_m(\sum_\tau 1_{\{pos^\tau = k\}})$$

and by the same token as above this is

$$= m(pos = k).$$

The second statement of our theorem is proved in the same way. □

3.3. σ-Symmetrically Distributed Random Variables. σ-Independence.

Definition 3.8. The real-valued RVs $f_1, \ldots, f_t$ on the DPS (Ω, m) are said to be ***σ-symmetrically distributed*** if for every $\sigma = (\sigma_1, \ldots, \sigma_t) \in \{-1, 1\}^t$ and for every permutation τ of $\{1, \ldots, t\}$ the RVs

$$\sigma_{\tau(1)} f_{\tau(1)}, \ldots, \sigma_{\tau(t)} f_{\tau(t)}$$

have the same joint distribution as $f_1, \ldots, f_t$.

Example 3.9. Let $f_1, \ldots, f_t$ be real-valued IID RVs on the DPS (Ω, m) and let $-f_1$ have the same distribution as f_1. Then $f_1, \ldots, f_t$ are obviously σ-symmetrically distributed.

Exercise. 3.10. Reformulate definition 3.8. in an analogous fashion as it was done for definition 3.1. by definition 3.2. and proposition 3.3. Formulate and prove the analogon to exercise 3.6.

Definition 3.11. The real-valued RVs $f_1, \ldots, f_t$ on the DPS (Ω, m) are said to be σ- *independent* if for m-almost every ω the t-tuple $(f_1(\omega), \ldots, f_t(\omega))$ is σ-independent, i.e. if for any choice of $\alpha_1, \ldots, \alpha_t \in \{-1, 0, 1\}$, not all $\alpha_u = 0$, we have

$$\sum_{u=1}^{t} \alpha_u f_u(\omega) \neq 0 \quad m\text{-a.e.}$$

Exercise 3.12. Prove that $f_1, \ldots, f_t$ are σ-independent iff their joint distribution p fulfills

$$p(\cup_\alpha H_\alpha) = 0$$

where $\alpha = (\alpha_1, \ldots, \alpha_t)$ runs over $\{-1, 0, 1\}^t \backslash \{(0, \ldots, 0)\}$ and the $3^t - 1$ hyperplanes H_α are defined by

$$H_\alpha = \{x \mid x = (x_1, \ldots, r_t) \in \mathbf{R}^t, \quad \sum_{u=1}^{t} \alpha_u x_u = 0\}.$$

The inconvenience mentioned in the introduction to this section arises as follows:

If $f_1, \ldots, f_t$ are IID RVs on a DPS (Ω, m), their joint distribution p has the form $q \times \cdots \times q$ (t factors) where q is a DPD on $\mathbf{R}$ (proposition III.3.20). Now if e.g. $q_y > 0$ for some $y \in \mathbf{R}$, then clearly $p_{(y,\ldots,y)} > 0$ and hence, with the above notation, $p(H_{(1,-1,0,\ldots,0)}) > 0$. Thus IID random variables are never σ-independent in discrete stochastics. If, however, we deal with continuous distributions q which assign probability 0 to every single real number, then every hyperplane in $\mathbf{R}^t$ is a nullset for $p = q \times \cdots \times q$ and q-distributed IID RVs turn out to be σ-independent in the obvious sense generalizing definition 3.11.

3.4. The Probabilistic Arcsin Law of Erik Sparre Andersen.

Theorem 3.13. Let $f_1, \dots, f_t$ real-valued σ-symmetric and σ-independent RVs on the DPS (Ω, m). Let the RVs max, pos, Max, Pos be defined as in theorem 3.7. Then

$$pos = Pos\; m\text{-a.e.}$$
$$max = Max\; m\text{-a.e.}$$

and all these four RVs have the same distribution, namely, $a^{(t)}$, that is

$$m(pos = k) = m(max = k) = \binom{2k}{k}\binom{2(t-k)}{t-k}\frac{1}{2^{2t}} \quad (k = 0, \dots, t).$$

PROOF. Only the last statement is less than obvious. For any $\sigma = (\sigma_1, \dots, \sigma_t) \in \{-1, 1\}^t$ and any permutation τ of $\{1, \dots, t\}$ define the RV

$$max^{\sigma,\tau}(\omega) = max^{(f_1(\omega), \dots, f_t(\omega))}(\sigma, \tau),$$

where $max(\sigma, \tau)$ is defined as in §2 subsection 3.
By σ-symmetry, we get

$$\begin{aligned} m(max = k) &= \frac{1}{2^t t!}\sum_{\sigma,\tau} m(max^{\sigma,\tau} = k) \\ &= \frac{1}{2^t t!} E_m\Big(\sum_{\sigma,\tau} 1_{\{max^{\sigma,\tau} = k\}}\Big) \end{aligned}$$

The sum in the bracket counts, for every $\omega \in \Omega$, for how many among the σ, τ

$$max(f_1(\omega), \dots, f_t(\omega))(\sigma, \tau) = k$$

holds. If ω avoids a suitable m-nullset, the real numbers $f_1(\omega), \dots, f_t(\omega)$ are σ-independent, hence by theorem 2.4. (combinatorial arcsin law), the said sum equals the constant value

$$\binom{2k}{k}\binom{2(t-k)}{t-k}\frac{1}{2^t t!}.$$

Thus

$$m(max = k) = a_k^{(t)} = \binom{2k}{k}\binom{2(t-k)}{t-k}\frac{1}{2^t t}$$

follows. □

4. Fluctuations of Random Walks

In this section we investigate the RVs max and pos in a situation which blatantly violates the hypothesis of σ-independence which was so essential for the arcsin results in §2 and §3: random walk with increments ± 1. We obtain results of arcsin type again, partly thanks to the fact that our random walks walk with step width 1 only and hence display sort of continuity: if you walk from integer a to integer b, you visit every integer between a and b at least once.

4.1. A Minimal Model for Random Walk.

We set the stage for the subsequent subsections by constructing sort of a minimal model for random walks with step width 1. We put, for an integer $t > 0$,

$$\begin{aligned}
\Omega^{(t)} &= \{-1,1\}^t = \{\sigma = (\sigma_1,\ldots,\sigma_t) \mid \sigma_1,\ldots,\sigma_t = \pm 1\} \\
m^{(t)} &= \text{equidistribution over} \quad \Omega^{(t)} \\
X_u(\sigma) &= \sigma_u \qquad (1 \le u \le t) \\
S_0(\sigma) &= 0 \\
S_u(\sigma) &= X_1(\sigma) + \ldots + X_u(\sigma) \qquad (1 \le u \le t) \\
max^{(t)}(\sigma) &= max(\sigma) = \min\{u | 0 \le u \le t, S_u(\sigma) = \max_{0 \le v \le t} S_v(\sigma)\} \\
pos^{(t)}(\sigma) &= pos(\sigma) = |\{u | 0 \le u \le t, S_u(\sigma) > 0\}| \\
p(\sigma) &= |\{u | 1 \le u \le t\}, \sigma_u = 1\}| \\
n(\sigma) &= |\{u \mid 1 \le u \le t, \sigma_u = -1\}| (\sigma = (\sigma,\ldots,\sigma_t) \in \Omega^{(t)})
\end{aligned}$$

The models for $t+1$ extends the model for t in an obvious way by splitting paths into two and halving their probabilities. We will make use of this later.

4.2. Desiré André's Reflection Principle.

The famous reflection principle (André [1887]) may be formulated in different ways. Here we choose the following form:

Theorem 4.1. (Reflection principle). Let $a > 0$ be an integer. Then for every integer $b < a$

$$m^{(t)}(S_t = b, S_u = a \quad \text{for at least one} \quad u \le t)$$

$$= m^{(t)}(S_t = a + (a - b)).$$

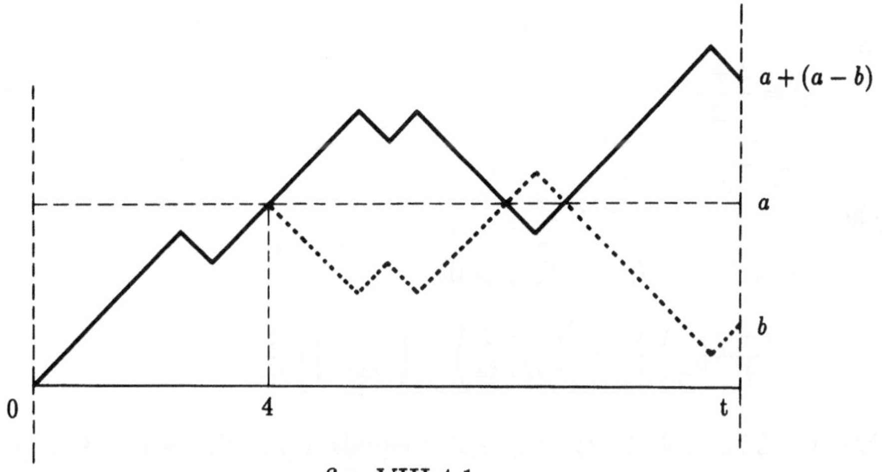

fig. VIII.4.1

PROOF. $a+(a-b)$ is b mirrored on a. The idea of our subsequent proof can be visualized as shown in figure VIII.4.1.

Verbally: every path with increments ± 1 which starts at 0 and ends (after t steps) at b and visits a at least once (say, after $u \leq t$ steps for the first time) corresponds in a one-to-one fashion to the path which coincides with the given path until u and coincides with its image obtained by mirroring it at a for the remaining steps; the new path goes from 0 to $a+(a-b)$ and visits a after u steps for the first time. As $m^{(t)}$ is equidistribution over $\Omega^{(t)}$, this bijection entails equality of probabilities for the two path sets in question. □

Another variant of the reflection principle sounds as follows:
There are precisely as many ± 1-paths visiting c at time s and b at time t, hitting $a \in \mathbf{Z} \backslash [c, b]$ at least once inbetween, as there are paths from $a+(a-c)$ to b during the time interval from s to t.

The proof if this variant is very similar to the proof of theorem 4.1. and is left to the reader.

4.3. The Ballot Theorem.

One may ask after the probability that the winner of a one-man-one-vote election constantly leads during the counting of votes. The answer is given by

Theorem 4.2. (Ballot theorem) Let, for any integer $t > 0, r$ be a natural number and $p, q \geq 0$ integers such that

$$t = p + q$$

$$r = p - q$$

i.e.

$$p = \frac{t+r}{2}$$
$$q = \frac{t-r}{2}.$$

Then

$$m^{(t)}(S_t = r, S_1, \ldots, S_{t-1} > 0)$$
$$= \frac{p-q}{t}\binom{t}{p}\frac{1}{2^t} = \left[\binom{t-1}{\frac{t+r-2}{2}} - \binom{t-1}{\frac{t+r}{2}}\right]\frac{1}{2^t}$$

Proof. The probability in question equals, up to the factor $\frac{1}{2^t} = \frac{1}{|\Omega^{(t)}|}$, the number of those ± 1-paths which start at 0 and end at r after t steps without ever returning to 0. Every such path is at 1 after one step. By our variant of the reflection principle (with $a = 0, b = r, c = 1, s = 1$) this equals the number of those paths which go in $t-1$ steps from 1 to r, minus the number of all paths which go from -1 to r in $t-1$ steps. A path with $t-1$ steps belongs to the former of these two classes iff it goes up $p-1$ times and down q times. And it belongs to the latter class iff it increases p times and decreases $q-1$ times. Thus (we use the equalities $t-1 = (p-1)+q = p+(q-1), r-1 = (p-1)-q$) the number in question turns out to be

$$\begin{aligned}\binom{t-1}{p-1} - \binom{t-1}{q-1} &= \frac{(t-1)!}{(p-1)!q!} - \frac{(t-1)!}{p!(q-1)!} \\ &= \frac{p-q}{t}\cdot\frac{t!}{p!q!} \\ &= \frac{p-q}{t}\binom{t}{p}\end{aligned}$$

Moreover $p-1 = \frac{t+r}{2} - 1 = \frac{t+r-2}{2}$, $\binom{t-1}{q-1} = \binom{t-1}{(t-1)-(q-1)} = \binom{t-1}{p} = \binom{t-1}{\frac{t+r}{2}}$. This does it. □

As a consequence of the Ballot theorem, the (conditional) probability of a permanent lead during vote counting, of a winner with majority r, is

$$\frac{p-q}{t} = \frac{r}{t}.$$

4.4. The Arcsin Laws for Random Walks.

Obviously, in our random walk model, t and S_t always have the same parity, i.e.

> S_t can only be even if t is even.
> S_t can only be odd if t is odd.

More precisely

> If t is even, the possible values for S_t are
> $-t, -t+2, \dots, -2, 0, 2, \dots, t-2, t$
> If t is odd, the possible values for S_t are
> $-t, -t+2, \dots, -1, 1, \dots, t-2, t$

Proposition 4.3. In our model,

$$m^{(t)}(S_1 > 0, \dots, S_t > 0) = \begin{cases} \dfrac{1}{2}\dbinom{2s}{2}\dfrac{1}{2^t} & \text{for} \quad t = 2s \\ \dbinom{2s}{s}\dfrac{1}{2^t} & \text{for} \quad t = 2s+1 \end{cases}$$

PROOF. We first consider the case $t = 2s$ and split the probability in question according to the possible values of S_t:

$$m^{(t)}(S_1, \dots, S_t > 0) = \sum_{k=1}^{s} m^{(t)}(S_1, \dots, S_{t-1} > 0, S_t = 2k)$$

By the Ballot Theorem, we may continue

$$= \frac{1}{2^t}\sum_{k=1}^{s}\left[\binom{2s-1}{s+k-1} - \binom{2s-1}{s+k}\right]$$

As $\binom{2s-1}{2s} = 0$, our sum telescopes down to

$$= \frac{1}{2^t}\binom{2s-1}{s} = \frac{1}{2^t}\cdot\frac{1}{2}\binom{2s}{s}.$$

If $t = 2s+1, S_{t-1} = S_{2s} > 0$ implies $S_{2s} \geq 2$, hence $S_t > 0$ follows automatically. Thus

$$\begin{aligned} m^{(t)}(S_1, \dots, S_t > 0) &= m^{(t)}(S_1, \dots, S_{t-1} > 0) \\ &= 2m^{(t-1)}(S_1, \dots, S_{t-1} > 0) \\ &= \frac{1}{2^t}\binom{2s}{s} \end{aligned}$$

by our previous result. □

Passing from > 0 to ≥ 0, we get even simpler formulas:

Proposition 4.4. In our model,

$$m^{(t)}(S_1 \geq 0, \ldots, S_t \geq 0) = \begin{cases} \binom{2s}{s}\frac{1}{2^t} & \text{for } t = 2s \\ \\ \frac{1}{2}\binom{2s}{s}\frac{1}{2^t} & \text{for } t = 2s-1 \end{cases}$$

PROOF. A path with t steps and all $S_1, \ldots, S_t > 0$ becomes a path with $t-1$ steps and all vertices ≥ 0 if we skip increment (step) no. 1, and this procedure is bijective. Thus

$$m^{(t)}(S_1, \ldots, S_t > 0) = \frac{1}{2}m^{(t-1)}(S_1, \ldots, S_{t-1} \geq 0)$$

or, equivalently,

$$m^{(t)}(S_1, \ldots, S_t \geq 0) = 2 \cdot m^{(t+1)}(S_1, \ldots, S_{t+1} > 0)$$

Now if $t = 2s$ is even, proposition 4.3 allows us to continue with

$$= 2 \cdot \binom{2s}{s}\frac{1}{2^{t+1}} = \binom{2s}{s}\frac{1}{2^t}.$$

If $t = 2s-1$ is odd, then S_t is odd and hence ≥ 1 if it is ≥ 0. No matter which increment no. $t+1$ we add, we still remain in the nonnegative half of $\mathbf{R}$. Thus

$$\begin{aligned} m^{(t)}(S_1, \ldots, S_t \geq 0) &= m^{(t+1)}(S_1, \ldots, S_{t+1} \geq 0) \\ &= \binom{2s}{s}\frac{1}{2^{t+1}} = \frac{1}{2}\binom{2s}{s} = \frac{1}{2}\binom{2s}{s}\frac{1}{2^t}. \end{aligned}$$

We now can prove sort of finite arcsin formulas in a way quite similar to the proof of theorem 2.4. The necessity to distinguish between odd and even makes the result and the proof clumsier, however. □

Theorem 4.5. (arcsin law for random walks.)
In our model, we have

1. in case $t = 2s$

$$m^{(t)}(pos = k) = m^{(t)}(\max = k) = \begin{cases} \binom{2s}{s}\frac{1}{2^t} \text{ for } k = 0 \\ \frac{1}{2}\binom{2j}{j}\binom{2(s-j)}{s-j}\frac{1}{2^t} \\ \quad \text{for } 0 < k \leq t\ k = 2j \\ \quad \text{or } k = 2j+1 \end{cases}$$

2. in case $t = 2s - 1$

$$m^{(t)}(pos = k) = m^{(t)}(\max = k) = \begin{cases} \frac{1}{2}\binom{2s}{s}\frac{1}{2^t} \quad \text{for} \quad k \neq 0 \\ \frac{1}{4}\binom{2j}{j}\binom{2(s-j)}{s-j}\frac{1}{2^t} \\ \text{for } 0 < k < t, k = 2j \\ \binom{2j}{j}\binom{2(s-j-1)}{s-j-1}\frac{1}{2^t} \\ \text{for } 0 < k < t,\ k = 2j+1 \\ \binom{2(s-1)}{s-1}\frac{1}{2^t} \text{ for } k = t \end{cases}$$

PROOF. As the equivalence principle (thorem 3.7.) doesn't need the hypothesis of σ-independence, it clearly applies here: *pos* and *max* have the same distribution. In the sequel, we will deal essentially with *max* only.

1. Let $t = 2s$.

a) If $k = 0$,

$$\begin{aligned} m^{(t)}(\max = k) &= m^{(t)}(\max = 0) \\ &= m^{(t)}(S_1, \ldots, S_t \leq 0) \\ &= \binom{2s}{s}\frac{1}{2^t} \end{aligned}$$

by proposition 4.4.

b) If $0 < k \leq t$, we argue as follows:

$$\begin{aligned} m^{(t)}(max = k) &= m^{(t)}(S_o, \ldots, S_{k-1} < S_k; S_{k+1}, \ldots, S_t \leq S_t) \\ &= m^{(t)}(S_k - S_o, \ldots, S_k - S_{k-1} > 0; \\ &\qquad S_{k+1} - S_k, \ldots, S_t - S_k \leq 0) \end{aligned}$$

As $m^{(t)}$ is a (very special) product probability distribution on $\Omega^{(t)}$, the two groups of RVs

$$S_k - S_o, \ldots, S_k - S_{k-1} \text{and} \quad S_{k+1} - S_k, \ldots, S_t - S_k$$

are independent (exercise) and we may continue

$$\begin{aligned} m^{(t)}(max = k) &= m^{(t)}(S_t, S_k - S_1, \ldots, S_k - S_{k-1} > 0) \\ &\quad \cdot m^{(t)}(S_{k+1} - S_k, \ldots, S_t - S_k \leq 0) \\ &= m^{(k)}(S_k, S_k - S_1, \ldots, S_k - S_{k-1} > 0) \\ &\quad \cdot m^{(t-l)}(S_1, \ldots, S_{t-k} \leq 0) \\ &= m^{(k)}(max^{(k)} = k) \cdot m^{(t-k)}(max^{(t-k)} = 0) \\ &= m^{(k)}(S_1, \ldots, S_k > 0) \\ &\quad \cdot m^{(t-k)}(S_1, \ldots, S_{t-k} \leq 0) \end{aligned}$$

Now we have to distinguish cases according to the parity of k; as t is even, it equals the parity of $t-k$. Thus if both k and $t-k$ are even, propositions 4.2 and 4.3 yield, with $k = 2j, t-k = 2(s-j)$,

$$\begin{aligned} m^{(t)}(max = k) &= \frac{1}{2}\binom{2j}{j}\frac{1}{2^k} \cdot [\binom{2(s-j)}{s-j}\frac{1}{2^{t-k}} \\ &= \frac{1}{2}\binom{2j}{j}\binom{2(s-j)}{s-j}\frac{1}{2^t} \end{aligned}$$

If both k and $t-k$ are odd, with $k = 2j+1, t-k = 2(s-j)-1$, we obtain

$$\begin{aligned} m^{(t)}(max = k) &= \binom{2j}{j}\frac{1}{2^k} \cdot \frac{1}{2}\binom{2(s-j)}{s-j}\frac{1}{2^{t-k}} \\ &= \frac{1}{2}\binom{2j}{j}\binom{2(s-j)}{s-j} \end{aligned}$$

Thus the case $t = 2s$ is settled.

2. Let $t = 2s - 1$. Here the proof goes in the same way, up to the fact that the parities of k and $t-k$ are always opposite because t is odd. The details are left to the reader. □

Our results carry over in an obvious fashion to the following more general situation: $X_1, \ldots, X_t$ are independent ± 1-valued RVs on a DPS (Ω, m), and

$$S_0 \equiv 0 \quad , S_u = X_1 + \ldots + X_u \qquad (u = 1, \ldots, t);$$

the RVs max and pos are derived from $S_0, S_1, \ldots, S_t$ in the same fashion as in our previous minimal model. The details are left to the reader.

5. The Andersen-Spitzer Formula

In this section we prove a comprehensive formula of Andersen [1953/54] and Spitzer [1956] which relates the sequence of the distributions of the RVs

P_n = number of strictly positive ones among the first n partial sums

to the sequence of the probabilities

a_n = probability of the n-th partial sum to be strictly positive

under the IID assumption about the underlying RVs. We remind the reader that we can't have a nontrivial infinite sequence of IID RVs on a DPS (see the last remark in ch. III). It will, however turn out that we obtain the distribution of $P_1, \ldots, P_n$ already from $a_1, \ldots, a_n$, which would enable us to work with a sequence of underlying DPS's. We will not carry out this in detail.

As in the previous sections, we will start with a purely combinatorial result of Bohnenblust, from which we subsequently draw probabilistic conclusions.

5.1. Bohnenblust's Lemma.

has never been published by its inventor – it is quoted under his name e.g. in Spitzer [1956]. It deals with cycle decompositions of permutations and with the concave hull of paths.

We start with a preparatory lemma. Let

(1) $c_1, \ldots, c_t$

be real numbers, of which we shall consider cyclic permutations. In order to do this conveniently, we extend (1) periodically to obtain an infinite sequence,

$$c_1, \ldots, c_t,\ c_1, \ldots, c_t, \ldots,$$

of which, however, we will really use the first $2t$ terms only. Define

$$\begin{aligned} S_0(k) &= 0 \\ S_u(k) &= c_k + c_{k+1} + \ldots + c_{k+u-1} \quad (u = 0, \ldots, t;\ k = 1, 2, \ldots) \end{aligned}$$

All $S_t(k)$ have the same value $c_1 + \ldots + c_t$. Let us put

$$\begin{aligned} c'_u &= c_u - \tfrac{1}{t}(c_1 + \ldots + c_t) && (u = 1, 2, \ldots) \\ S'_0(k) &= 0 \\ S'_u(k) &= c'_k + c'_{k+1} + \ldots + c'_{k+u-1} && (u = 0, \ldots, t;\ k = 1, 2, \ldots,) \end{aligned}$$

Clearly

$$S'_u(k) = S_u(k) - \frac{u}{t}(c_1 + \ldots + c_t)$$

and in particular

$$S'_t(k) = 0 \qquad (k = 1, 2, \ldots)$$

Let us assume that $c_1, \ldots, c_t$ *rationally independent:* linear combination with rational coefficients not all zero can't bring them to zero. This implies that, while $c'_1 + \ldots + c'_t = 0$, we can't combine less than all t of the c'_u to 0 with rational coefficients not all zero (exercise for the reader!). This in turn implies that for every $k = 1, 2, \ldots$ the values $S'_1(k), \ldots, S'_t(k)$ are pairwise different. Let us now enumerate $S'_1(1), \ldots, S'_t(1)$ in decreasing order:

$$s_0 > s_1 > \ldots s_{t-1}.$$

It is easy to see that $S'_k(1) = s_j$ means nothing else but the following: there are precisely j strictly positive terms in the finite sequence $S'_1(k), \ldots, S'_t(k)$. Now what does $S'_u(k) > 0$ mean for the original partial sums $S_u(k)$? It means, geometrically, that the point $(u, S_u(k))$ lies above the straight segment joining (0,0) and $(t, S_t(k))$ in the plane. This observation completes the proof of our preparatory

Lemma 5.1. Let t be a natural number and $c_1, \ldots, c_t$ rationally independent reals. Then for every $j \in \{0, 1, \ldots, t-1\}$ there is exactly one cyclic permutation τ_j of $\{1, \ldots, t\}$ such that, with the notation

$$S_u(\tau) = c_{\tau(1)} + \ldots + c_{\tau(u)} \qquad (u = 0, 1, \ldots, t;\ \tau \text{ a permutation of } \{1, \ldots, t\})$$

precisely j of the $S_u(\tau_j)$ $(u = 1, \ldots, t-1)$ lie above the straight segment joining (0,0) and $(t, S_t(\tau_j))$ in the plane.

We shall now get rid of the restriction to cyclic permutations by cycle decomposition of arbitrary permutations. And the latter ones will have their path-geometric counterpart in what we will call the *concave hull* of a path.

Let $c_1, \ldots, c_t$ be rationally independent reals and τ any permutation of $\{1, \ldots, t\}$. For every τ we plot the path with the increments $c_{\tau(1)}, \ldots, c_{\tau(t)}$, i.e.

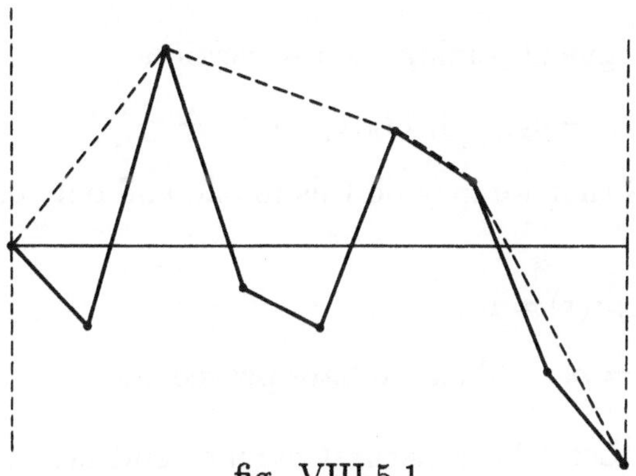

fig. VIII.5.1

with the vertices (0,0), $(1, c_{\tau(1)})$, $(2, c_{\tau(1)} + c_{\tau(2)}), \ldots, (t, c_{\tau(1)} + \ldots + c_{\tau(t)}) = (t, c_1 + \ldots + c_t)$. The dotted line in figure VIII.5.1 indicates the *concave hull* of this path.

Let $0, u_1(\tau), u_2(\tau), \ldots, u_{r(\tau)}(\tau) = t$ be the indices of coincidence between the given path and its concave hull (marked in bold in the picture), and define a new permutation ϱ_τ by its cycle representation

$$\varrho_\tau = (\tau(1), \ldots, \tau(u_1(\tau)) \circ (\tau(u_1(\tau) + 1), \ldots, \tau(u_2(\tau)) \circ \ldots \\ \circ(\tau(u_{r(\tau)-1}(\tau) + 1), \ldots, t).$$

Let us now show that the mapping $\tau \to \varrho_\tau$ is bijective. For this it will be sufficient to prove surjectivity: every permutation ϱ of $\{1, \ldots, t\}$ can be obtained as $\varrho = \varrho_\tau$ from some τ. But this is simple: take the cycle representation of ϱ, say,

$$\varrho = (\mu_1, \ldots, \mu_{s_1}) \circ (\mu_{s_1+1}, \ldots, \mu_{s_2}) \circ \ldots \circ$$

$$(\mu_{s_{n-1}+1}, \ldots, \mu_t),$$

where $\mu_1, \ldots, \mu_t$ are the numbers $1, \ldots, t$ written in a certain order. This representation of ϱ is unique up to the order in which we write the cycles, and up to cyclic reordering of the indexes within each cycle. Let us now firstly assume we have chosen the order of the cycles in such a fashion that

$$c_{\mu_1} + \ldots + c_{\mu_{s_1}} > c_{\mu_{s_1+1}} + \ldots + c_{\mu_{s_2}} > \ldots > c_{\mu_{s_{n-1}+1}} + \ldots + c_{\mu_t}$$

And secondly let us assume that we have chosen the arrangement of the members within each cycle in such a fashion that the vertices belonging to that cycle remain strictly below the segment joining the first with the last of them, that is, the vertices

$$(s_{\nu-1} + 1, c_{\mu_1} + \ldots + c_{\mu_{s_{\nu-1}+1}}), \ldots, (s_\nu - 1, c_{\mu_1} + \ldots + c_{\mu_{s_\nu - 1}})$$

remain below the segment joining the two vertices

$$(s_{\nu-1},\ c_{\mu_1}+\ldots+c_{\mu_{s_{\nu-1}}}),\ (s_\nu, c_{\mu_1}+\ldots+c_{\mu_{s_\nu}}).$$

Lemma 5.1. tells us that we may do this in one and only one way. If we now define

$$\tau(1)=\mu_1,\ldots,\tau(t)=\mu_t,$$

we obviously get $\varrho=\varrho_\tau$. – Thus we have proved the

Proposition 5.2. Let t be a natural number and $c_1,\ldots,c_t$ rationally independent reals. Then the mapping $\tau\to\varrho_\tau$ described above is a bijection among all permutations of $\{1,\ldots,t\}$.

Let us now get rid of the restriction to rationally independent t-tuples of reals:

Lemma 5.3. Let t be a natural number and $c_1,\ldots,c_t$ any reals. Then there are rationally independent reals $c'_1,\ldots,c'_t$ such that

$$J\subseteq\{1,\ldots,t\},\ \sum_{u\in J}c_u>0\Longrightarrow\sum_{u\in J}c'_u>0$$

$$J\subseteq\{1,\ldots,t\},\ \sum_{u\in J}c_u\le 0\Longrightarrow\sum_{u\in J}c'_u\le 0.$$

Proof. Choose $\epsilon>0$ such that $t\epsilon<\min\{|\sum_{u\in J}c_u|\,|J\subseteq\{1,\ldots,t\},\ \sum_{u\in J}c_u>0\}$ and put $U_\epsilon=\{(c'_1,\ldots,c'_t)|c'_1,\ldots,c'_t$ rationally independent and $c_1-\epsilon\le c'_1\le c_1,\ldots,c_t-\epsilon\le c'_t\le c_t\}$. Then the conclusion of the lemma obviously holds for any $(c'_1,\ldots,c'_t)\in U_\epsilon$. The set $U_\epsilon\subseteq\mathbf{R}^+$ is nonempty because the rationally dependent t-tuples are contained in a countable union of hyperplanes $\subseteq\mathbf{R}^t$. □

We now finally arrive at Bohnenblust's Lemma. For any reals $c_1,\ldots,c_t$, and for any permutation τ of $\{1,\ldots,t\}$ let us define, as earlier,

$$\begin{aligned}\max(\tau) &= \text{position of the first maximum}\\ &= min\{u|0\le u\le t,\ c_{\tau(1)}+\ldots+c_{\tau(u)}\ge c_{\tau(1)}+\ldots+c_{\tau(v)}\\ &\quad (0\le u\le t)\}\\ pos(\tau) &= \text{number of strictly positive partial sums}\\ &= |\{u|1\le u\le t,\ c_{\tau(1)}+\ldots+c_{\tau(u)}>0\}|\end{aligned}$$

and, now newly,

$$pz(\tau)=\text{sum of the lengths of all strictly positive cycles in }\tau,$$

where a cycle $(\mu_{s_{\nu-1}+1}, \ldots, \mu_{s_\nu})$ in the cycle representation of τ is called *positive* if $c_{\mu_{s_{\nu-1}+1}} + \ldots + c_{\mu_{s_\nu}} > 0$.

Theorem 5.4. (Bohnenblust's Lemma) Let t be a natural number and $c_1, \ldots, c_t$ reals. Then for every $k \in \{0, 1, \ldots, t\}$ we have

$$
\begin{aligned}
|\{\tau | pos(\tau)\}| &= |\{\tau | max(\tau) = k\}| \\
&= |\{\tau | pz(\tau) = k\}|.
\end{aligned}
$$

PROOF. In order to calculate, for a given (τ), the values $max(\tau)$, $pos(\tau)$, $pz(\tau)$, we have only to check, for certain sets $J \subseteq \{1, \ldots, t\}$, whether $\sum_{u \in J} c_{\tau(u)}$ is > 0 or not. Thus we may assume $c_1, \ldots, c_t$ to be rationally independent, by lemma 5.3. The first one of our above equalities is a statement from the combinatorial equivalence principle (theorem 1.1.). In order to prove the second equality, we make use of the bijection $\tau \leftrightarrow \varrho_\tau$ established in proposition 5.2., and observe that, for every τ, we collect precisely the positive cycles when going from 0 until $max(\tau)$. This does it. □

5.2. The Andersen-Spitzer Formula.

We will no incorporate the above combinatorial results in a probabilistic framework. Let t be a natural number and $f_1, \ldots, f_t$ IID real valued RVs on some DPS (Ω, m). Let p denote the distribution of f_1 (and thus also of f_2 etc.) Put

$$
\begin{array}{lcll}
S_0 & = & 0 & \\
S_u & = & f_1 + \ldots + f_u & (1 \leq u \leq t) \\
pos(\omega) & = & |\{u | 0 \leq u \leq t,\ S_u(\omega) > 0\}| & (\omega \in \Omega) \\
q_t & = & \text{distribution of } pos_t & \\
a_u & = & m(S_u > 0). &
\end{array}
$$

The distribution of $S_0, S_1, \ldots, S_t$ and pos_t, and the value of a_u depend upon p alone. We may calculate and interprete probabilistically these for arbitrarily large values of u, t by passing the models with t sufficiently large. – It is our aim to prove

Theorem 5.5. (Andersen [1953/54], Spitzer [1956]) Let p be a DPD in $\mathbf{R}$, and let $q_o, q_1, \ldots$, and $a_o, a_1, \ldots$ be derived from p as above. Define

$$
\begin{aligned}
\chi_t(\lambda) &= \textstyle\sum_{k=0}^{\infty} e^{-\lambda k} q_t(\{k\}) \ (= E_m[e^{-\lambda pos_t}]) \quad (t = 0, 1, \ldots) \\
\varrho_n(\lambda) &= (1 - a_n) + e^{-\lambda n} a_n \ (= E_m[e^{-\lambda n 1_{]0,\infty[}(S_n)}]) \ (n = 1, 2, \ldots)
\end{aligned}
$$

as functions of the real variable $\lambda > 0$. – Then

$$\sum_{t=0}^{\infty} \chi_t(\lambda) z^t = e^{\sum_{n=1}^{\infty} \frac{\varrho_n(\lambda)}{n} z^n} \qquad \text{(Andersen-Spitzer-Formula)}$$

holds for $\lambda > 0$ and $|z| < 1$, the power series in question being absolutely convergent for these values of λ and z.

PROOF. Let us remark beforehand that $\chi_t(\lambda)$ and $\varrho_n(\lambda)$ are nothing but the Laplacians of the RVs pos_t resp. $1_{]0,\infty[}(S_n)$. We shall, however, not make explicit use of the theory of Laplace transforms here. – In applying our previous combinatorial results, we shall emphasize their dependence upon a chosen t-tuple $c = (c_1, \dots, c_t)$ of reals by writing out an affix c or $(c_1, \dots, c_t)$ explicitly. That is, we shall write

$$\begin{array}{lll} pos_t^c(\tau) & or & pos_t^{(c_1,\dots,c_t)}(\tau) \text{ instead of } pos_t \\ pz_t^c(\tau) & or & pz^{(c_1,\dots,c_t)}(\tau) \text{ instead of } pz_t \\ max_t^c(\tau) & = & min\{u | 0 \le u \le t, c_{\tau(1)} + \dots + c_{\tau(u)} \ge c_{\tau(1)} + \dots + c_{\tau(v)} \\ & & (0 \le v \le t)\} \end{array}$$

With $c = (f_1(\omega), \dots, f_t(\omega))$, i.e. ($id$ denotes the identical permutation)

$$\begin{array}{lll} max_t(\omega) & = & max_t^{(f_1(\omega),\dots,f_t(\omega))}(id) \\ pos_t(\omega) & = & pos_t^{(f_1(\omega),\dots,f_t(\omega))}(id) \\ pz_t(\omega) & = & pz_t^{(f_1(\omega),\dots,f_t(\omega))}(id) \end{array}$$

we now fit our previous combinatorics into the conceptual framework of our theorem. Bohnenblust's Lemma (theorem 5.4) now implies (details as an exercise for the reader) that for every t the RVs pos_t, max_t, pz_t have the same distribution. Let us now calculate, for an arbitrary permutation τ of $\{1, \dots, t\}$ and its cycle representation

$$\tau = (\mu_1, \dots, \mu_{s_1}) \circ \dots \circ (\mu_{s_{r-1}}, \dots, \mu_{s_t}),$$

the value of

$$(2) \quad E_m\left(e^{\lambda\, pz_t^{(f_1(\omega),\dots,f_t(\omega))}(\tau)}\right).$$

Using the abbreviations

$$\begin{array}{llll} \theta(x) & = & 1_{]0,\infty[}(x) & (x \in \mathbf{R}) \\ I_\varrho & = & \{\mu_{s_{\varrho-1}+1}, \dots, \mu_{s_\varrho}\} & (\varrho = 1, \dots, r;\ s_0 = 0, s_r = t), \end{array}$$

we obtain

$$pz_t^{(f_1,\ldots,f_t)}(\tau) = \sum_{\varrho=1}^{r} |I_\varrho| \theta \left(\sum_{u \in I_\varrho} f_{\mu_\varrho} \right).$$

Clearly the RVs $\sum_{u\in I_\varrho} f_{\mu_\varrho}$ ($\varrho = 1,\ldots,r$) are independent since they are obtained from the independent RVs $f_1,\ldots,f_t$ by formation of r disjoint index groups. Thus

$$\begin{aligned} E_m\left[e^{-\lambda\, pz_t^{(f_1,\ldots,f_t)}(\tau)}\right] &= E_m\left[e^{-\lambda \sum_{\varrho=1}^{r} |I_\varrho|\theta(\sum_{u\in I_\varrho} f_u)}\right] \\ &= E_m\left[\Pi_{\varrho=1}^{r} e^{-\lambda|I_\varrho|\theta(\sum_{u\in I_\varrho} f_u)}\right] \\ (3) \qquad &= \Pi_{\varrho=1}^{r} E_m\left[e^{-\lambda|I_\varrho|\theta(\sum_{u\in I_\varrho} f_u)}\right] \end{aligned}$$

Now, as $f_1,\ldots,f_t$ are IID, (2) doesn't really depend upon τ and may thus also be evaluated as

$$E_m\left[e^{-\lambda\, pz_t^{(f_1,\ldots,f_t)}(id)}\right] = E_m\left[e^{-\lambda\, pz_t}\right],$$

which, in view of the distribution equality of pz_t and pos_t, is

$$= E_m\left[e^{-\lambda pos_t}\right] = \chi_t(\lambda).$$

Our IID hypothesis likewise implies that the value of factor no. ϱ in (3) depends on $|I_\varrho|$ only. We may thus continue with

$$\chi_t(\lambda) = \Pi_{\varrho=1}^{r} E_m\left[e^{\lambda|I_\varrho|\theta(S_{|I_\varrho|})}\right]$$

$$(4) \qquad = \Pi_{\varrho=1}^{r}\, \varrho_{|I_\varrho|}(\lambda).$$

Now, for any choice of nonnegative integers $\ell_1, \ell_2, \ldots$ such that

$$t = \sum_{j=1}^{\infty} j\, \ell_j$$

(clearly only finitely many of the ℓ_j can be > 0), there are precisely

$$\frac{t!}{\Pi_{j=1}^{\infty}(j^{\ell_j}\, \ell_j!)}$$

permutations τ in whose cycle representation ℓ_j cycles of length j occur ($j = 1, 2, \ldots$). For each of these permutations, (4) equals

$$\Pi_{j=1}^{\infty} \varrho_j(\lambda)^{\ell_j}.$$

Averaging over all $t!$ permutations, we get

$$\begin{aligned} \chi_t(\lambda) &= \frac{1}{t!} \sum_{t=1\cdot\ell_1+2\cdot\ell_2+\ldots} t\, \Pi_{j=1}^{\infty} \frac{1}{\ell_j!} \left(\frac{\varrho_j(\lambda)}{j}\right)^{\ell_j} \\ &= \sum_{t=1\cdot\ell_1+2\cdot\ell_2+\ldots} \Pi_{j=1}^{\infty} \frac{1}{\ell_j!} \left(\frac{\varrho_j(\lambda)}{j}\right)^{\ell_j}. \end{aligned}$$

Our proof now ends with bare power series calculations: for $z \in \mathbb{C}$, $|z| < 1$ the following power series are absolutely convergent and fulfil

$$\begin{aligned} \sum_{t=0}^{\infty} \chi_t(\lambda) z^t &= \sum_{t=0}^{\infty} \sum_{t=1\cdot\ell_1+2\cdot\ell_2+\ldots} \Pi_{j=1}^{\infty} \frac{1}{\ell_j!} \left(\frac{\varrho_j(\lambda) z^j}{j}\right)^{\ell_j} \\ e^{\sum_{n=1}^{\infty} \frac{\varrho_n(\lambda)}{n} z^n} &= \Pi_{n=1}^{\infty} e^{\frac{\varrho_n(\lambda)}{n} z^n} \\ &= \Pi_{j=1}^{\infty} e^{\frac{\varrho_j(\lambda)}{j} z^j} \\ &= \Pi_{j=1}^{\infty} \sum_{\ell_j=0}^{\infty} \frac{1}{\ell_j!} \left(\frac{\varrho_j(\lambda)}{j} z^j\right)^{\ell_j} \end{aligned}$$

□

After rearranging terms we have the desired formula.

6. Outlook

The fluctuation story told in the preceding five sections is no more than a cross-section of a vast theory. I believe this cross-section to be in a way representative, although the original literature does, as a rule, not follow the presentation scheme

do combinatorics first and fill in probability afterwards

chosen here. Andersen [1953/54], Spitzer [1956], Wendel [1958][1960] are, in my opinion, the most important original papers to be read. My own presentation largely follows Jacobs [1969] plus older mimeographed texts by myself, and the very substantial Diplomarbeit Strehl [1970], which allows to carry over large parts of the theory to RVs with values in $\mathbb{R}^d$, $d > 1$. See also Dinges [1965].

IX. Optimal Strategies in Casinoes: Red and Black

Imagine yourself being blackmailed to pay 1.000.000 \$ at five o'clock next morning. It is late evening, all banks are closed, your automatic bank account is empty. But you have 1 \$ in your pocket, and a casino is open. So what will you do? You will try to win the 1.000.000 \$ by gambling in that casino. And you will try to maximize the probability of your survival by a proper choice of your betting strategy.

One obvious betting device sounds as follows:

> *Desperado's Strategy (DESP):* at every stage of the game, stake all the money you have, until you have at least 1.000.000 \$. Then you stop and walk out.

The probability of desperado's survival depends, of course, upon the rules of the game and the probability parameters of the aleatoric device employed by the casino. If the rules imply that in the case of "no success" the whole stake is lost, our desperado will survive only in case of uninterrupted success. It the success probability is p and if a minimum of n successes yield 1.000.000 \$, the probability of survival is (assuming independent trials) p^n; it's horribly tiny. You can do a little bit better employing the

> *Bold Strategy (BOLD):* follow Desperado's Strategy until you have $a \geq \frac{1}{2} \cdot 1.000.000$ \$ in your pocket for the first time. Then stake only 1.000.000 minus a, i.e. the precise amount you still need. So if $a > 500.000$ and you lose, you have still $a - 500.000$ in your pocket and can go further ahead. Whenever you are below 500.000 you play Desperado (and get broke in case of "no success"), and whenever you are above 500.000, you stake only the amount you still need.

Clearly, this modification of "Desperado" adds some new possibilities of survival and thus improves your chance of survival.

> *Question:* Can you do still better than by playing BOLD?
> *Answer:* For a certain class of casinos, the answer is: BOLD is optimal.

In this chapter, this answer will be made precise, and proved (theorem 3.3). The reader will thereby gain insight into a typical branch of a vast theory which was systematically displayed for the first time in the famous book

[1965] Dubins, L., and L. Savage,
How to gamble if you must.

Dubins-Savage [1965] contains a considerable machinery based on the notion of a finitely additive set function, but the essentials of the theory can be reconstructed within the usual measure-theoretical framework (see e.g. Sudderth [1969][1969a][1971]). In the present chapter we will get along with finite probability spaces. We essentially follow Hansen-Walz [1971].

1. Strategies and Their Probability of Success

We will consider casinos characterizable by bare two parameters:

p = probability of success in a single trial
r = inverse winning rate, i.e.: if the gambler stakes e in a single trial, he gets $\frac{e}{r}$ in case of success. In the case "no success", the stake is forfeited.

Thus $r \leq 1$ means $\frac{e}{r} \geq e$, i.e. the winner gets at least his stake back. We will always assume $r \leq 1$ in this chapter. $r = \frac{1}{2}$ means $\frac{e}{r} = 2e$, that is, in the case of success, the stake earns once more its own value; this particular casino is called *Red and Black*. $p = \frac{1}{2}$ means that the probability structure of the casino amounts to the tossing of a fair coin; $r \geq \frac{1}{2}$ then means that the winners stake e earns one more e, but he has to pay $\frac{2r-1}{r}e$ tax out of it; this casino is therefore called the "taxed coin". We will show later (theorem 3.3) that BOLD is optimal if $p \leq \frac{1}{2}$ and $r \geq \frac{1}{2}$, plus some extras. This includes the case of Red and Black with $p \leq \frac{1}{2}$. Stakes will be subject to the following

Staking Rule: A gambler with a fortune a in his pocket may stake any amount $o \leq e \leq a$.

Thus a gambler staking e out of his fortune a will be left with the new fortune

$$a - e \quad \text{in the case of "failure"}$$

$$a - e + \frac{e}{r} = a + \frac{1-r}{r}e \text{ in the case of success}$$

after his trial. If we use the abbreviations

$$\bar{p} = 1 - p$$
$$\bar{r} = 1 - r$$

we may rewrite the new fortune as

$a - e$ in the case of failure: probability $\bar{p}$
$a + \frac{\bar{r}}{r}e$ in the case of success: probability p

We will, for every natural number n, employ the sample space $\{0,1\}^n$, endowed with the obvious product probability distribution, as a model for n independent trials; 1 stands for success and 0 for failure. Thus e.g. for $n = 5, m_{10010} = p^2\bar{p}^3$.

Further, we will normalize the goal to be 1 (instead of 1.000.000).

A *strategy* for such a model with fixed duration n is a function $\varphi(a,k)$, defined for reals $a \geq 0$ and integers $k \geq 0$ and fulfilling $\varphi(a,k) \leq a$ everywhere. This function tells the gambler:

if your fortune is a, and if still k trials are ahead, stake $e = \varphi(a,k)$

We may now formalize our verbal strategies mentioned in the introduction to this chapter:

$$\text{DESP:} \quad \varphi^{DESP}(a,k) = \begin{cases} a & \text{if} \quad a < 1 \\ 0 & \text{if} \quad a \geq 1 \end{cases}$$

$$\text{BOLD:} \quad \varphi^{BOLD}(a,k) = \begin{cases} a & \text{if} \quad a < \frac{1}{2} \\ 1-a & \text{if} \quad \frac{1}{2} \leq a < 1 \\ 0 & \text{if} \quad a \geq 1 \end{cases}$$

These function (strategies) do not depend upon k, but we will compare them with others allowed to depend upon k. The specification $\varphi(a,k) = 0$ for $a \geq 1$ means, of course, "stop and walk out after you have reached your goal".

Let us see how this works for small values of n. We shall denote the probability of survival after n trials, for a given strategy, a given initial fortune a, and a given casino (no notation), by $P_n^{\varphi}(a)$, and plot $P_n^{BOLD}(a) = P^{\varphi^{BOLD}}(a)$ for Red and Black and $n = 0,1,2,3$ (figure IX.1.1).

It is clear how this goes on: if you are allowed $n+1$ trials,

$0 \leq a < \frac{1}{2}$ makes you survive iff you win the first trial (probability P, resulting fortune $2a$) and then survive with n further trials: survival with probability $pP_n^{BOLD}(2a)$

casino: Red and Black

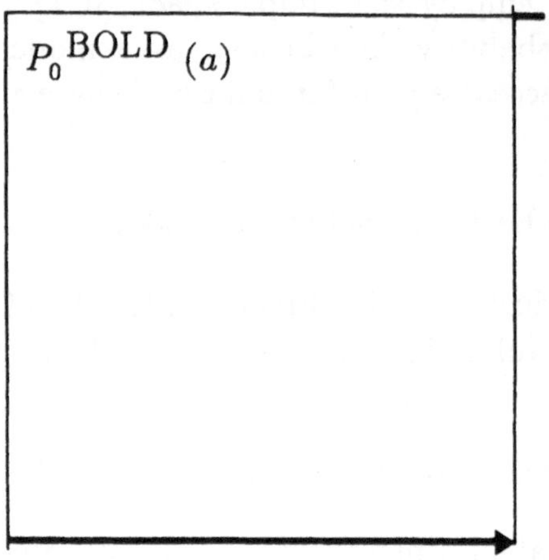

BOLD: $n = 0$

no trial
either you have it and survive.
or you don't have it and die.

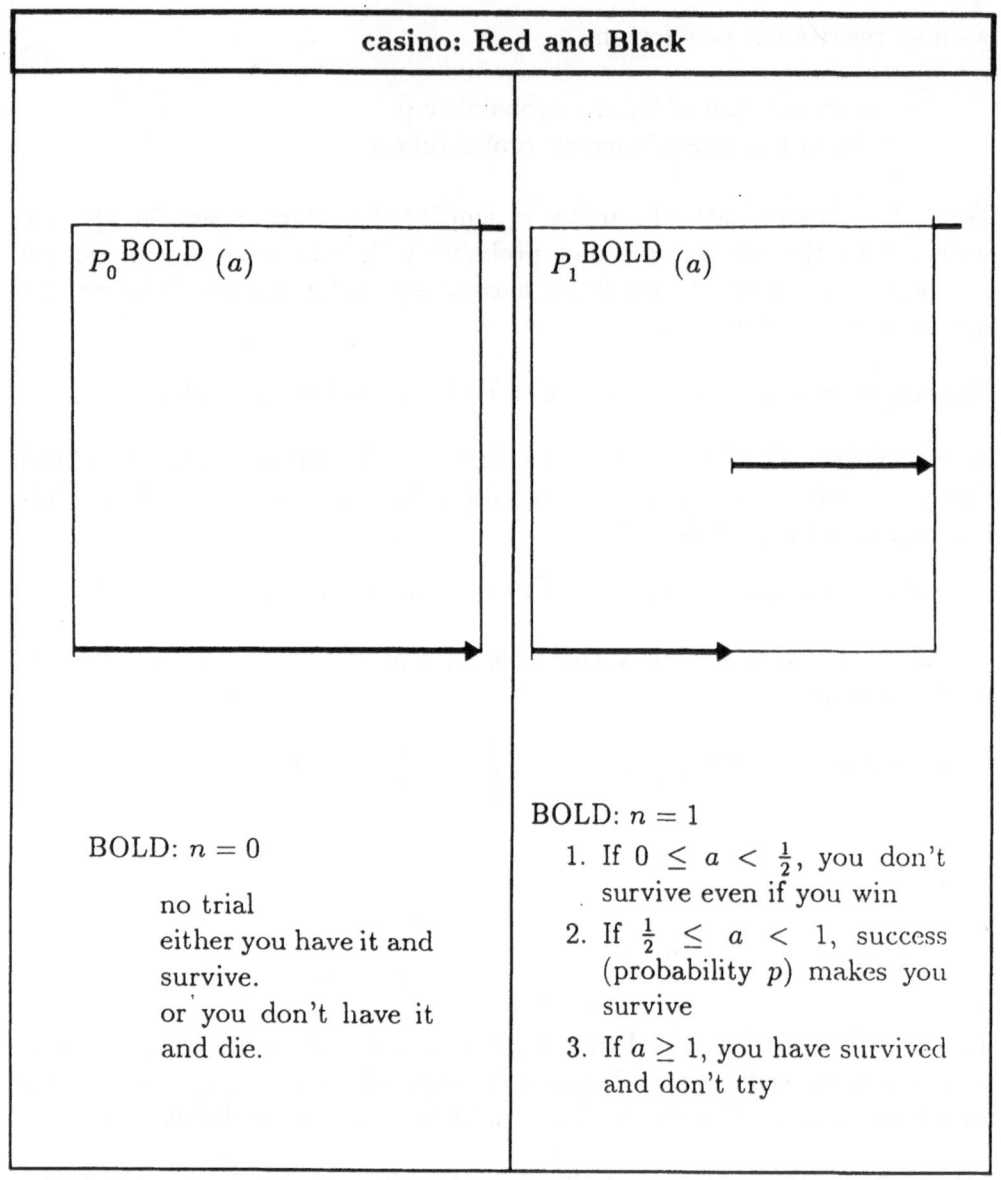

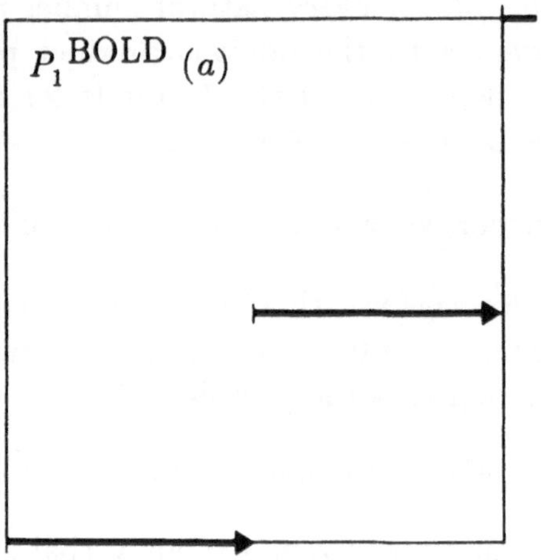

BOLD: $n = 1$

1. If $0 \leq a < \frac{1}{2}$, you don't survive even if you win
2. If $\frac{1}{2} \leq a < 1$, success (probability p) makes you survive
3. If $a \geq 1$, you have survived and don't try

fig. IX.1.1

$\frac{1}{2} \leq a$ makes you survive iff you win the first trial (probability p, resulting fortune 1) or you lose it (probability $\bar{p}$, resulting fortune $a - (1 - a) = 2a - 1 \in [0, \frac{1}{2}[$) and then survive with n further trials: survival with probability $p = \bar{p}P_n^{BOLD}(2a - 1)$.

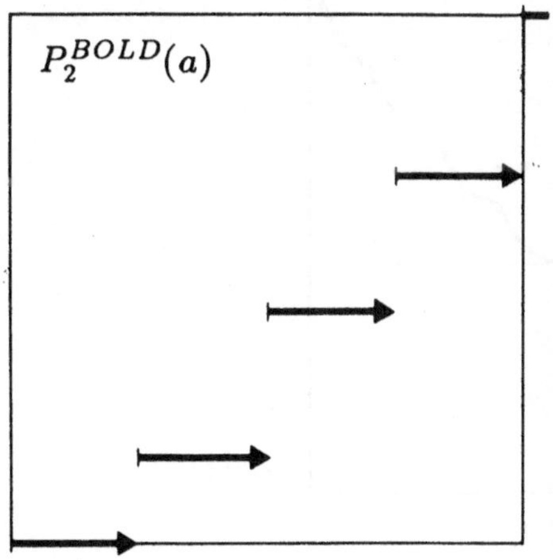

BOLD: $n = 2$

1. If $0 \leq a < \frac{1}{4}$, you can't survive even if you win in both trials
2. If $\frac{1}{4} \leq a < \frac{1}{2}$, you survive iff you win in both trials: probablity p^2
3. If $\frac{1}{2} \leq a < \frac{3}{4}$, you survive iff you win in the first trial: probability p
4. If $\frac{3}{4} \leq a < 1$, you survive even if you lose in the first trial but win the second: probability $p + \bar{p}p = p(1+p)$
5. If $a \geq 1$ you have survived and don't try

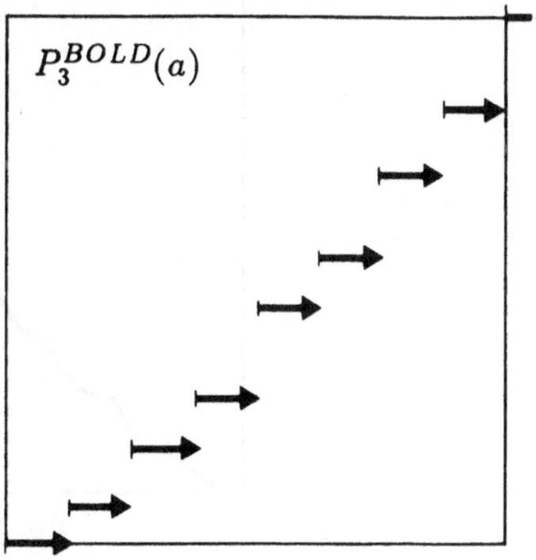

BOLD: $n = 3$

1. If $0 \leq a < \frac{1}{8}$, you can't survive even if you win thrice
2. If $\frac{1}{8} \leq a < \frac{1}{4}$, you survive iff you win in all three trials
3. If $\frac{1}{4} \leq a < \frac{3}{8}$, you survive iff you win in the first two trials
4. If $\frac{3}{8} \leq a < \frac{1}{2}$, you survive iff you win the first trial, and then at least one more trial
   etc.
5. If $\frac{7}{8} \leq a < 1$ you survive iff you win at least one trial.
6. If $a \geq 1$, you have survived and don't try.

fig. IX.1.2

The recursion formula which expresses this, is

$$P_{n+1}^{BOLD}(a) = \begin{cases} pP_n^{BOLD}(2a) & \text{for} \quad 0 \leq a < \frac{1}{2} \\ p + \bar{p}P_n^{BOLD}(2a-1) & \text{for} \quad \frac{1}{2} \leq a \leq 1 \\ 1 & \text{for} \quad a \geq 1 \end{cases}$$

and has the following visual impact:

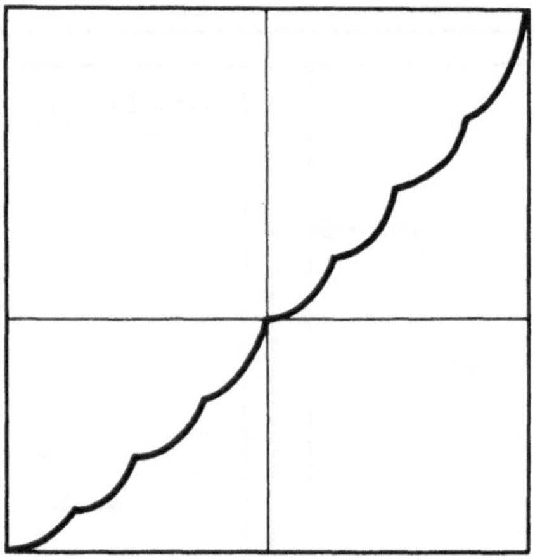

fig. IX.1.3

In order to plot the graph of $P_{n+1}^{BOLD}(a)$, take two copies of the graph of $P_n^{BOLD}(a)$ and squeeze

the first copy into the rectangle (1)
the second copy into the rectangle (2)

in fig. IX.1.3

It is likewise obvious how this generalizes to an arbitrary casino (parameters $0 < p < 1, 0 < r < 1$) and an arbitrary strategy φ:

If you start with a stake $\varphi(a, n+1)$ and win the first trial, you arrive at a new fortune $a + \frac{\bar{r}}{r}\varphi(1, n+1)$ and survive iff φ makes you survive in n trials, starting from there: probability

$$pP_n^{\varphi}(a + \frac{\bar{r}}{r}\varphi(a, n+1))$$

If you lose the first trial, you are left with a new fortune $a - \varphi(a, n+1) \geq 0$ and survive with probability

$$\bar{p}P_n^{\varphi}(a - \varphi(a, n+1))$$

This yields the recursion formula

$$P_{n+1}^{\varphi}(a) = pP_n^{\varphi}(a + \frac{\bar{r}}{r}\varphi(a, n+1)) + \bar{p}P_n^{\varphi}(a - \varphi(a, n+1))$$
$$(n = 0, 1, \ldots)$$

Clearly $P_0^{\varphi}(a)$ is the same for all φ because φ doesn't really prescribe any action:

$$P_0^{\varphi}(a) = \begin{cases} 0 & \text{for } 0 \leq a < 1 \\ 1 & \text{for } a \geq 1 \end{cases}$$

Definition 1.1. Let $0 < p, r < 1$. A strategy φ is said to be *optimal* for the casino described by the parameters p, r and for n trials if

$$P_n^{\varphi}(a) \geq P_n^{\psi}(a) \qquad (a \geq 0)$$

for all strategies ψ (remember that p, r enter into the computation of $P_n^{\varphi}, P_n^{\psi}$).

We will show (theorems 2.3, 3.3):

1. BOLD is always optimal if $n \leq 2$
2. For $n \geq 3$ BOLD is optimal iff $p \leq \frac{1}{2}$ and $r \geq \frac{1}{2}$. This will include the cases $p \leq \frac{1}{2}, r = \frac{1}{2}$ ("Red and Black") and $p = \frac{1}{2}, r \geq \frac{1}{2}$ ("taxed coin").

2. Some Properties of BOLD

We will write $\varphi^{BOLD}(a)$ instead of $\varphi^{BOLD}(a, k)$ since there is no dependence upon k in BOLD, and define

$$\varphi^{BOLD}(a) = \begin{cases} a & \text{for} \quad 0 \leq a \leq r \\ \frac{r}{\bar{r}}(1-a) & \text{for} \quad r \leq a \leq 1 \\ 0 & \text{for} \quad a \geq 1 \end{cases}$$

In §1 we had considered the special case $r = \frac{1}{2}$.

We will write

$$P_n \quad \text{instead of} \quad P_n^{BOLD}$$

throughout this and the next section. Obviously, the recursion formula derived in §1 for the special case $r = \frac{1}{2}$ ("Red and Black") generalizes as follows:

$$P_{n+1}(a) = \begin{cases} pP_n(\frac{a}{r}) & \text{for} \quad 0 \leq a \leq r \\ p + \bar{p}P_n(\frac{a-r}{\bar{r}}) & \text{for} \quad r \leq a \leq 1 \\ 1 & \text{for} \quad a \geq 1 \end{cases}$$

and the visual interpretation is analogous. Obviously $P_{n+1}(a) = p + \bar{p}P_n(\frac{a-r}{\bar{r}})$ also holds for $a \geq 1$ since this implies $\frac{a-r}{\bar{r}} \geq \frac{1-r}{1-r} = 1$.

In view of the above visual interpretation, the obvious inequality $P_0 \leq P_1$ perpetuates herself:

$$P_0(a) \leq P_1(a) \leq P_2(a) \leq \ldots \quad (a \geq 0)$$

and likewise we see that

every $P_n(a)$ is an isotone function of $a \geq 0$.

Lemma 2.1. If $0 \le e \le a, \quad a + \frac{\bar{r}}{r}e \ge 1$, then

$$(*) \qquad pP_n(a + \frac{\bar{r}}{r}e) + \bar{p}P_n(a - e) \le P_{n+1}(a) \quad (n = 0, 1, \ldots)$$

PROOF. $r \le 1, a \ge e$ and $a + \frac{\bar{r}}{r}e \ge 1$ imply

1. $a = ra + \bar{r}a \ge ra + \bar{r}e = r(a + \frac{\bar{r}}{r}e) \ge r$ and hence $P_n(a + \frac{\bar{r}}{r}e) = 1, \quad P_{n+1}(a) = p + \bar{p}P_n(\frac{a-r}{\bar{r}})$
2. $a - e = \frac{1}{\bar{r}}\bar{r}a - \frac{r}{\bar{r}} \cdot \frac{\bar{r}}{r}e = \frac{1}{\bar{r}}(a - ra) - \frac{r}{\bar{r}} \cdot \frac{\bar{r}}{r}e = \frac{a}{\bar{r}} - \frac{r}{\bar{r}}(a + \frac{\bar{r}}{r}e) \le \frac{a-r}{\bar{r}}$

The isotony of P_n now yields

$$pP_n(a + \frac{\bar{r}}{r}e) + \bar{p}P_n(a - e) \le p + \bar{p}P_n(\frac{a-r}{\bar{r}}) = P_{n+1}(a)$$

□

Lemma 2.2. If $0 \le e \le a, n = 0$ or $n = 1$, then

$$(*) \qquad pP_n(a + \frac{\bar{r}}{r}e) + \bar{p}P_n(a - e) \le P_{n+1}(a)$$

PROOF. Our previous lemma settled the case $a + \frac{\bar{r}}{r}e \ge 1$. If $a + \frac{\bar{r}}{r}e < 1$, then $a < 1$, hence $P_0(a + \frac{\bar{r}}{r}e) = 0 = P_0(a - e)$, and thus

$$pP_0(a + \frac{\bar{r}}{r}e) + \bar{p}P_0(a - e) = 0 \le P_1(a),$$

settling the case $n = 0$. For $n = 1$ we distinguish two cases:

Case I: $a < r$. Then $P_1(a - e) = 0$ and

$$a + \frac{\bar{r}}{r}e = \frac{ra + \bar{r}e}{r} \le \frac{ra + \bar{r}a}{r} = \frac{a}{r}$$

hence

$$pP_1(a + \frac{\bar{r}}{r}e) + \bar{p}P_1(a - e) = pP_1(a + \frac{\bar{r}}{r}e) \le pP_1(\frac{a}{r}) = P_2(a)$$

Case II: $a \ge r$. Then $P_2(a) \ge p$. From $a + \frac{\bar{r}}{r}e < 1$ we infer $P_1(a + \frac{\bar{r}}{r}e) \le p$ and $a - e \le a + \frac{\bar{r}}{r}e < 1$, hence $P_1(a - e) \le p$. This yields

$$pP_1(a + \frac{\bar{r}}{r}e) + \bar{p}P_1(a - e) \le pp + \bar{p}p = p \le P_2(a).$$

□

This allows us to prove already our first optimality result:

Theorem 2.3. For $n = 0, 1, 2$ BOLD is optimal.

PROOF. For $n = 0$ all P_0^φ are equal, hence BOLD is trivially optimal in this case. For $n = 0, 1$ we now deduce optimality of BOLD for $n + 1$ from optimality of BOLD for n as follows:

$$\begin{aligned} P_{n+1}^\varphi(a) &= \bar{p}P_n^\varphi(a - \varphi(a, n+1)) + pP_n^\varphi(a + \frac{\bar{r}}{r}\varphi(a, n+1)) \\ &\le \bar{p}P_n(a - \varphi(a, n+1)) + pP_n(a + \frac{\bar{r}}{r}\varphi(a, n+1)), \end{aligned}$$

as BOLD is assumed to be optimal for n. Applying lemma 2.2. with $e = \varphi(a, n+1) \le a$, we may continue with

$$\le P_{n+1}(a),$$

proving the theorem. □

From this proof we read the

Remark 2.4. Whenever p, r are such that $(*)$ holds for all $0 \le e \le a$ and all $n = 0, 1, \ldots$, optimality of BOLD follows for all $n = 0, 1, \ldots$. In order to prove optimality of BOLD for $p \le \frac{1}{2} \le r$, as announced at the end of §1, we have nothing to do but to prove $(*)$ for $0 \le e \le a, \quad, n = 0, 1, \ldots$ under the hypothesis $p \le \frac{1}{2} \le r$. As lemma 2.1. settles the case $a + \frac{\bar{r}}{r}e \ge 1$ once and for all, we may assume $a + \frac{\bar{r}}{r}e < 1$ in the sequel whenever we like.

3. The Optimality of BOLD for $p \le \frac{1}{2} \le r$

We have only to establish $(*)$ for $0 \le e \le a$ and all $n = 0, 1, \ldots$.

Lemma 3.1. Let $n \in \{0, 1, \ldots\}$, $p \le \frac{1}{2} \le r$ and

$$(*) \qquad pP_n(a + \frac{\bar{r}}{r}e) + \bar{p}P_n(a - e) \le P_{n+1}(a) \quad (0 \le e \le a)$$

Then for all α, β, γ such that $0 \le \alpha \le \gamma$, $0 \le \beta \le \gamma$, we have

(1) $p[P_n(\alpha) + P_n(\beta)] \le P_{n+1}((r - \bar{r})\gamma + \bar{r}(\alpha + \beta))$

(2) $\bar{p}[P_n(\alpha) + P_n(\beta)] \le (\bar{p} - p) + P_{n+1}((r - \bar{r})\gamma + \bar{r}(\alpha + \beta))$

PROOF. We may and shall assume $0 \le \beta \le \alpha \le \gamma$. Put

$$a = r\alpha + \bar{r}\beta \quad , \quad e = r(\alpha - \beta)$$

This yields, after an easy calculation,

$$a + \frac{\bar{r}}{r}e = \alpha \quad , \quad a - e = \beta$$

Now $(*)$ implies

(3) $\quad pP_n(\alpha) + \bar{p}P_n(\beta) \leq P_{n+1}(r\alpha + \bar{r}\beta)$

As $r \geq \frac{1}{2}$ implies $r \geq \bar{r}$, we get

(4) $\quad r\alpha + \bar{r}\beta = (r - \bar{r})\alpha + \bar{r}(\alpha + \beta) \leq (r - \bar{r})\gamma + \bar{r}(\alpha + \beta)$

and as P_{n+1} is an isotone function, we obtain

$$pP_n(\alpha) + \bar{p}P_n(\beta) \leq P_{n+1}((r - \bar{r})\gamma + \bar{r}(\alpha + \beta))$$

As $p \leq \bar{p}$ (by $p \leq \frac{1}{2}$), the left member here is $\geq p[P_n(\alpha) + P_n(\beta)]$, which proves (1). In order to get (2), we have only to observe

$$\begin{aligned} \bar{p}[P_n(\alpha) + P_n(\beta)] &= (\bar{p} - p)P_n(\alpha) + [pP_n(\alpha) + \bar{p}P_n(\beta)] \\ &\leq (\bar{p} - p) + P_{n+1}(r\alpha + \bar{r}\beta)(by(3)) \\ &\leq (\bar{p} - p) + P_{n+1}((r - \bar{r})\gamma + \bar{r}(\alpha + \beta)) \end{aligned}$$

by (4) and the isotony of P_{n+1}. □

PROOF. of $(*)$ by induction: The case $n = 0$ has been settled by lemma 2.2. We assume $(*)$ to be true for n and establish its validity for $n + 1$ as follows:

Let $0 \leq e \leq a$. The case $a + \frac{\bar{r}}{r}e \geq 1$ has been settled by lemma 2.1. Assume now $a + \frac{\bar{r}}{r}e < 1$. We distinguish four cases:

Case I	:	$a + \frac{\bar{r}}{r}e \leq r$
Case II	:	$a \leq r \leq a + \frac{\bar{r}}{r}e$
Case III	:	$a - e \leq r \leq a$
Case IV	:	$r \leq a - e$

This is obviously exhaustive.

Case I: $a + \frac{\bar{r}}{r}e \leq r$.– We have $a \leq r$ and hence, by the recursion formular and the induction hypothesis

$$pP_{n+1}(a + \tfrac{\bar{r}}{r}e) + \bar{p}P_{n+1}(a - e)$$

$$= ppP_n(\tfrac{a + \frac{\bar{r}}{r}e)}{r} + \bar{p}pP_n(\tfrac{a-e}{r})$$

$$= p[pP_n(\tfrac{a}{r} + \tfrac{\bar{r}}{r}\tfrac{e}{r}) + \bar{p}P_n(\tfrac{a}{r} - \tfrac{e}{r})]$$

$$\leq pP_{n+1}(\tfrac{a}{r}) = P_{n+2}(a)$$

Case II: $a \leq r \leq a + \frac{\bar{r}}{r}e$.– The recursion formula yields

$$\begin{aligned} pP_{n+1}(a + \tfrac{\bar{r}}{r}e) &= \bar{p}P_{n+1}(a - e) \\ &= p\left[p + \bar{p}P_n\left(\tfrac{(a+\frac{\bar{r}}{r}e)r}{\bar{r}}\right)\right] + \bar{p}pP_n(\tfrac{a-e}{r}) \\ &= p^2 + \bar{p}p\left[P_n\left(\tfrac{(a+\frac{\bar{r}}{r}e)-r}{r}\right) + P_n(\tfrac{a-e}{r})\right] \end{aligned}$$

We have

$$a = ra + \bar{r}a = r(a + \frac{\bar{r}}{r}a) \geq r(a + \frac{\bar{r}}{r}e) \geq r^2.$$

If we put

$$\begin{aligned} \gamma &= \frac{a - r^2}{\bar{r}r} \\ \alpha &= \frac{(a + \frac{\bar{r}}{r}e) - r}{\bar{r}} \\ \beta &= \frac{a - e}{r}, \end{aligned}$$

we have, by lemma 3.1,

$$\begin{aligned} \gamma &\geq 0 \\ \alpha &= \frac{(a + \frac{\bar{r}}{r}e) - r}{\bar{r}} \leq \frac{(a + \frac{\bar{r}}{r})a - r}{\bar{r}} \\ &= \frac{ra + \bar{r}a - r^2}{r\bar{r}} = \frac{a - r^2}{r\bar{r}} = \gamma \\ \beta &= \frac{a - e}{r} = \frac{\frac{\bar{r}}{\bar{r}}a - e}{r} = \frac{\frac{1-r}{\bar{r}}a - e}{r} \\ &= \frac{\frac{a}{\bar{r}} - \frac{r}{\bar{r}}(a + \frac{\bar{r}}{r}e)}{r} \leq \frac{\frac{a}{\bar{r}} - \frac{r}{\bar{r}}r}{r} = \frac{a - r^2}{r\bar{r}} = \gamma \\ (r - \bar{r})\gamma + \bar{r}(\alpha + \beta) &= \frac{r - \bar{r}}{r\bar{r}}(a - r^2) + (a + \frac{\bar{r}}{r}e - r) + \frac{\bar{r}}{r}(a - e) \\ &= \frac{a}{\bar{r}}\left(\frac{r - \bar{r}}{r} + \bar{r} + \frac{\bar{r}^2}{r}\right) - \frac{r^2}{\bar{r}}\left(\frac{r - \bar{r}}{r} + \frac{\bar{r}}{r}\right) \\ &= \frac{a}{\bar{r}}\left(\frac{r - \bar{r} + r\bar{r} + \bar{r}^2}{r}\right) - \frac{r^2}{\bar{r}} \\ &= \frac{a - r^2}{\bar{r}} \end{aligned}$$

(as $r - \bar{r} + r\bar{r} + \bar{r}^2 = r - \bar{r} + \bar{r}(r + \bar{r}) = r - \bar{r} + \bar{r} = r$).

We may now apply (1) and obtain

$$p\left[P_n(\alpha)+P_1(\beta)\right] \le P_{n+1}\left(\frac{a-r^2}{\bar r}\right)$$

and thus, by the recursion formula

$$\begin{aligned} & pP_{n+1}\left(a+\frac{\bar r}{r}e\right)+\bar pP_{n+1}(a-e) \\ = {} & p\left[p+\bar pP_n\left(\frac{(a+\frac{\bar r}{r}e)-r}{\bar r}\right)\right]+\bar ppP_n\left(\frac{a-e}{r}\right) \\ = {} & p^2+\bar pp\left[P_n(\alpha)+P_n(\beta)\right] \\ \le {} & p^2+\bar pP_{n+1}\left(\frac{a-r^2}{\bar r}\right) \end{aligned}$$

Now $\frac{a-r^2}{\bar r} \le \frac{r-r^2}{\bar r} = \frac{r\bar r}{\bar r} = r$, and thus the recursion formula yields

$$\begin{aligned} = {} & p^2+\bar ppP_n\left(\frac{a-r^2}{\bar r r}\right) \\ = {} & p\left[p+\bar pP_n\left(\frac{\frac{a}{r}-r}{\bar r}\right)\right] \end{aligned}$$

As $a \ge r^2$, and thus $r \le \frac{a}{r} \le 1$, this expression is

$$pP_{n+1}(\frac{a}{r}) = P_{n+2}(a),$$

again by the recursion formula.

Case III: $a-e \le r \le a$. – We start, as in case II, with

$$\begin{aligned} & pP_{n+1}\left(a+\frac{\bar r}{r}e\right)+\bar pP_{n+1}(a-e) \\ = {} & p^2+\bar pp\left[P_n\left(\frac{(a+\frac{\bar r}{r}e)-r}{\bar r}\right)+P_n\left(\frac{a-e}{r}\right)\right] \end{aligned}$$

We shall apply lemma 3.1. with

$$\begin{aligned} \gamma &= 1 \\ \alpha &= \frac{(a+\frac{\bar r}{r}e)-r}{\bar r} \\ \beta &= \frac{a-e}{r} \end{aligned}$$

and thus verify

$$\begin{aligned} & (r-\bar r)\gamma+\bar r(\alpha+\beta) \\ = {} & r-\bar r+\bar r\left(\frac{(a+\frac{\bar r}{r}e)-r}{\bar r}+\frac{a-e}{r}\right) \end{aligned}$$

$$\begin{aligned}
&= r - \bar{r} + a + \frac{\bar{r}}{r}e - r + \frac{\bar{r}}{r}a - \frac{\bar{r}}{r}e \\
&= a + \frac{\bar{r}}{r}a - \bar{r} = \frac{ra + \bar{r}a}{r} - \bar{r} = \frac{a}{r} - \bar{r} = \frac{a}{r} - 1 + r \\
&= \frac{a - r}{r} + r
\end{aligned}$$

Lemma 3.1. (formula (2)) and the recursion formula yield

$$\begin{aligned}
& pP_{n+1}(a + \frac{\bar{r}}{r}e) + \bar{p}P_{n+1}(a - e) \\
\leq\ & p^2 + p\left[(\bar{p} - p) + P_{n+1}(\frac{a - r}{r} + r)\right] \\
=\ & p\bar{p} + pP_{n+1}(\frac{a - r}{r} + r) \\
=\ & p\bar{p} + p\left[p + \bar{p}P_n(\frac{a - r}{r\bar{r}})\right] \\
=\ & p + p\bar{p}P_n(\frac{a - r}{r\bar{r}})
\end{aligned}$$

But

$$\begin{aligned}
\frac{a - r}{\bar{r}} &= \frac{r}{\bar{r}}(ra + \bar{r}a) - \frac{r}{\bar{r}} = a - e + e + \frac{r}{\bar{r}}a - \frac{r}{\bar{r}} \\
&= (a - e) + \frac{r}{\bar{r}}(a + \frac{\bar{r}}{r}e) - \frac{r}{\bar{r}} \leq r + \frac{r}{\bar{r}} - \frac{r}{\bar{r}} = r
\end{aligned}$$

allows us to continue

$$p + p\bar{p}P_n(\frac{a - r}{r\bar{r}}) = p + \bar{p}P_{n+1}(\frac{a - r}{\bar{r}}) = P_{n+2}(a)$$

Case IV: $a - e \geq r$.– We get

$$\begin{aligned}
& pP_{n+1}(a + \frac{\bar{r}}{r}e) + \bar{p}P_{n+1}(a - e) \\
=\ & p\left[p + \bar{p}P_n\left(\frac{(a + \frac{\bar{r}}{r}e) - r}{\bar{r}}\right)\right] + \bar{p}\left[p + \bar{p}P_n(\frac{a - e - r}{\bar{r}})\right] \\
=\ & p + \bar{p}\left[pP_n\left(\frac{a - r}{\bar{r}} + \frac{\bar{r}}{r} \cdot \frac{e}{\bar{r}}\right) + \bar{p}P_n\left(\frac{a - r}{\bar{r}} - \frac{e}{\bar{r}}\right)\right] \\
\leq\ & p + \bar{p}P_{n+1}(\frac{a - r}{\bar{r}}) = P_{n+2}(a)
\end{aligned}$$

Now $(*)$ is established also for $n + 1$, and our proof is complete. □

We have thus proved

Theorem 3.3. If $p \leq \frac{1}{2} \leq r$, then BOLD is optimal for all $n = 0, 1, \ldots$.

4. Non-Optimality of BOLD if $p \le \frac{1}{2} \le r$ Fails

In this section we shall define a particular strategy φ with the following property: if either $p > \frac{1}{2}$ or $r < \frac{1}{2}$ (or both), then for every $n = 3, 4, \ldots$ there is some a_n such that $0 < a < 1$ and

$$P_n^{\varphi}(a_n) > P_n^{BOLD}(a_n).$$

In fact, φ will be defined by

$$\varphi(a,n) = \begin{cases} 2\bar{r}r^2 & \text{for} \qquad a = r^2 + 2\bar{r}r^2, n = 3 \\ \varphi^{BOLD}(a,n) & \text{otherwise.} \end{cases}$$

and we will choose

$$a_n = r^{n-1}(a + 2\bar{r})$$

We begin with the observation (remember $0 \le r \le 1$)

$$\begin{aligned} \mid 2r - 1 \mid < 1 \Longrightarrow r^2 + 2\bar{r}r^2 &= r(1 - \bar{r}) + 2\bar{r}r^2 \\ &= r + r\bar{r}(2r - 1) < r + r\bar{r} \le 1 \end{aligned}$$

Consequently

(1) $a_n = r^{n-3}(r^2 = 2\bar{r}r^2) < r^{n-3}$

Next we prove

(2) $P_n^{\varphi}(a_n) = p^{n-3} P_3^{\varphi}(r^2 + 2rr^2)$

(3) $P_n^{BOLD}(a_n) = p^{n-3} P_3^{BOLD}(r^2 + 2\bar{r}r^2)$

thus reducing all cases $n > 3$ to the case $n = 3$.

We proceed by induction. For $n = 3$, (2) and (3) are obviously true. Assume (2) and (3) to be true for some $n \ge 3$. As $n + 1 > 3$, $\varphi(a, n + 1) = \varphi^{BOLD}(a, n + 1)$, and the recursion formula for BOLD applies:

$$\begin{aligned} P_{n+1}^{\varphi} &= pP_n^{\varphi}(\frac{a_{n-1}}{r}) \quad \text{by (1))} \\ &= pP_n^{\varphi}(a_n) = p^{n+1-3} P_3^{\varphi}(r^2 + 2\bar{r}r^2) \\ P_{n+1}^{BOLD}(a_{n+1}) &= p^{n+1-3} P_3^{BOLD}(r^2 + 2\bar{r}r^2) \end{aligned}$$

We now compute

$$\begin{aligned} P_3^{\varphi}(r^2 + 2\bar{r}r^2) &= pP_2^{\varphi}(r^2 + 2\bar{r}r^2 + \frac{\bar{r}}{r} 2\bar{r}r^2) + \bar{p}P_2^{\varphi}(r^2 + 2\bar{r}r^2 - 2\bar{r}r^2) \\ & \qquad (\text{stake } 2\bar{r}r^2) \\ &= pP_2^{\varphi}(r^2 + 2\bar{r}r(r + \bar{r})) + \bar{p}P_2^{\varphi}(r^2) \end{aligned}$$

$$
\begin{aligned}
&= pP_2^{\varphi}(r^2+2\bar{r}r)+\bar{p}P_2^{\varphi}(r^2)\\
&= pP_2^{\varphi}(r(r+\bar{r}+\bar{r}))+\bar{p}P_2^{\varphi}(r^2)\\
&= pP_2^{\varphi}(r(1+\bar{r}))+\bar{p}P_2^{\varphi}(r^2)\\
(&= pP_2^{\varphi}(r+\bar{r}r)+\bar{p}P_2^{\varphi}(r^2)))\\
&= pP_2^{BOLD}(r+\bar{r}r)+\bar{p}P_2^{BOLD}(r^2)\\
&= p(p+\bar{p}P_1(r))+\bar{p}pP_1(r)\\
&= p^2+2p^2\bar{p}
\end{aligned}
$$

Assume now $r < \frac{1}{2}$, hence $2r-1<0$, and thus

$$r^2+2\bar{r}r^2 = r + r\bar{r}(2r-1) < r$$

It follows that

$$
\begin{aligned}
P_3^{BOLD}(r^2+2\bar{r}r^2) &= pP_2^{BOLD}(r+2\bar{r}r)\\
&= p(p+\bar{P}_1^{P_1^{BOLD}}(2r))\\
&= p^2+\bar{p}p^2 < p^2+2\bar{p}p^2 = P_3^{\varphi}(r^2+2\bar{r}r^2)
\end{aligned}
$$

Thus φ is better than BOLD for $n=3$ and $a_3 = r^2+2\bar{r}r^2$.

Next let $r \geq \frac{1}{2}$ but $p > \frac{1}{2}$. Then $2r-1 \geq 0$ and hence

$$r^2+2\bar{r}r^2 = r+r\bar{r}(2r-1) \geq r$$

$$2r-1 = r+(r-1) < r < 1.$$

Recursion yields

$$
\begin{aligned}
P_3^{BOLD}(r^2+2\bar{r}r^2) &= p+\bar{p}P_2^{BOLD}(r(2r-1))\\
&= p+\bar{p}pP_1(2r-1)\\
&= p
\end{aligned}
$$

As $2p>1$, we get

$$p = p(p+\bar{p}) = p^2+p\bar{p} < p^2+p\bar{p}2p = p^2+2\bar{p}p^2,$$

that is,

$$P_3^{BOLD}(a_3) < P_3^{\phi}(a_3).$$

$$
\begin{aligned}
& \ldots pF_0^{\pi}(r - 2^{\pi}r) + P_0^{\pi}(r^{\pi}) \\
&= p[F_0^{\pi}(r \ldots) + \ldots] + pP_0^{\pi}(r^{\pi})] \\
&\quad [\illegible] \\
&\quad [\illegible] \\
&= \ldots [\illegible] \\
&= [\illegible]
\end{aligned}
$$

Assume also $p < \frac{1}{2}$, hence, $2p - 1 < 0$, and therefore

$$\ldots = r + (2p - 1)r^2 < r.$$

It follows that

$$
\begin{aligned}
& F^{BOLD}(\ldots + 2r^2) = pF^{BOLD}(x + 2r)\,[\illegible] \\
& \quad [\illegible] \\
& \quad = \ldots p^2 < \ldots [\illegible]
\end{aligned}
$$

That is, is better than BOLD for $r < \frac{1}{2}$ and $2x = r + 2r^2$ …

Consider $r = \frac{1}{2}$ and p … Then … < 0 and hence

… $(2p - 1) < \ldots$

$$2p - 1 \ldots + (r - 3) < r < 1.$$

Inequality yields

$$
\begin{aligned}
F^{BOLD}(\ldots) &= p + F_0^{BOLD}(x)(2p - 1)\ldots \\
&\quad + \ldots pR(2p - 1) \\
&= \gamma
\end{aligned}
$$

As $2p < 1$, we get

$$\ldots(p - q) = \ldots + 2p^4$$

that means

$$F^{\ldots}(x) < F_1^{\pi}(x)$$

X. Foundational Problems

In the preceding chapters we have deloped stochastics as a mathematical theory with a particular motivation: to cope with random phenomena in qualitative and, above all, in quantitative terms. We had more or less left aside a few questions of principle, such as

> what is random, after all?
> what is the spiritual meaning of probability?

We could do this because, as the scientific experience of four centuries has shown, most of the tasks given to stochastics by the empirical sciences can be performed without touching upon such questions in a more than intuitive way. There are, however, a few mathematical theories which have been developed in order to answer such questions of principle in a precise fashion. In this chapter we will report on three such theories, namely,

1. the theory of randomness, after Mises [1919], Kolmogorov [1965], Martin-Löf [1966], Schnorr [1970] [1970a] [1977].
2. the theory of subjective probabilities and related topics.
3. the theory of belief functions, after Dempster [1967], [1968], Shafer [1976].

We will leave aside considerations of a more philosophical character which are treated e.g. in Reichenbach [1942], Carnap [1950], Cohen [1977], and we will not go into too many technical details in the sequel.

1. The Theory of Randomness

If we toss a coin (0 = head, 1 = tail) 1000 times, it is extremely improbable to obtain a rather regular 0-1-word of length 1000 like

0000 00
1111 11
0000 ...... 01 (only one 1)

According to the weak law of large numbers we expect to get a sequence with less than 400 or more than 600 zeros with a probability $\frac{\sigma^2}{1000\cdot(\frac{1}{10})^2} \leq \frac{1}{4}\cdot\frac{1}{10} = \frac{1}{40}$ only, but even so nobody would believe that

0101 01

would ever turn out by sheer random, although it contains exactly 500 symbols 0 and 500 symbols 1. Random yields irregular sequences like

010101011110011001101000011011 00010010001101000011001001101000
11011010001111010011001110111011110001

which follow no apparent rule. Such examples, however striking, are, of course, no substitute for a precise mathematical definition of the notion of a *random 0-1-sequence.* It is the purpose of the theory of randomness to proffer such a definition and to show that the random sequences so defined have the properties which one would expect a) intuitively, and b) according to the "almost everywhere" theorems of probability theory.

It will be enough for our present purpose, to consider only two-symbol sequences here; the extension to more general situations would be a bare technical matter.

First general observation: we should not try to define the concept of randomness for *finite* 0-1-sequences, if the definition is to reflect the idea of irregularity ("no rule"), because every finite 0-1-sequence *is* sort of a rule – we may learn it by heart, and we may reproduce it at will, without making use of a random mechanism. Thus the theory to be explained here will be a theory about *infinite* 0-1-sequences and will lead to a decomposition of the set $\Omega = \{0,1\}^{\mathbf{N}}$ of all 0-1-sequences into a set R of sequences called *random*, and a remainder N of sequences called *non-random.* We mention in passing that there is also a theory of *degrees of randomness*, and actually such an idea is crucial for the ideas brought forth by Kolmogorov [1965]. But let us first begin with a historical sketch.

1.1. Von Mises' Kollektive.

The mathematical theory of randomness begins with Mises [1919]. Richard von Mises (1883-1953) called an infinite 0-1-sequance

$$\omega = \omega_1\omega_2\omega_3\ldots \quad (\omega_1,\omega_2\ldots \in \{0,1\})$$

a *Kollektiv* (with $p = \frac{1}{2}$) if

$$\lim_{n\to\infty} \frac{1}{n}(\omega_1 + \ldots + \omega_n) = \frac{1}{2},$$

and if the same "fifty-fifty" property holds for every subsequence

$$\omega_{k_1}\omega_{k_2}\omega_{k_3}\ldots$$

$$0 \leq k_1 < k_2 < k_3 < \dots$$

of ω for which the choice of the $k_1, k_2, \dots$ has been made "according to some rule". This verbiage reflects von Mises' ("frequentist") idea that probabilities are limits of relative frequencies. It was soon discovered that it wasn't a rigorous mathematical definition because the condition "according to some rule" had not been precisely stated. During the thirties a theory of formal languages was developed in mathematical logic; on this basis Abraham Wald (1902-1950) was able to transform von Mises' idea into a precise definition (Wald [1936][1937]); an easy application of the strong law of large numbers showed the existence of many Kollektive. The new theory suffered, however, a severe blow when Jean Ville constructed an example of a Kollektiv ω which didn't do what one would expect from a 0-1-sequence resulting from a random experiment: it showed a preference for 1, fulfilling

$$\frac{1}{n}(\omega_1 + \omega_1 + \omega_n) \geq \frac{1}{2} \quad (n = 1, 2, \dots);$$

this violates the loglog theorem which implies that the average in question is $> \frac{1}{2}$ for some n, and $< \frac{1}{2}$ for some others, almost surely (Ville [1939]). The matter was discussed intensely by important probabilists at a meeting near Genève in 1937, and the theory of Kollektive fell into oblivion for about 20 years thereafter. To this day, Kolmogorov's [1933] measure-theoretical foundation of probability theory, has proved to be the overwhelmingly successful tool for probability theory, and practically all probabilists were only too glad to work within this frame and to pay no attention to von Mises' idea. For a detailed historical discussion see Krengel [1990].

1.2. The Proposal of Kolmogorov.

It is remarkable that sort of a revival of von Mises' ideas was originated by Andrej Nikolajevic Kolmogorov (1903-1987) himself. In his seminal paper Kolmogorov [1965] he proposed to define randomness via computational complexity, this giving a precise meaning to the ancient idea, that

> random is beyond rational explanation (παϱαλογος,
> Aristotle phys.196-197)

or, a bit more specific

> a 0-1-sequence should be called random if you can't learn it by
> heart

We will explain one of the proposals made in Kolmogorov [1965] here, making intuitive use of some notions of the theory of Turing machines.

A Turing machine, adapted to our present purposes, is an automaton A which produces ***finite*** 0-1-strings w upon insertion of ***programs*** which we assume to be formulated as finite 0-1-strings themselves. We shall employ the standard notation $\{0,1\}^*$ for the set $\{\Box\} \cup \bigcup_{n=1}^{\infty}\{0,1\}^n$ of all finite 0-1-strings, including the empty string $\Box$. Different Turing machines will perform this task in different ways. There is e.g. a Turing machine C which is usually called the "copying machine" and which produces w after the program w has been inserted. Another Turing machine may produce specific sequences w in particularly simple fashion. We might e.g. imagine, for any given $w \in \{0,1\}^*$, a machine especially "tailored" for w: it produces w from the program 0. We may express this also by saying that this machine "has learned w by heart". This is especially plausible if w is extremely regular, such as $w = 00\ldots0$; the program 0 is then nothing but an expression of our brief thought "take 0 and repeat it". All this can be made precise within a formal theory of Turing machines. The reader should learn from this sketch that different Turing machines produce the same $w \in \{0,1\}^*$ from different programs $p \in \{0,1\}^*$, and in particular from programs p of different lengths $|p|$ ($|p| = n \iff p \in \{0,1\}^n$). We don't exclude the possibility that A can't produce w at all, or that A produces the same w from many different programs. Let us now define for any Turing machine A, and any $w \in \{0,1\}$, the ***program complexity*** of w for A as

$$K_A(w) = \text{length of the shortest program which makes } A \text{ produce } w$$

($= \infty$ iff A can't produce w at all). Thus e.g. the copying machine C yields

$$K_C(w) = |w| \qquad (w \in \{0,1\}^*)$$

One of the basic results of Turing theory is the existence (and explicit construction) of at least one ***universal Turing machine***. Such a universal machine U simulates any other Turing machine A if we insert an appropriate simulation program $s(A) \in \{0,1\}^*$ as a prefix. That is: if A produces w from program p, then U produces w from program $s(A)p$ (p preceded by $s(A)$). This implies

$$K_U(w) \leq K_A(w) + |s(A)|$$

We may apply this especially to $A = V =$ another universal Turing machine, and thus see: the program complexities for different universal Turing machines differ by at most a constant, namely, the length of the prefix program which makes the one universal machine simulate the other.

Let us now choose a specific universal Turing machine U and write

$$K(w) \text{ for } K_U(w) \qquad (w \in \{0,1\}^*).$$

Proposition 1.1. For any integers $n > 0, d \geq 0$ these are at least

$$2^n(1 - \frac{1}{2^d}).$$

0-1-strings w of length n such that

$$K(w) \geq n - d.$$

In particular (take $d = 0$) there is at least one $w \in \{0,1\}^*$ with $K(w) \geq n$.

PROOF. Put $c = n - d$. $K(w) < c$ means: there is a program p with $|p| < c$ such that U produces w from p. There are 2^e programs of length e and thus

$$1 + 2 + \ldots + 2^{c-1} = \begin{cases} 0 & \text{for } c = 0 \\ 1 & \text{for } c = 1 \\ \frac{2^c - 1}{2 - 1} & \text{for } c > 1 \end{cases}$$

programs of length $< c$. The number of all 0-1-strings w with length n and $K(w) < n - d$ is therefore $< 2^{n-d}$. Consequently there are at least $2^n - 2^{n-d} = 2^n(1 - \frac{1}{2^d})$ 0-1-strings with length n and $K(w) \geq n - d$. □

Let us now turn to *infinite* 0-1-sequences

$$\omega = \omega_1\omega_2 \ldots \in \{0,1\}^{\mathbb{N}}.$$

We might imagine, at least for some ω like $000\ldots$, that a Turing machine A prints out ω as far as we want, when given a suitable program p. For the program complexities of the initial sections $\omega_1 \ldots \omega_n$ $(n = 1, 2, \ldots)$ of w this means

$$K(\omega_1 \ldots \omega_n) \leq K_A(\omega_1 \ldots \omega_n) + |s(A)|$$

where the length $|s(A)|$ of the simulation program $s(A)$ is independent of ω and n. But $K_A(\omega_1 \ldots \omega_n) \leq |p|$, and thus we see:

> if an $\omega \in \{0,1\}^{\mathbb{N}}$ can be produced by a Turing machine upon the insertion of one single program, then the program complexities of the initial sections of ω remain bounded.

This, along with the above proposition, motivates the following

Definition 1.2. (Kolmogorov [1965]). We call an infinite 0-1-sequence $\omega = \omega_1\omega_2 \ldots$ *quasi-random* if there is a constant $d(= d(w))$ such that

(1) $K(\omega_1 \ldots \omega_n) \geq n - d \quad (n = 1, 2, \ldots).$

As we have seen, no Turing machine can print out quasi-random sequences because (1) forces the complexities of the initial sections to be unbounded. But can we obtain quasi-random sequences by other means, or at least prove their existence?

1.3. Per Martin-Löf's Objection.

It turned out immediately that the answer was "no", and thus 3.2 was not yet the appropriate definition of randomness. Making use of an old (pigeon-hole-type) device of Borel [1920], Per Martin-Löf [1966] (*1942) was able to prove

Proposition 1.3. There are no quasi-random sequences. More specifically: if $d_1, d_2, \ldots$ is a "Turing-computable" sequence of non-negative integers such that

$$\sum_{n=1}^{\infty} \frac{1}{2^{d_n}} = \infty$$

(that is, the d_n must not "grow too fast": $d_n = [\log_2 n] + 1$ would do it, but $d_n = n$ not). Then for every $\omega = \omega_1 \omega_2 \ldots \in \{0,1\}^{\mathbb{N}}$.

$$K(\omega_1 \ldots \omega_n) < n - d_n$$

happens infinitely often.

1.4. Per Martin-Löf's Proposal.

Per Martin-Löf [1966] now made another proposal for a definition of the randomness of an infinite 0-1-sequence:

> call an infinite 0-1-sequence *random* if it " survives all randomness tests."

The crucial point was, of course, to give a precise meaning to the notion "randomness test", and subsequently to prove the existence of (many) random sequences in the above sense.

A test in the sense of mathematical statistics is given as a critical region, a certain subset of the underlying basic set: if you are in that critical region, you have failed the test (the null hypothesis is declined). Martin-Löf [1966] adapts this notion to the present purpose, taking $\Omega = \{0,1\}^{\mathbb{N}}$ as the basic space and considering so-called *cylinder sets* of various orders $n \in \mathbb{N}$: if $n \in \mathbb{N}$ and $\omega_1 \ldots \omega_n \in \{0,1\}^n$, then

$$[\omega_1 \ldots \omega_n] = \{\omega | \omega_1 \ldots \omega_n \eta_{n+1} \eta_{n+2} \ldots j \eta_{n+1}, \eta_{n+2}, \ldots \in \{0,1\}\}$$

is the cylinder set of order n associated to the 0-1-string $\omega_1 \ldots \omega_n$ of length n: it consists of all 0-1-extensions (to the right) of that string. A subset E of Ω is said to be of order n if it can be represented as a union of cylinders of order n and if n is the smallest number with this property. Every $[\omega_1 \ldots \omega_n]$ is of order n while we may write it as $[\omega_1 \ldots \omega_n] = [\omega_1 \ldots \omega_n 0] \cup [\omega_1 \ldots \omega_n 1]$ as well. If $E \subseteq \Omega$ is of order n, and a union of k cylinders of order n, we define $m(E) = \frac{k}{2^n}$. It is easily checked that the system $\mathcal{F}$ of all sets of finite order is stable under finite unions, intersections and differences, and contains Ω (which is of order 0). $m : \mathcal{F} \to \mathbf{R}$ takes values in $[0,1], m(\emptyset) = 0,\ m(\Omega) = 1$, and is (finitely) ***additive***: $m(E \cup F) = m(E) + m(F) \quad (E, F \in \mathcal{F},\ E \cap F = \emptyset)$. We mention in passing that *measure theory* would allow us to extend m to a σ-additive set function ≥ 0 on the σ-field generated by $\mathcal{F}$ (ch. III, §3), but we will not make use of this possibility.

A *randomness test* in the sense of Martin-Löf [1966], is a descending sequence

$$K_1 \supseteq K_2 \supseteq \ldots$$

of sets from $\mathcal{F}$ such that

1. $m(K_n) \leq \frac{1}{2^1},\ m(K_2) \leq \frac{1}{2^2}, \ldots,$
2. $K_1, K_2, \ldots$ can be produced by a Turing machine after the insertion of one single program – we will not specify in detail what this precisely means, asking the reader to rely on his intuitive understanding here.

If some $\omega \in \Omega$ belongs to K_n, we will say that it *fails this test at stage n*. $K_1 \cap K_2 \cap \ldots$ are the *total failures* for our test, and it is intuitively clear that the set of all such total failures is a "m-nullset".

To be a bit more specific, we might imagine that K_r consists of all those ω which fulfil

$$\left|\frac{1}{N}(\omega_1 + \ldots + \omega_N) - \frac{1}{2}\right| > \epsilon$$

for some large $N = N_m$ which allows to estimate $m(K_r) \leq \frac{1}{2^r}$ by Tschebyshev's inequality; clearly this K_r is of order $\leq N$, hence in $\mathcal{F}$, and it is intuitively clear that it can be described by a Turing machine, even in a uniform way for all r.

Many other randomness tests will certainly exist, and a result from Turing theory – an analogon to the existence of a universal Turing machine – proves the existence of at least one *universal randomness test*

$$U_1 \supseteq U_2 \supseteq \ldots$$

– universal in the sense that for every other randomness test

$$K_1 \supseteq K_2 \supseteq \dots$$

there is a constant c (depending on $K_1, K_2, \dots$) such that

$$K_{r+c} \subseteq U_r \quad (r = 1, 2, \dots)$$

This implies

$$\bigcap_r K_r \subseteq \bigcap_r U_r,$$

that is, the total failures of any randomness test are among the total failures of our universal test, and for any other universal test the total failures would be exactly the same.

Definition 1.4. An infinite 0-1-sequence $\omega = \omega_1\omega_2\dots$ is said to be a *random sequence* if it is not a failure for any randomness tesst, that is if it passes or "survives" every randomness test.

Again we see, looking at some universal test, that "m-almost all" 0-1-sequences are random: many many such sequences exist.

On the other hand it is intuitively clear that no single random sequence $\omega = \omega_1\omega_2\dots$ can be printed by a Turing machine upon insertion of a single finite program, because every such program producing $\omega = \omega_1\omega_2\dots$ would also produce

$$K_r = [\omega_1 \dots \omega_r]$$

with

$$m(K_r) = \frac{1}{2^r} \quad (r = 1, 2, \dots),$$

and ω would certainly be a total failure (the only one) for this randomness test.

Our above sketch of a randomness test checking the law of large numbers can be paralleled by other tests checking other laws of probability theory such as the loglog theorem (ch. IV §6) etc. It may even be expected that some tests check laws that have not yet been explicitly discovered. As a random ω fails none of all these tests, we may sloppily state

> a random 0-1-sequence fulfils all laws of probability theory which can be checked by randomness tests – those laws that are already discovered, and also those that will be discovered in the future (or even never).

The key results of Turing theory which lead to this conclusion, are nothing but a bit more sophisticated analogues of the primitive fact that there is a mechanical device producing all finite symbol strings that can be formed from the usual european alphabet – including Wycliffe's Bible, Goethe's Faust etc. Only: the device doesn't recognize the value, the beauty, the interest of all those strings – it does not even distinguish the meaningful from the meaningless.

1.5. Schnorr's Rehabilitation of von Mises' Kollektive.

In a sequence of important papers, Claus-Peter Schnorr (*1943) took up the ideas of Per Martin-Löf and extended them in various directions: Schnorr [1970] [1970a] [1977]. He e.g. defined *degrees of randomness* for 0-1-sequences and proved that sequences with a given degree of randomness are Turing constructible. As a byproduct of his investigations, he was able to show that von Mises' old idea, if suitably generalized, yields random sequences in Schnorr's sense: a selection rule in the sense of Mises [1919], Wald [1936][1937] may be interpreted as a "constructible measure preserving transformation" of the space of all 0-1-sequences; there are more such transformations than those thus obtained, and if a 0-1-sequence remains "fifty-fifty" under *all* of them, it is random (Schnorr [1970]).

1.6. Random Numbers.

The practical production of 0-1-sequences which are "sufficiently random" for everyday business takes place on a much simpler level than the one discussed in purely theoretical papers. In some cases, physical random processes such as electron emission are used, but what one normally encounters in commercial computers are number theoretical algorithms. A standard reference for these is Knuth [1969] vol. 2 ch. 3 (see also Hlawka-Firneis-Zinterhof [1981], Zielinski [1972]). And one of the standard methods of generating random symbols is the *linear congruential method* based on four integers ≥ 0

ω_0 = the starting value
a = the multiplier
c = the increment
m = the modulus $(m > \max\{\omega_0, a, c, \})$.

One then defines recursively

$$\omega_{n+1} = (a\omega_2 + c) \bmod m.$$

This sequence will, of course, be periodic and thus by no means really random, but by proper choice – characterizable by number-theoretical conditions – of our four integers one can achieve large periods and a good "equidistribution" of the resulting sequence $\omega_0, \omega_1, \ldots \in \{0, 1, \ldots, m-1\}$ so that it can serve as a substitute for veritable random. For $m = 10$, the following sequence of digits $0, 1, \ldots, 9$ is obtained

1698523404231289764965392410891183523154274919340176938033639364142947758800329968570625740298607
99485439544334654342794301242913213285122778564951811855409821518787160959795225872087836764359454
23575396525544759112685580860337579636985166071719295892087433458904411521520604403702093934973385
04580799341110619965655423061516346302093786247913492588528832269782049777533187085089058075529519
4215215287463129441560336323988204280548475744140534391630760359188409494541962365603343998592444
30986002800636855374673910150177972170796645381539784029091409955944931736299493905375933303495004
0433079684647144634931925943123602672820855334370600832055660446398056304546277106987411951698818
2767884113307513226549494587371582105036728236839134411433807739083461844168148015712375040473675
6474891844250266201216015522244278811077657046404078543330658688140498280133144190057840435675524
0306593661089854597100569930348261902244081728525866311348798350978741945180422591096091204936574
0231304743717262067052461248011824214048504345288813032513300754671220688047332725717434089850630
7136076144257843838724439397826586598831860890468908192277125235889215753472727387755431089786670
2614446710470909591532805080756813251652787324919789660361283223726842592535516656906580423374595
63908250271551874727746761960592632322711352735974567016796605318852960626624812526388846208663504
9044225843059427158405755767839036742303909929258947694721342943448324920705318570102301158526437
3976548397358704772035059395585725131123802410388999407050003771745613771385671244962159456368567
9363659548545310249848312452753185718432133483766512962143643184748633552412861331033642339309810
2667295237263545281738689055204783059665307188958743895309839983845546138957101959731276667113033
1706959372911469829871258563654545897374334763498433082455419756643481498608918356653547415238576
2886281260507077640287250337862976645557040076268300305063294011818590514278260374488667771127075
3880892977966521672206244301494575522609576101505570795612254308272325128300899168650993806295964
9598452419619030296764076046477324677597897958724036013686010011158661769830784838385080965401213
6839721135035823843836295123381780456893272216214611864499317877145823758224876924519946617888052
0270103744102153430270241791389227156039657433408889967648333011379306280616883336048509699523412
5881007241187298717680689662947185859673802632447126762419675434269738192774255309815050064124762
5155286337917983231638651155560308781375775055192344929779687259210556891234120174875804823974664
8447747548962544945636414839175772490357974391192677396392642522336132022582386791972546495057937
25634940960431000081563244726263888888814701762356082273864661968871485150609629258347331919715867
7880045558430131028245532446832934454236581530957111951676692825395399482604267653960669631776650
4447683925383117484897684432484603210654453793122267575656559094081719249649631833419806898163599
5136188481992183918899482432814120990496092802412640953204931493362388796339021983271712466811573
3922216606690187723440492094618500916701777775048709934648083012774030304513785158140262561159611
6709976969700664445155788004408207151250004639089001068727358953159122754838387111729767573844647
6336020837004412078319872302611355481668815810916454234690190886591539906964920442047324083700386
1386869757602394347698557672946421523900530347072522503484454712919751518633834247379883792462754
5494391350779068531968910390304764639148191584891938785905972907175970004765316225529440915778361
03128197066679820437152208230292006945334637777830798490750362565666636875047769074752441819398142
2289587809548007189189026568623822846736111390790608379915152887281329911924326149093879148707172
9484417835570113988009080058542398358054528422546173230818988564708538951931294513463387512427686
0006928121180108237756579187105757320261120713813767796821038975250017990073468575722194530253676
2720583751122886066699181833315168734388077895103354496327085001414349267997429750605130295495272
2435201570488451334419757061098634672089033759198890011935684007963172990373538244850219967303963
2448184461161053344613321073158562857149879594892563652521193626301615470877297045406021196866053
2829333547882686331289999281171606748013699332997324514224843753772661119432525716465816692256275
8319211725543509972571982341906923049591770399169569672772693160904827467798597175892983507696431
6654321578454159032878890973521535786229467140772749309582936425472726282094181445649291962758137
1177972693001380205561556035234915451025152422227916028472231843694886891394125322990228012020425
92402822871046547741745774238730621900418399287545010699338896002508393487690226573959057745128674
9590923590238506127452012414313955949647517743129981521406711482754039770195056300444338947918317
67964629882990556852607918691226355545715917016078401206307638831599578871256166715253177518992633
0447773928808854569819529642029852932307791971878739538009232103844826313595077834208043932312113
3988894134606855481593400417494094015049969605762172647050229782704723919402466779709527654843907
3383551245478513761360155550286656350581020982031437374323076860561610531542648950763869104434777
3167596096374524988796533020896414885171881133706163740033437233589145406743511477478469207888739
6985688286900106486510144387325001549915450541318776649889850656966294023251559806271765722933698
2622403046470452800847030581331306275340636122485956163923715047961595684960981957129499745624141
3817469187735509584133848276994167005451180862432871941752100959711074225410676760204727347375090
92769573766299383485068815478160797170228479434088856148837200933246403735039239814369455146228227
8383270147680515481030051011149251747518497071428319532756519670169636306055447674036365371076859
17097644008150567406101440948405050238354555883327173286509250608553574317837585541772741472989203
94026556549237239279762707881317267588122425622474886313481385589698445720364961785575435839882214
45594455081029335941014417675115146267448286479333251447042636035583081559547667475364907878964049
0920411466338662166835958537702797077170447574607060510127675891555486558516218521476319196519803
6415302913392737253784916026623806359587210675177727704130116729368582876844436265999153830459449
8822321956624089197070837611856857986290462444026793426197706634683798846602113745949182574040937
2219862153308965087056825941338221357101068810507638347590956018500506685848331555969321858112975
4074710663975516593542835004161068071324254551153780143250958127507286307231240998730782143010969
4128119787047937421571328714473475421900886345592583825132576022930304818495857368077619910449441
9620204192182098505141640620057621341912826887124394603453790177168069755718613164001347141310345
8921661401796775260514767063809413258235333759242691338713845834675390089448563756514985254919630
4448085230526672381247699577063479717720455011766917225674624309132270338488549849104906581634054
9492531397617452997141213950937632788416957583845870571922101933869657869018075929280079749195655
1102649798194083653373736893601522000594152071760937126005979822652214845773816985186308082433690

7382372896608529936261902516888531925375280363961560046885673566204619444583375548008519570397923644
9514633057281271739909593257518668664960446973106488753020473787388979655742355558168785471179206 0
8668508287323829350520101118696940757076650791231342342130114792661119090484242383073315103171090
9027171760917143571211196624940135636705825634462999901430035304983306541556548170765801256871506 2
7235225039050850176066647399375732979435141655060176362397073893360635705173019665401271179862938
0911805792905069185011865434483888176940934510020043397106278073700121381657171127303954735653161
2588879459646941443819549438136940262566985534909233855579260066905188671464006154717522062040593 6
4293222084031821312872056181340452911737825210771212750149525197667562208808813605485720544205653
7830630317519267287737885911477551177513148666801844677067954007716031993872923555710345809778 0799
2481085387344252725375625156581883344549553454696757234806353723997021117457974833295093566102372
8282445316815242104311492134476548527669175917013707413228638162449604712851664683930497770037924
7954317306349705030605557374640721929232032961832848305707601765516746430761004962384672630770311
5906559738846613721725740398612502332225184717474822584984944888781421695591092802199760261869334
1295767051702906724386389909976195505536022456272844287484304750062858273460840124233736000048609
2417197862220097885437674847384652690880089495095772463597672532889551237248832302617838634291102
2988039479609600151658221231106112695169504187168770731137724141042921802060910876371028378511337
6896345444235319239309537058797930395882165497958649854813690112729431383344355773388696609736390
4769739861248447414639606828806362002794796225193819982247391493188923784671233595297038125086985
5402730495584296497876973567639101915841144569470213124161287907573108018912539261025005794679591
9300464782611737909499289682758764382564132604074977195926940673806741718001647280710860458074378
5025298976418488175627646567572770617978776158801696758808806632918343112459017826555408046020083 8
7258509401641331294154113357839667400661769438644607552445085590494561165791075558093895524997271 0
0140614136588445802500273308318796107240127986621961072594403407811693883649105141755067054735506
0633771856599274854155040127290215649614403707712857234836664892856669131154615777453415750491512
8447663133349705990399875562876857445245343499762621399645064570409811495082926621424788412129250
8973266699639048468907576674094354130038139319061760041852460511421527976862563052164074248118579
4292173917505618633433373553787843137635515698054221504373489747088497525499051899864987612342352
3409745298282977006666883598606745030829925288409983945069474848780858571547778919513227568894681 1
9368109315914517530254638270501785134390619438497434460591603272685755379453129309510479826932402
4724528230836536503481932573644325313674881513692213865029955370351604175047993315497593819837575
758515901

If we transform it into a 0-1-sequence by coding all even digits $(0, 2, 4, 6, 8)$ into 0 and all odd ones into 1, we obtain, representing 1 graphically by ■ and 0 by a blank

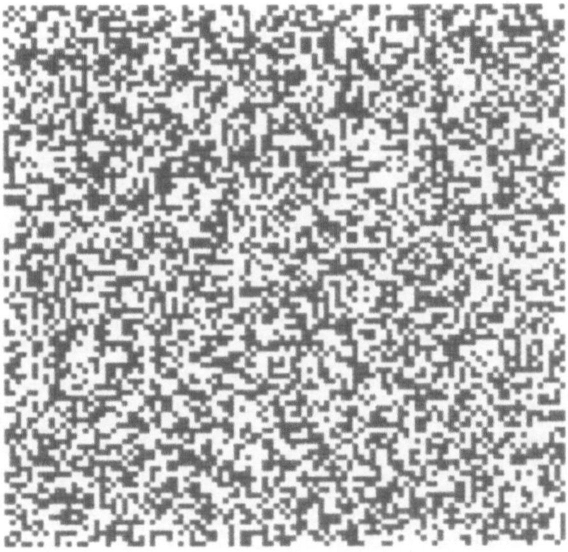

Also the decimal expansions of e and π are considerably random. Transforming them as we just did, we get the pictures for e:

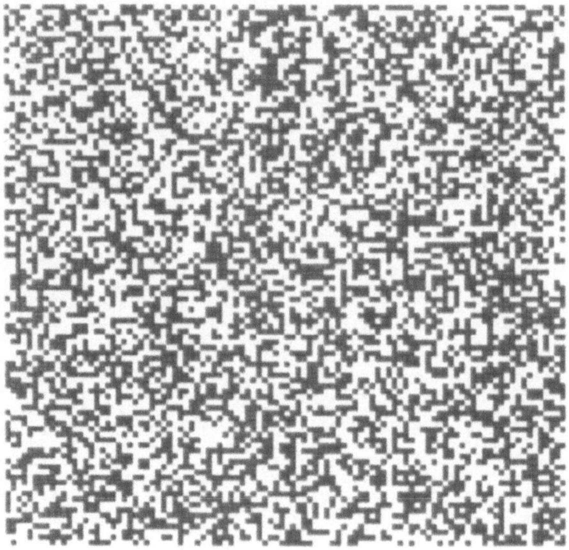

for π:

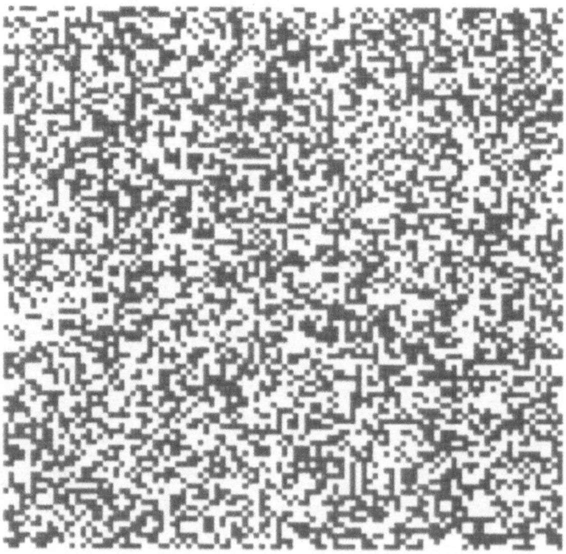

It is, of course, tempting to visualize other number-theoretically defined sets in this fashion. If we e.g. "spiral" a tape showing ■ at the prime number places and a blank else, we obtain

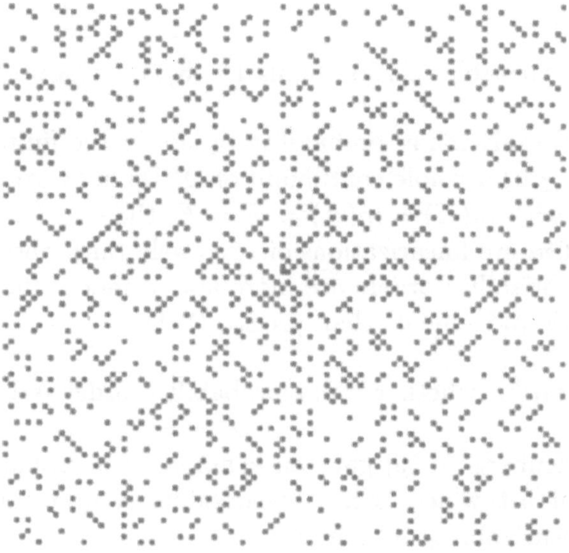

2. Subjective Probabilities

It has been argued that a DPS (Ω, m) with all its subtle arrangement of real numbers, does not adequately reflect the subjective fuzziness of probability assignments. As an answer to objections of this sort, some probabilists have developed a theory of so-called ***subjective probabilities***. In this theory, one doesn't try to precisely say ***how probable*** a certain event is, but rather to define a consistent way of calling some events ***more probable*** than others. The heart of this theory (quoted here after Fishburn [1986]) is the

Definition 2.1. Let Ω be a finite nonempty set. A binary relation $\prec$ on $\mathcal{P}(\Omega)$ is called a ***probability order*** if it has the following properties:

1. comparability: for any $E, F \subseteq \Omega$ either $E \prec F$ or $F \prec E$ or both
2. transitivity: $E, F, G \subseteq \Omega,\ E \prec F, F \prec G \Longrightarrow E \prec G$
3. additivity: $E, F, G \subseteq \Omega, (E \cup F) \cap G = \emptyset \Longrightarrow [E \cup G \prec F \cup G \text{ iff } E \prec F]$

Let $\prec$ be a binary relation on $\mathcal{P}(\Omega)$ and m a DPD on Ω; we say that m is a *linearization* of $\prec$ if for any $E, F \subseteq \Omega$

$$(*) \quad E \prec F \Longleftrightarrow m(E) \leq m(F).$$

If there exists a linearization of a relation $\prec$ in $\mathcal{P}(\Omega)$, $\prec$ is called *linearizable*.

Clearly every linearizable relation fulfils 1) - 3), i.e. is a probability order.

There are only finitely many probability orders in the finite set $\mathcal{P}(\Omega)$ while there is (for $|\Omega| \geq 2$) a whole continuum of DPDs in Ω. Actually, if $\prec$ has a linearization m with $m(E) = m(F) \Rightarrow E = F$, then all DPDs from a whole neighborhood of m are linearizations of $\prec$ as well. Such observations confirm the view that probability orders are more apt to model the fuzziness of our daily probabilistic judgements than precise DPDs.

The obvious question whether every probability order is linearizable has a negative answer:
Kraft-Pratt-Seidenberg [1959] gave an example of a probability ordering $\prec$ in a five-element set which is not linearizable. A necessary and sufficient condition for linearizability is given in

Theorem 2.2. (Kraft-Pratt-Seidenberg [1959]). Let $\Omega \neq \emptyset$ be a finite set. Then a probability order $\prec$ of $\mathcal{P}(\Omega)$ is linearizable if and only if it is

4) *strongly additive* in the following sense: if $A_1, \ldots, A_m, B_1, \ldots, B_m \in \mathcal{P}(\Omega)$ fulfil

$$(**) \quad \sum_{k=1}^{m} 1_{A_k} = \sum_{k=1}^{m} 1_{B_k}$$

then

$$(***) \quad A_1 \prec B_1, \ldots, A_m \prec B_m \Longrightarrow B_1 \prec A_1, \ldots, B_m \prec A_m$$

The necessity of 4. is obvious: if m linearizes $\prec$, take expectations (for m) in $(**)$, finding $\sum_k m(A_k) = \sum_k m(B_k)$ and get $(***)$ immediately. The sufficency of 4. is harder; the proof in Kraft-Pratt-Seidenberg [1959] uses techniques from the theory of linear inequalities.

We will not got to further lengths on subjective probabilities here. Fishburn [1986] is recommended as a starting point for further reading.

3. Belief ("bel") Functions

The usual probability theory as is has been explained – within the limits of discreteness – in the preceding chapters, has sometimes been criticized in the following way:

> in usual discrete probability theory, you can't start dealing with probabilities before you have fully established a certain DPS (Ω, m); this is true too much for daily practice. Take the example of a criminal process in court: the State Attorney starts with some suspicion – clearly he deals with probabilities; the witnesses make their statements, some very clear, some confused – in any case a matter of probability; the jury makes a decision supposed to be true "beyond reasonable doubt" – still a probabilistic statement. The probability space in which all these probabilities fit together can only be constructed after the process is over; but the decisions have to be make before that.

Probability theory can easily cope with such objections by offering product space constructions:

> some first probability space $(\Omega^{(0)}, p^{(0)})$ reflects what we believe to know in the beginning, say, of our abovementioned legal process. Witness no. 1 brings in a new basic set $\Omega^{(1)}$: every $\omega_0 \in \Omega^{(0)}$ is split into several $\omega_0\omega_1$ with ω_1: We may plot this in the form of a tree (figure X.3.1) and construct a DPD in $\Omega^{(0)} \times \Omega^{(1)}$ from $p^{(0)}$ and a transition matrix $P^{(1)} = (P^{(1)}_{\omega_0\omega_1})_{\omega_0 \in \Omega^{(0)}, \omega_1 \in \Omega^{(1)}}$. We may continue in this fashion, obtaining the final DPS step by steps the process goes on.

One might also defend the use of classical discrete probability theory in such situations by saying that we constantly presuppose only the *existence* of the final DPS, but *discover* its true shape only in successive stages as the informations drop in.

Nevertheless the abovementioned objections have their own vital traditions, often motivated by the problem of finding justice in court (see e.g. Cohen [1977]), and have led to an interesting way of constructing probability spaces from so-called *belief functions* ("bel functions"). I will try to make the reader acquainted with the basic ideas of this development here; more details and careful discussions can be found in Shafer[1976].

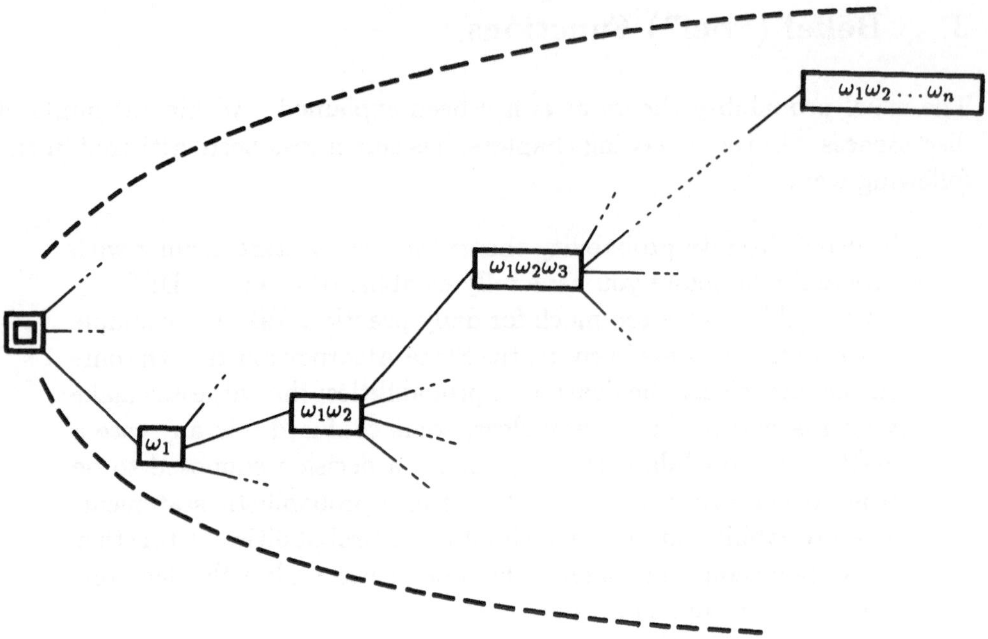

fig X.3.1

In the sequel, we will confine ourselves to *finite* basic sets D. While usual probability theory ascribe probabilities $p_j \geq 0$ the ***points*** j of D such that

$$\sum_{j \in D} p_j = 1,$$

and derive an additive set function $p : \mathcal{P}(\Omega) \to [0,1]$ from this by

$$p(E) = \sum_{j \in E} p_j \qquad (E \subseteq \Omega),$$

belief functions ascribe probabilities $b_E \geq 0$ to the ***subsets*** E of Ω such that

$$\begin{array}{rcl} b_\emptyset & = & 0 \\ \sum_{E \subseteq \Omega} b_E & = & 1; \end{array}$$

we may then derive a set function

$$m_b : \mathcal{P}(\Omega) \to [0,1]$$

from b by

$$m_b(f) = \sum_{E \subseteq F} b_E \qquad (F \subseteq \Omega).$$

It is easy to recover b from m_b by so-called *Moebius inversion:*

$$\begin{aligned}
\sum_{E\subseteq F}(-1)^{|F\setminus E|}m_b(E) &= \sum_{E\subseteq F}(-1)^{|F\setminus E|}\sum_{A\subseteq E} b_A \\
&= \sum_A b_A \sum_{A\subseteq E\subseteq F}(-1)^{|F\setminus E|}1^{|E\setminus A|} \\
&= \sum_A b_A(1-1)^{|F\setminus A|} \\
&= b_F
\end{aligned}$$

(we have used the binomial theorem here; $(1-1)^{|F\setminus A|}=0$ unless $F=A$).

It is easily seen that m_b is additive iff b is >0 only on one-element sets $\subseteq \Omega$, i.e.

$$b_E > 0 \Longrightarrow |E| = 1;$$

thus classical DPDs fit into the new frame as special cases.

Belief functions allow us to express intuitions like

> I believe the true $j \in D$ to lie in E, and believe it with strength b_E, but I don't want to make any statement about *where* in E the truth lies.

Let us illustrate this by a few examples.

Example 3.1. The *empty* bel function e is defined by

$$\begin{aligned}
e_\Omega &= 1 \\
e_E &= 0 \quad (e\subseteq\Omega,\ E\neq\Omega).
\end{aligned}$$

It expresses the triviality that the true j is in Ω, and nothing more.

Example 3.2. A bel function is called *simple* if there is a $\Omega_0\subseteq\Omega$, $\Omega_0\neq\Omega$, and an $0\le\alpha\le 1$ such that

$$\begin{aligned}
b_{\Omega_0} &= \alpha \\
b_\Omega &= 1-\alpha \\
b_E &= 0 \text{ else}
\end{aligned}$$

Such a b expresses the idea that one believes, with strength α, the truth to lie in Ω_0, and no further statement should be made. The empty bel function appears as the simple bel function with $\alpha = 0$.

Let us imagine that two witnesses shine up in court with their bel functions $b^{(1)}, b^{(2)}$, and that the judge or the jury wants to combine $b^{(1)}$ and $b^{(2)}$ into a new bel function which reflects the situation resulting from the two evidences have been given.

We will certainly say that the two witnesses (resp. their bel functions $b^{(1)}, b^{(2)}$) are *incompatible* if

$$b_E^{(1)} > 0,\ b_F^{(2)} > 0 \Longrightarrow E \cap F = \emptyset$$

or, equivalently,

$$\left[\bigcup_{b_E^{(1)}>0} E\right] \cap \left[\bigcup_{b_F^{(2)}>0} F\right] = \emptyset;$$

in fact witness no.1 gives the true j no chance to lie outside the set in the left [ ], and witness no. 2 does the same with the right [ ]. – If this is the case, the obvious composite of $b^{(1)}$ and $b^{(2)}$ is the *empty* bel function e. If $b^{(1)}$ and $b^{(2)}$ are *compatible*, then there are $E, F \subseteq D$ such that

$$E \cap F \neq \emptyset, \text{ and } b_E^{(1)} \cdot b_F^{(2)} > 0$$

We thus have

$$(1) \quad \sum_{E \cap F \neq \emptyset} b_E^{(1)} b_F^{(2)} > 0,$$

and may define a new bel function b by

$$(DR) \quad \begin{aligned} b_\emptyset &= 0 \\ b_G &= \frac{\sum_{E\cap F=G} b_E^{(1)} b_F^{(2)}}{\sum_{E\cap F\neq\emptyset} b_E^{(1)} b_F^{(2)}} \quad \text{for } G \neq \emptyset \end{aligned}$$

In fact the denominator here is > 0 and guarantees

$$\sum_{G \subseteq D} b_G = 1.$$

We shall now write

$$b = b^{(1)} \oplus b^{(2)}$$

and call b the *(Dempster) composite* of $b^{(1)}$ and $b^{(2)}$; (DR) is called *Dempster's Rule* (Dempster [1967][1968]). Since compatibility is equivalent to (1) and since obviously

$$\begin{aligned} \sum_{E\cap F\neq\emptyset} b_E^{(1)} b_F^{(2)} &\leq \sum_{E,F} b_E^{(1)} b_F^{(2)} = \left(\sum_D b_E^{(1)}\right) \cdot \left(\sum_F b_F^{(2)}\right) \\ &= 1 \cdot 1 = 1, \end{aligned}$$

it is a plausible idea to interpret

$$\sum_{E\cap F\neq 0} b_E^{(1)} b_F^{(2)}$$

as sort of *degree of compatibility* between $b^{(1)}$ and $b^{(2)}$. Shafer [1976] elaborates thoroughly on this idea.

Let us look at a few examples in order to see how Dempster's Rule works. It is obvious that $\oplus$ is *commutative.*

Example 3.3. The empty bel function e (example 3.1.) is a *neutral element* for $\oplus$: as $e_D = 1$, e is compatible with any bel function b and

$$(e \oplus b)_G = \frac{b_G}{\sum_F b_F} = b_G \qquad (G \neq \emptyset)$$

Example 3.4. Let $b^{(1)}$, $b^{(2)}$ be simple, but not empty, say with $0 < \alpha_1, \alpha_2 \leq 1$,

$$b_{\Omega_i}^{(i)} = \alpha_i, \; b_\Omega^{(i)} = 1 - \alpha_i \quad (i = 1, 2)$$

If $\Omega_1 \cap \Omega_2 = \emptyset$, $\alpha_1 = \alpha_2 = 1$, then $b^{(1)}$ and $b^{(2)}$ are incompatible. If $\alpha_1 < 1$, $\alpha_2 = 1$, then

$$\sum_{E\cap F\neq 0} b_E^{(1)} b_F^{(2)} = \sum_{E\cap\Omega_2\neq\emptyset} b_E^{(1)} = \begin{cases} 1-\alpha_1 & \text{if} \quad \Omega_1\cap\Omega_2 = \emptyset \\ 1 & \text{if} \quad \Omega_1\cap\Omega_2 \neq \emptyset \end{cases}$$

Thus we obtain

$$(b^{(1)} \oplus b^{(2)})_{\Omega_2} = \frac{1}{1-\alpha_1} b_\Omega^{(1)} = \frac{1-\alpha_1}{1-\alpha_1} = 1$$

i.e. $b^{(1)} \oplus b^{(2)} = b^{(2)}$ if $\Omega_1 \cap \Omega_2 = \emptyset$, and $b^{(1)} \oplus b^{(2)} = b^{(1)}$ if $\Omega_1 \cap \Omega_2 \neq \emptyset$. – If $\alpha_1 < 1$, $\alpha_2 < 1$, we have

$$\sum_{E\cap F\neq\emptyset} b_E^{(1)} b_F^{(2)} = \begin{cases} 1-\alpha_1\alpha_2 & \text{if} \quad \Omega_1\cap\Omega_2 = \emptyset \\ 1 & \text{if} \quad \Omega_1\cap\Omega_2 \neq \emptyset \end{cases}$$

and consequently, for $\Omega_1 \cap \Omega_2 = \emptyset$

$$\begin{aligned} (b^{(1)} \oplus b^{(2)})_{\Omega_1} &= \alpha_1(1-\alpha)_2 \\ (b^{(1)} \oplus b^{(2)})_{\Omega_2} &= (1-\alpha_1)\alpha_2 \\ (b^{(1)} \oplus b^{(2)})_{\Omega} &= (1-\alpha_1)(1-\alpha_2) \end{aligned}$$

and $(b^{(1)} \oplus b^{(2)})_E = 0$ else. In the case $\Omega_1 \cap \Omega_2 \neq \emptyset$ we get

$$(b^{(1)} \oplus b^{(2)})_G = \begin{cases} \alpha_1\alpha_2 & \text{if} \quad G = \Omega_1 \cap \Omega_2 \\ \alpha_1(1-\alpha_2) & \text{if} \quad G = \Omega_2 \\ (1-\alpha_1)\alpha_2 & \text{if} \quad G = \Omega_1 \\ (1-\alpha_1)(1-\alpha_2) & \text{if} \quad G = \Omega \\ 0 & \text{else.} \end{cases}$$

Exercise. 3.5. Prove: if $b^{(1)}$ and $b^{(2)}$ are both concentrated on one-element subsets of Ω (and hence are identifiable with DPDs in Ω), so is $b^{(1)} \oplus b^{(2)}$ provided $b^{(1)}$ and $b^{(1)}$ are compatible.

The obvious question whether $\oplus$ is associative has no simple answer, largely due to compatibility problems; see Shafer [1976] pp. 62 ff.

Not every bel function can be represented as a $b^{(1)} \oplus \ldots \oplus b^{(n)}$ with simple belief functions $b^{(1)}, \ldots, b^{(n)}$, as the reader may verify solving

Exercise 3.6. Let $|\Omega| = 3$, say $\Omega = \{a, b, c, \}$ and

$$b_{\{a,b\}} = b_{\{a,c\}} = b_\Omega = \frac{1}{3}$$

Show that b cannot be represented as a Dempster composite of a finite number of simple bel functions.

We conclude this brief introduction to bel functions with a few examples from the book Shafer [1976], to which the reader is referred for further studies.

Example 3.7. ("Guilty-Innocent"). At a trial the question is whether the defendant is guilty (g) or innocent (i). Let us thus work with the very simple $\Omega = \{g, i\}$ and see what happens if both the Judge and the State Attorney have a bel function on $\mathcal{P}(\Omega)$ each, and combine them in order to come to a conclusion. Classical juridical opinion would suggest that the Judge's bel function a be given by

$$\begin{aligned} a_{\{i\}} &= 1 \\ a_E &= 0 \quad \textit{else} \end{aligned}$$

But this bel function yields $a \oplus b = a$ with every bel function b; in fact $E \cap F \neq \emptyset$, $a_E b_F > 0$ iff $E = \{i\} \subseteq F$, and in this case we get $G = E \cap F = \{i\}$, $\sum_{G=E\cap F} a_E b_F = \sum_{i \in F} b_F = \sum_{E\cap F\neq\emptyset} a_E b_F$, i.e. $(a \oplus b)_G = 1$ iff $G = \{i\}$, and $= 0$ else: $a \oplus b = a$.

One could also prescribe that the Judge show his impartiality by choosing $a = e =$ the empty bel function. As we have seen in example 3.1., this would lead to $a \oplus b = e \oplus b = b$ for every bel function b; thus the State Attorney would carry the day. If, thirdly, the Judge's bel function a is given by

$$a_{\{i\}} = 0.9,\ a_\Omega = 0.1$$

and the State Attorney's bel function b by

$$b_{\{i\}} = 0.1,\ b_{\{g\}} = 0.9,$$

we get

$$\begin{aligned}\sum_{E\cap F\neq\emptyset} a_E b_F &= (0.9)\cdot(0.1)+0.1 = 0.19\\ (a\oplus b)_{\{i\}} &= \frac{0.1}{0.19} = \frac{10}{19} > \frac{1}{2}\\ (a\oplus b)_{\{g\}} &= \frac{(0.1)\cdot(0.9)}{0.19} = \frac{0.09}{0.19} = \frac{9}{19} < \frac{1}{2}.\end{aligned}$$

Example 3.8. ("The Burglary at the Sweet Shop"). Sherlock Holmes is investigating the burglary of a sweetshop. By examining the opened safe, he concludes with degree α of certainty nearly 1 that the thief was left-handed. By a different evidence, Mr. Holmes is able to conclude, with a degree β of certainty very close to 1 that the thief was an insider. He combines now two bel functions on

$$\Omega = \{LI, LO, RI, RO\}$$

(L = left-handed, R = right-handed, I = insider, O = outsider, of course), namely

the bel function a associated with the first evidence:

$$\begin{aligned}a_{\{LI,LO\}} &= \alpha,\ a_\Omega = 1-\alpha\\ a_E &= 0 \quad \text{else}\end{aligned}$$

the bel function b associated with the second evidence:

$$\begin{aligned}b_{\{LI,RI\}} &= \beta,\ b_\Omega = 1-\beta\\ b_E &= 0 \quad \text{else}\end{aligned}$$

We obtain $a \oplus b$ in the following way:

$$\begin{aligned}\sum_{E\cap F\neq 0} a_E b_F &= 1\\ (a\oplus b)_{\{LI\}} &= \alpha\beta \text{ etc.}\end{aligned}$$

This value $(a\oplus b)_{\{LI\}} = \alpha\beta$ is again very close to 1 while $a_{\{LI\}} = 0$, $b_{\{LI\}} = 0$. Now the clerk of the sweetshop is an insider, of course, and, as it turns out, is left-handed. Conclusion: with a very high degree $\alpha\beta$ of certainty, he was the thief.

APPENDIX A: The Marriage Theorem

This very important combinatorial theorem was discovered and proved independently by Ph. Hall [1935] and W. Maak [1935]. It soon turned out that it is equivalent to a graph theoretical theorem of D. König [1916]. Weyl [1949] contributed the interpretation in terms of "marriage". The simple induction proof given below is from Halmos-Vaughan [1950].

Theorem (Marriage Theorem). Let W, M be two nonempty finite sets. For every $w \in W$ let $F(w)$ be a subset of M (interpretation: the elements of W are "women", the elements of M "men", $F(w)$ is the set of all "friends" of w ($w \in W$)). Then the following two statements are equivalent:

MM: there exists a monogamic marriage f of all women such that every woman marries one of her friends; that is

a) $f : W \to M$ is one-to-one

b) $f(w) \in F(w) \quad (w \in W)$

PC ("party condition"): on every party given by some women for their friends, there are no less males than females; that is

$$P \subseteq W \Longrightarrow |P| \leq |\bigcup_{w \in P} F(w)|$$

(the power of a set S is denoted by $|S|$ as usual).

PROOF. $MM \Longrightarrow PC$ is obvious: on every party there are at least the husbands of the inviting ladies, and no husband is married to two different ladies. – Proof of $PC \Longrightarrow MM$ by induction over $|W|$:

1) $|W| = 1$. – PC implies only $P(w)$ to be nonempty: w may marry anyone of her friends.

2) Assume $|W| > 1$ and $PC \Longrightarrow MM$ true in all situations with less than $|W|$ women.

Case I: There is a party $P_0 \subseteq W$, $\emptyset \neq P_0 \neq W$ with $|\bigcup_{w \in P_0} F(w)| = |P_0|$.– By induction hypothesis we may marry the $w \in P_0$ to the $m \in \bigcup_{w \in P_0} F(w) = M_0$ in a monogamous fashion: $f_0 W_0 = M_0$. Send these couples on honeymoon trips. The remaining women and men now constitute the new situation $W_1 = W \backslash W_0$, $M_1 = M \backslash M_0$, $F_1(w) = F(w) \backslash M_0$ ($w \in W_1$) with $|W_1| < |W|$ (as $W_0 \neq \emptyset$). But PC is fulfilled here again: a $P_1 \subseteq W_1$ with $|\bigcup_{w \in P_1} F_1(w)| < |P_1|$ would lead to the contradiction $|\bigcup_{w \in P_0 \cup P_1} F(w)|$- $= |M_0| + |\bigcup_{w \in P_1} F_1(w)| < |P_0| + |P_1| = |P_0 \cup P_1|$. – Thus we may, by induction hypothesis, marry $f_1 : W_1 \to M_1$. f_0 and f_1 combine into a marriage $f : W \to M$.

Case II: For every party $\emptyset \neq P \subseteq W$ we have $|P| < |\bigcup_{w \in P} F(w)|$. – Choose any $w_o \in W$, marry it to any $m_0 \in F(w_0)$ and define $W_1 = W \backslash \{w_0\}$, $M_1 = M \backslash \{m_0\}$, $F_1(w) = F(w) \backslash \{m_0\}$ $(w \in W_1)$. This new situation, with $|W_0| < |W|$, now fulfils PC again: $\emptyset \neq P \subseteq W_1 \Longrightarrow |\bigcup_{w \in P} F_1(w)| \geq |\bigcup_{w \in P} F(w)| - 1 \geq |P|$. Thus the induction hypothesis allows us to marry also the $w \in W_1$. □

Exercise. Show that in case $|F(w)| \geq r > 0$. $(w \in W)$ there are at least $r!$ different marriages in case $r \leq |W|$, and at least $r!/(r - |W|)!$ ones in case $r > |W|$.

APPENDIX B: Markovian Semigroups

Let D be a finite nonempty set. In Ch.II §4 we proved that for every stochastic $D \times D$-matrix P the sequence $I, P, P^2, P^3, \dots$ is (exponentially) asymptotic periodic. The method employed there may be characterized as a "method of invariant subsets", namely, of D.

In this appendix, we will obtain the same, and some much more general results, by a method which may be called "the semigroup method". It applies generally to compact abelian semigroups – here of stochastic $D \times D$-matrices. It is a descendant of the Jacobs-de Leeuw-Glicksberg method in operator ergodic theory (see e.g. Krengel [1985]) which, in my opinion, deserves the attention of probabilists, even on the elementary level of this book.

Let again $V(\subseteq \mathbf{R}^D)$ denote the set of all probability vector over D, and $W(\subseteq \mathbf{R}^{D \times D})$ the set of all stochastic $D \times D$-matrices. We recall from ch.II §4 that W is convex, compact, and a semigroup (i.e. stable under matrix multiplication). We will focus attention to abelian subsemigroups of W such as the set $\{I, P, P^2, \dots, \}$ of all iterates of a single stochastic matrix P ("cyclic case") or sets $\{P^{(t)} | t \geq 0\}$ where $(P^{(t)})_{t \geq 0}$ is a one-parameter subsemigroup of W (appendix C). A first obvious remark:

> the closure (in $\mathbf{R}^{D \times D}$) of an abelian subsemigroup of W is a *compact* abelian subsemigroup of W.

Actually, *compact* abelian subsemigroups of W will be the main object of our investigations here. Let G be such a semigroup.

Definition B1. A compact nonempty subset H of G is called an *ideal* in G if

$HQ \subseteq H$ for all $Q \in G$, i.e. iff
$P \in H, Q \in G \Longrightarrow PQ \in H$

Let $Id(G)$ denote the set of all ideals in G.

Clearly, G is in $Id(G)$.

Proposition B2. The intersection of all ideals in G is an ideal in G: the *minimal ideal* $\underline{G}$ of G.

PROOF. Obviously $H, K \in Id(G) \Rightarrow HK = \{PQ|P \in H, Q \in K\} \in Id(G)$, $HK \subseteq H \cap K$. Thus the intersection of any two ideals in G contains an ideal in G. The proposition now follows essentially by Cantor's intersection theorem. □

Proposition B3. The minimal ideal $\underline{G}$ of G contains exactly one idempotent, i.e. a stochastic matrix P fulfilling $\underline{P}\underline{P} = \underline{P}$; $\underline{P}$ is the unique neutral element of $\underline{G}$, i.e. it fulfils $\underline{P}\underline{Q} = \underline{Q}$ for every $\underline{Q} \in \underline{G}$, and it is the only element of $\underline{G}$ with this property. Actually, $\underline{G}$ is a compact subgroup of G.

PROOF. For any $R \in G$ we have $\underline{G}R = \underline{G}$, as $\underline{G}R$ clearly is an ideal in G again, and $\underline{G}$ is minimal. Thus for every $Q \in \underline{G}$ we may find some $P \in \underline{G}$ such that $Q^2P = Q$ (put $R = Q^2$). If we now put $\underline{P} = Q^2P^2$, we obtain $\underline{P}\underline{P} = Q^2P^2Q^2P^2 = QPQP = Q^2P^2 = \underline{P}$.
However we choose an idempotent $\underline{P}$ in $\underline{G}$, the following holds:
as $\underline{G}\underline{P} = \underline{G}$, every $\underline{Q} \in \underline{G}$ may be written $\underline{R}\underline{P}$ with some $\underline{R} \in \underline{G}$; now $\underline{Q}\underline{P} = \underline{R}\underline{P}\underline{P} = \underline{R}\underline{P} = \underline{Q}$ follows, i.e. $\underline{P}$ is a neutral element of $\underline{G}$. If $\underline{P}'$ is another neutral element of $\underline{G}$, we obtain $\underline{P} = \underline{P}'\underline{P} = \underline{P}'$. — $\underline{G}$ is a group with neutral element $\underline{P}$, as $\underline{G}\underline{Q} = \underline{G}$ ($\underline{Q} \in \underline{G}$) is tantamount to division within G. □

Theorem B4. (Splitting Theorem). Let G be a compact abelian subsemigroup of W and $\underline{P}$ the only idempotent (hence neutral element) of its minimal ideal $\underline{G}$. Then the linear subspaces $\mathbf{R} = \mathbf{R}^D\underline{P}, \mathbf{F} = \mathbf{R}^D(I - \underline{P})$ of $\mathbf{R}^D$ form a direct decomposition of $\mathbf{R}^D$, i.e. every vector $x \in \mathbf{R}^D$ has a unique decomposition

(2) $\quad x = \underline{x} + f$ with $\underline{x} \in \mathbf{R}$, $f \in \mathbf{F}$.

Both $\mathbf{R}$ and $\mathbf{F}$ are G-invariant, i.e. every $P \in G$ yields $\mathbf{R}P \subseteq \mathbf{R}, \mathbf{F}P \subseteq \mathbf{F}$. They can be characterized as follows:

(3) $\quad \mathbf{R} = \{x | x \in \mathbf{R}^D$ and for every $P \in G$.
there is a $Q \in G$ such that $xPQ = x$ ("reversibility") $\}$

(4) $\mathbf{F} = \{x | x \in \mathbf{R}^D$ and there is at least one $P \in G$ such that $xP = 0\}$

After restriction to $\mathbf{R}, G$ is a *group* of nonsingular linear transformations.

PROOF. For any $x \in \mathbf{R}^D$ put $\underline{x} = x\underline{P}$, $f = x(I - \underline{P}) = x - \underline{x}$ in order to obtain a decomposition (1). If $x = \underline{x}' + f'$ is another decomposition with $\underline{x}' \in \mathbf{R}$, $f' \in \mathbf{F}$, $\mathbf{R} \ni \underline{x} - \underline{x}' = f' - f \in \mathbf{F}$ follows. But as $(I - \underline{P})\underline{P} = \underline{P} - \underline{P}\underline{P} = \underline{P} - \underline{P} = 0$, every vector from $\mathbf{R}$, while remaining fixed under $\underline{P}$, goes into $0 \in \mathbf{R}^D$ if it belongs also to $\mathbf{F}$, that is $\mathbf{R} \cap \mathbf{F} = \{0\}$, and $\underline{x} = \underline{x}'$, $f = f'$ follows.—
If $P \in G$, then $P\underline{G} = \underline{G} :\subseteq$ follows because $\underline{G}$ is an ideal, and $=$ then follows because $P\underline{G}$ is an ideal, and $\underline{G}$ is minimal. — G-invariance of $\mathbf{R}$: if $\underline{x} \in \mathbf{R}$, we get, for any $P \in G$, $\underline{x}P = \underline{x}\underline{P}P = (xP)\underline{P} \in \mathbf{R}$. — G-invariance of $\mathbf{F}$: if $f \in \mathbf{F}$, we may write $f = x - x\underline{P}$ for some $x \in \mathbf{R}^D$ and get for any $P \in G$, $fP = (xP) - (xP)\underline{P} \in \mathbf{F}$. — The group property of $\underline{G}$ entails the group property of G within $\mathbf{R}$ (exercise). – Characterization of $\mathbf{R}$: if $\underline{x} \in \mathbf{R}$, $P \in G$, we derive from $\underline{G}P = \underline{G}$ the existence of a $\underline{Q} \in \underline{G}$ that $P\underline{Q} = \underline{P}$, that is $\underline{x}P\underline{Q} = \underline{x}\underline{P} = \underline{x}$. This proves $\subseteq$ in (2). In order to establish $\supseteq$, take any $x \in \mathbf{R}^D$ with the "reversibility" property described in (2), split it into $x = \underline{x} + f$, $\underline{x} \in \mathbf{R}$, $f \in \mathbf{F}$, and find Q such that $x\underline{P}Q = x$. As $f\underline{P} = 0$, we conclude $x = x\underline{P}Q = \underline{x}\underline{P}Q + f\underline{P}Q = \underline{x}\underline{P}Q \in \mathbf{R}$.– Characterization of $\mathbf{F}$: if $f \in \mathbf{F}$, then $f\underline{P} = 0$ follows, proving $\subseteq$ in (3). In order to establish $\supseteq$, let $x \in \mathbf{R}^D$, $P \in G$ be such that $xP = 0$. Split $x = \underline{x} + f$, $\underline{x} \in \mathbf{R}$, $f \in \mathbf{F}$ and find $Q \in G$ such that $\underline{x}PQ = \underline{x}$. We conclude $0 = xP = xPQ = \underline{x}PQ + fPQ = \underline{x} + fPQ \Longrightarrow \underline{x} \in \mathbf{R} \cap \mathbf{F} \Longrightarrow \underline{x} = 0 \Longrightarrow x = f \in \mathbf{F}$. –
We now focus attention to the "reversible" subspace $\mathbf{R}$ of $\mathbf{R}^D$ and to the action of G within $\mathbf{R}$. In fact, G and $\underline{G}$ coincide within $\mathbf{R}$, as $G\underline{P} = \underline{G}$. Thus G acts, within $\mathbf{R}$, as a group. – The theorem is proved. □

Theorem B5. $\mathbf{R}$ is a vector sublattice of $\mathbf{R}^D$.

PROOF. As all vector lattice operations can be obtained from the operation $x \to x_+$ via linear operations, it suffices to prove

$$r \in R \Longrightarrow r_+ \in R.$$

Now if $r \in \mathbf{R}$ and $P \in G$, we may find $Q \in G$ such that $rPQ = r$, by (2). By $r = r_+ - r_-$, this leads to

$$r = rPQ = r_+PQ - r_-PQ.$$

This is a representation of r as a difference of two nonnegative vectors (here the nonnegativitiy of all $P, Q \in G$ finally comes into action). From $r_+PQ =$

$r + r_- PQ \geq r$, $r_+ PQ \geq 0$ we conclude $r_+ PQ \geq r \vee 0 = r_+$. But the total mass $\langle r_+ \rangle$ of $r_+ \geq 0$ is preserved under the action of PQ, hence $r_+ PQ = r_+$ follows: $r_+ \in \mathbf{R}$. □

Next we concentrate on the compact convex set $\underline{V} = V \cap \mathbf{R}$ and its extremal points. Recall the definition of the *support* or *carrier* of a vector $x \in \mathbf{R}^D$:

$$supp(x) = \{j | j \in D,\ x_j \neq 0\}.$$

Theorem B6. Let r, r' be two different extremal points of $\underline{V}$. Then

$$supp(r) \cap supp(r') = \emptyset.$$

PROOF. $\neq \emptyset$ would be tantamount to $r \wedge r' \neq 0$. As $\mathbf{R}$ is a vector sublattice of $\mathbf{R}^D$, $\underline{r} = r \wedge r'$ belongs to $\mathbf{R}$ again. As $r \neq r'$, but both of them, being in V, have the same total mass 1, $r \neq \underline{r} \neq r'$ follows.

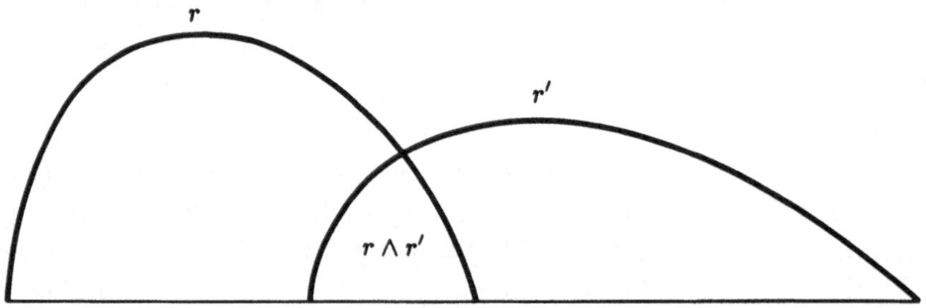

fig.B.1

It even follows that $\underline{r}$ is neither a multiple of r nor one of r'.

This leads now to a contradiction to the extremality either of r or of r' in $\underline{V}$: the two chunks $r \wedge r'$ and $r - (r \wedge r')$, into which r is splitted, are ≥ 0, and $\neq 0$, and yield, after renormalization to total masses 1, a convex-linear representation of r which contradicts the extremality of r in $\underline{V}$, unless $r \wedge r'$ is a nonnegative multiple of r, which can, however, not be the case, as we have shown. □

From the finiteness of D we now infer the

Theorem B7. $\underline{V}$ has only finitely many extremal points $r^{(1)}, \ldots, r^{(m)}$. They have pairwise disjoint supports, and they form a linear basis of $\mathbf{R}$.

PROOF. Only the last statement has still to be proved. As $\mathbf{R}$ is a vector lattice, it suffices to represent an arbitrary $r \geq 0$ from $\mathbf{R}$ as a linear combination of $r^{(1)}, \ldots, r^{(m)}$. Essentially the same argument as the one used in the proof of theorem B6 now shows that

$$r \wedge r^{(i)} \text{ is a nonnegative multiple of } r^{(i)}$$

In fact, if not, $r \wedge r^{(i)}$ and $r^{(i)} - (r \wedge r^{(i)})$ would constitute two "chunks" $\neq 0$ of $r^{(i)}$ which would contradict the extremality of $r^{(i)}$ in $\underline{V}$. □

Theorem B8. Every $P \in G$ permutes the extremal points of $\underline{V}$.

PROOF. As $\mathbf{R}P = \mathbf{R}$ and $VP \subseteq V$, $\underline{V}P \subseteq \underline{V}$ follows. If some $r^{(i)}P$ would not be an extremal point of $\underline{V}$, we would take a representation $r^{(i)}P = \alpha u + (1-\alpha)v$, $0 < \alpha < 1$, $u, v \in \underline{V}$, $u \neq v$ forbidden for extremal points, and then choose $Q \in G$ such that $r^{(i)}PQ = r^{(i)} = \alpha(uQ) + (1-\alpha)(vQ)$. As Q acts non-singularly in $\mathbf{R}$ and leaves $\underline{V}$ invariant, this turns out as a representation of $r^{(i)}$ incompatible with the extremality of $r^{(i)}$ in $\underline{V}$. □

We now turn to special types of semigroups.

I) The cyclic case.

Let $P \in W$ and G the closure of $\{I, P, P^2, \ldots,\}$. Clearly G is a compact abelian subsemigroup of W. Apply the previous general theory and consider the corresponding splitting

$$\mathbf{R}^D = \mathbf{R} + \mathbf{F}$$

and the set

$$\underline{V} = V \cap \mathbf{R}$$

whose extremal points $r^{(1)}, \ldots, r^{(m)}$ form a linear basis of $\mathbf{R}$ and are permuted by P. Splitting this permutation into cycles, we see that $I, P, P^2, \ldots$ is a purely periodic sequence when restricted to the subspace $\mathbf{R}$ of $\mathbf{R}^D$. The action within $\mathbf{F}$ of this matrix sequence is easily analyzed as follows:

for every $f \in \mathbf{F}$ there is a $Q \in G$ such that $fQ = 0$. As Q is a limit of a sequence of powers of P, there is a $n(f)$ such that

$$\|fP^{n(f)}\| \leq \frac{1}{2}\|f\|.$$

For any $n \geq n(f)$ we have the more

$$\|fP^n\| \leq \frac{1}{2}\|f\|$$

as P contracts the norm. An obvious approximation argument working in the (norm) unit ball of F leads to the existence of some n_0 such that

$$\|fP^{n_0}\| \leq \frac{1}{2}\|f\| \quad (f \in \mathbf{F})$$

(exercise). Another obvious argument now shows that the sequence $\|fP^n\|$ ($n = 0, 1, \ldots$) tends to 0 exponentially fast, and uniformly for all f from a norm-bounded set (exercise).

Putting these results together, we obtain the (exponentially-) asymptotic periodicity of the sequence $I, P, P^2, \ldots$, that is, the same general result as in ch. II §4. The cyclicity results for subsets of D obtained there can also easily be deduced from the results obtained here by the semigroup method. I leave the details as an exercise to the reader.

II) The one-parameter case.

Let $(P^{(t)})_{t\geq 0}$ be a one-parameter sub-semigroup of W (see appendix C), and let G be the closure of $\{P^{(t)} | t \geq 0\}$. Clearly G is a compact abelian subsemigroup of W. Our general theory applies, yielding $\mathbf{R}^D = \mathbf{R} + \mathbf{F}$, $\underline{V} = V \cap R$ and the result that the finitely many extremal points of $\underline{V}$ form a linear basis of $\mathbf{R}$ and are permuted by every $P^{(t)}$, $t \geq 0$. Let $\tau^{(t)}$ denote that permutation. Obviously $\tau^{(s+t)} = \tau^{(s)} \circ \tau^{(t)} \quad (s, t \geq 0)$. Now $\tau^{(t)} = (\tau^{\frac{t}{m!}})^{m!}$ shows that all these permutations are $m!$-th powers within the symmetry group of m objects, and thus leave everything fixed. Conclusion: all $P^{(t)}$ act as identity mappings within $\mathbf{R}$. Consequence: no periodicity, nay, convergence. That is, there is some $\underline{P} \in W$ such that

$$\lim_{t\to\infty} P^{(t)} = \underline{P}.$$

III. The eigenvalues of stochastic matrices.

We conclude this appendix with some results concerning the eigenvalues of a stochastic matrix P. To this end, we embed $\mathbf{R}^D$ into the complex space $\mathbf{C}^D$, but as P is a real matrix, it leaves both the real subspace $\mathbf{R}^D$ of $\mathbf{C}^D$ and the purely imaginary subspace $i\mathbf{R}^D$ invariant. From the behavior of $I, P, P^2, \ldots$ within $\mathbf{R}^D$ we conclude: for any $z \in \mathbf{C}^D$ the sequence $z, zP, zP^2, \ldots$ remains (componentwise) *bounded*. Conclusion:

every (complex) eigenvalue λ of a
stochastic matrix P fulfils
$|\lambda| \leq 1$.

There is a least one eigenvalue 1 of modulus 1, and if we have a sequence

$$x, xP, xP^2, \ldots, \quad \in \mathbf{R}^D$$

of period p, then for

$$\lambda = e^{i(\frac{2\pi}{p})}$$

the vector

$$z = x + \lambda xP + \ldots + \lambda^{\varrho-1} xP$$

clearly is an eigenvector ($\neq 0!$) for the eigenvalue λ, and this λ is a *root of unity.*

Theorem B9. Let P be a stochastic $D \times D$-matrix and λ an eigenvalue of P. Then $|\lambda| \leq 1$; if $|\lambda| = 1$, then λ is a root of unity.

PROOF. Apply our previous results for the cyclic case and split

$$\mathbf{C}^D = \mathbf{R} + i\mathbf{R} + \mathbf{F} + i\mathbf{F}$$

accordingly. The vectors from $\mathbf{F} + i\mathbf{F}$ go to 0 exponentially fast, and those from $\mathbf{R} + i\mathbf{R}$ behave strictly periodically, under iterated application of P. Every eigenvector for an eigenvalue λ with $|\lambda| = 1$ must therefore be in $\mathbf{R} + i\mathbf{R}$, and λ can only be a root of unity. □

Results of this type can, of course, be obtained also by entirely different methods, see e.g. Fritz-Huppert-Willems [1979].

APPENDIX C: One-parameter semigroups of stochastic matrices

Although continuous-time Markov theory is not an objective of this book, I think the reader should have the opportunity of acquainting himself with one of the basic facts about the continuous-time analogon of the set of all iterates of a stochastic matrix: one-parameter semigroups of stochastic matrices.

Let D be a finite (state) set. A one-parameter semigroup of stochastic $D \times D$-matrices is a continuous curve

$$(P^{(t)})_{\mathbf{R} \ni t \geq 0}$$

in W (i.e. with $P^{(t)} \in W$) such that

(5) $\quad P^{(0)} = I, \;\; P^{(s+t)} = P^{(s)} P^{(t)} \qquad (s, t \geq 0)$

Clearly such semigroups are abelian. We prove the basic characterization theorem about them here: A $D \times D$-Matrix $G = (G_{jk})_{j,k \in D}$ is called a *generator* if

(6) $\quad G_{jk} \geq 0 \qquad (j \neq k)$

(7) $\sum_k G_{jk} = 0 \qquad (j \in D)$

Theorem C.1. A continuous one-parameter sub-family $(P^{(t)})_{\mathbf{R} \ni t \geq 0}$ of W is a one-parameter semigroup iff there is a generator G such that

(8) $$P^{(t)} = e^{tG} \left(= \sum_{n=0}^{\infty} \frac{t^n}{n!} G^n \right).$$

PROOF. We define e^{tG} by the exponential series as indicated; convergence may be handled conveniently via the matrix norm

$$\|M\| = \sup_{\|x\| \leq 1} \|xM\|$$

which clearly fulfils

$$\|MN\| \leq \|M\| \cdot \|N\| \quad (M, N \in \mathbf{R}^{D \times D}).$$

It is obvious that the usual convergence proofs apply here mutatis mutandis, and that

$$e^{M+N} = e^M e^N \quad (M, N \in \mathbf{R}^{D \times D})$$

holds (addition componentwise, of course.) Clearly $M \to e^M$ is a continuous mapping of matrices. It is also obvious that this continuity is uniform in $\{M \| \|M\| \leq K\}$, however we choose the bound $K > 0$. In particular, e^{tG} depends uniformly continuously upon G and t if we keep these variables bounded.

Let now G be any generator. By arbitrarily small modification we may pass to generators H fulfilling (2) even with $>$. Now the matrix

$$\bar{P}^{(t)} = e^{tH} = I + tH + t^2 \left(\frac{H^2}{2!} + t\frac{H^3}{3!} + \ldots \right)$$

has all entries > 0 if $t > 0$ is small enough: on the diagonal because I contributes 1 there, and off the diagonal because tH contributes something > 0 there and the remainder term can not outweigh these contributions for small $t > 0$. It follows that $\bar{P}^{(t)}$ has all entries > 0 for small t, and by

$$\bar{P}^{(t)} = \left(\bar{P}^{(\frac{t}{n})} \right)^n \quad (n = 1, 2, \ldots)$$

this follows also for $t > 0$ arbitarily large. By $H \to G$ and continuity we arrive at $P^{(t)}_{jk} \geq 0$ $(j, k \in D)$ for $P^{(t)} = e^{tG}$.

Moreover, all row sums of $P^{(t)}$ are 1, because we have a contribution of 1 to it from I, and (by (3)) contributions 0 from the remaining terms of the defining series. Thus $P^{(t)} \in W$ for all $t \geq 0$, and we have a one-parameter semigroup $\left(P^{(t)}\right)_{t \geq 0}$.

Conversely, for a given one-parameter semigroup $(P^{(t)})_{t\geq 0}$ in W, we want to establish a representation (4), with a suitable generator G. Now, if such a representation exists, G can be obtained from it as

$$(9) \quad G = \lim_{0<t\to 0} \frac{1}{t}(P^{(t)} - I)$$

In fact, (4) implies $\frac{1}{t}(P^{(t)} - I) = G + t\frac{G^2}{2!} + \ldots$. Let us thus try to *define* G by (5). Firstly, we have to establish the convergence of $\frac{1}{t}(P^{(t)} - I)$ for $0 < t \to 0$.

We calculate, for some $b > 0$ and some small $t > 0$, and n such that $nt \leq b < (n+1)t$,

$$\begin{aligned} P^{(b)} - I &= \left(P^{(t)} - P^{(b-t)}\right) + \left(P^{(b-t)} - P^{(b-2t)}\right) + \ldots \\ &\quad + \left(P^{(b-(n-1)t)} - P^{(b-nt)}\right) + \left(P^{(b-nt)} - I\right) \\ &= \tfrac{1}{t}\left(P^{(t)} - I\right) t \left[P^{(b-t)} + P^{(b-2t)} + \ldots + P^{(b-nt)}\right] \\ &\quad + \left(P^{(b-nt)} - I\right) \end{aligned}$$

As $s \to P^{(s)}$ is continuous, we see that $o < t \to 0$ entails $n \to \infty$ and

$$P^{(b-nt)} - I \to 0$$

$$t \sum_{k=1}^{n} P^{(n-kt)} \to \int_0^b P^{(s)} ds$$

(the integral over a matrix-valued continuous function is to be understood componentwise, of course). Now if b is small enough, the $P^{(s)}$ contributing to the integral are all close to I, ensuring a non-singular $A = \int_0^b P^{(s)} ds$ and thus the existence of

$$\lim_{0<t\to 0} \frac{1}{t}(P^{(t)} - I) = (P^{(b)} - I)A^{-1}.$$

We are now indeed entitled to *define*

$$G = \lim_{0<t\to 0} \frac{1}{t}(P^{(t)} - I).$$

As both $P^{(t)}$ and I have all row sums $= 1$, G has all row sums 0, and since all entries off the diagonal of $P^{(t)} - I$ are ≥ 0, so are all those entries of G. Thus G is in fact a generator.

For any $t, h > 0$,

$$\frac{1}{h}\left(P^{(t+h)} - P^{(t)}\right) = \frac{1}{h}\left(P^{(h)} - I\right) P^{(t)}$$
$$\frac{1}{-h}\left(P^{(t-h)} - P^{(t)}\right) = \frac{1}{h}\left(P^{(h)} - I\right) P^{(t-h)}$$

and we see that both expressions tend to $GP^{(t)}$ as $0 < h \to 0$. Thus for $t > 0$, $P^{(t)}$ is differentiable at t, fulfilling the ordinary differential equation (actually a system of such)

$$\frac{d}{dt}P^{(t)} = GP^{(t)}$$

whose only solution is, according to standard theories, Be^{tG}, where B is any constant matrix. If we prescribe $\lim_{0<t\to 0} P^{(t)} = I$, $B = I$ follows, and we finally arrive at

$$P^{(t)} = e^{tG},$$

thus completing the proof of the theorem. □

Bibliography

[1963] Aczél, J. und Z. Daróczy, Charakterisierung der Entropien positiver Ordnung und der Shannonschen Entropie, Acta. Math. Acad. Sci. Hungar. 14 (1963), 95-121

[1949] Andersen, E.S., On the number of positive sums of random variables, Skand. Aktuarietidsskrift 32 (1949), 27-36

[1950] Andersen, E.S., On the frequency of positive partial sums of a series of random variables, Mat. Tidsskrift B, 33-35 (1950)

[1953] Andersen, E.S., On sums of symmetrically dependent random variables, Skand. Aktuarietidsskrift 36 (1953) 123-138

[1953/54] Andersen, E.S., On the fluctuations of sums of random variables I,II, Math. Scand. 1 (1953), 263-283, 2 (1954), 125-223

[1962] Andersen, E.S. The equivalence principle in the theory of fluctuations of random variables, Colleq. Comb. Math. in Prob. Th. Aarhus 1962, 12-16

[1887] André, D., Solution directe du problème résolu par M. Bertrand, C.R. Acad. Sci. Paris 105 (1887), 436-437

[1991] Bauer, H., Wahrscheinlichkeitstheorie, 4. Aufl. Berlin (deGruyter) 1991, engl. transl. New York (HR&W) 1972

[1763] Bayes, Th., An essay towards solving a problem in the doctrine of chances, Phil. Trans. Roy. Soc. London 53 (1763), 376-398, reprinted in Biometrika 45 (1958), 293-315.

[1913] Bernoulli, Jakob, Ars Coniectandi,, Basel (Thurnisií) 1713, also in: Werke Bd. 3, Basel (Birkhäuser) 1975

[1975] Bernoulli, Jakob, Werke Bd. 3, Basel (Birkhäuser) 1975

[1941] Berry, A.C., The accuracy of the Gaussian approximations to the sum of independent variables, TAMS 49 (1941), 122-136

[1985] Billingsley, P., Probability and measure, New York (Wiley) 1985

[1946] Birkhoff, G., Tres observaciones sobre el algebra lineal, Univ. Nac. Tucumán Rev., Ser., A, 5, 147-150

[1909] Borel, E., Sur les probabilités dénombrables et leurs applications arithmétiques, Rend, Circ. Mat. Palermo 26 (1909), 247-271

[1920] Borel, E., Méthodes et problèmes de la théorie des fonctions, Paris (Gauthier-Villars) 1920

[1968] Breiman, L., Probability, Reading/Mass. (Addison-Wesley) 1968

[1828] Brown, Robert, A brief account of microscopic observations made in the months of June, July and August, 1827, on the particles contained in the pollen of plants [Clarkia pulchella], and on the general existence of active molecules in anorganic and in organic bodies, Private print 1828

[1950] Carnap, P., Logical Foundations of Probability, Chicago (UP) 1950

[1924] Chintschin, A.I. Ein Satz der Wahrscheinlichkeitsrechnung, Fund.Math. 6 (1924), 6-20

[1956] Chintschin, A.I. On the fundamental theorems of information theory, Uspehi Mat. Nauk 11 (1956), 17-75, engl. transl., Newtonville/Mass. 1956, dt. Übers. Berlin 1957

[1974] Chung, K.L., Elementary probability theory with stochastic processes, Berlin etc. (Springer) 1974

[1977] Cohen, L.J. The Provable and the Probable, Oxford (Clarendon) 1977

[1981] Csiszár, I. and J. Körner, Information Theory: Coding Theorems for Discrete Memoryless Systems, Budapest (Akad. Kiadó) 1981

[1975] Csörgö, M., and P. Revész, A new method to prove LIL of Strassen (2 parts), ZfW 31, 255-259, 261-269

[1986] Davis, M., The art of decision making, New York etc. (Springer) 1986

[1967] Dempster, A.P. Upper and lower probabilities induced by a multivalued mapping, AMS 38 (1967), 325-339

[1968] Dempster, A.P., A generalization of Bayesian inference, J. Roy. Stat. Soc. (B) 30 (1968), 205-247

[1965] Dinges, H., Zufällige Pfade mit vertauschbaren Schritten, ZfW 3 (1965), 328-374

[1979] Dinges, H., Einsteins Beitrag zur statistischen Theorie der Diffusion, 15pp. preprint Frankfurt 1979.

[1953] Doob, J.L., Stochastic Processes, New York (Wiley) 1953

[1965] Dubins, L., and L.J. Savage, How to gamble if you must, New York (Mc Graw Hill) 1965

[1907] Ehrenfest, P., und T. Ehrenfest, Über zwei bekannte Einwände gegen das Boltzmannsche H-Theorem, Ph. Z. 8 (1907), 311-314

[1906] Einstein, A., Zur Theorie der Brownschen Bewegung, Ann. Phys. IV 19 (1906), 371-381

[1987] Engel, A., Stochastik, Stuttgart (Klett) 1987

[1947] Erdös, P. and M. Kac, On the number of positive sums of independent random variables BAMS 53 (1947), 1011-1020

[1956] Esséen, C.G., A moment inaquality with an application to the central limit theorem. Skand. Aktuarietidsskr. 39 (1956), 160-170

[1958] Esséen, C.G., On mean central limit theorems, Kungl.Tekn. Högsk. Handl. Stockholm 121 (1958), 31 pp.

[1956] Faddeev, D.K., On the nation of entropy of a finite probability scheme, Uspehi. Mat. Nauk. 11 (1956), 227-231, dt. Übers. Berlin 1957

[1958] Feinstein, A, Foundations of information theory, New York 1958

[1957] Feller, W., An introduction to probability theory and its applications, New York (Wiley) 1957

[1983] Fierz, M., Girolamo Cardano (1501-1576), Boston-Basel- Stuttgart (Birkhäuser) 1983

[1986] Fishburn, P.C., The axioms of subjective probability, Stat. Sci. 1 (1986), 335-358

[1978] Fisher, J., R.A. Fisher, The Life of a Scientist New York (Wiley) 1978

[1979] Fritz, F.-J., B. Huppert und W. Willems, Stochastische Matrizen, Berlin etc. (Springer) 1979

[1968] Gallager, R.G. Information theory and reliable communication, New York (Wiley) 1968

[1984] Gericke, H., Mathematik in Antike und Orient, Berlin etc. (Springer) 1984

[1990] Gericke, H., Mathematik im Abendland, Berlin etc. (Springer) 1990

[1662] Graunt, I., Natural and political observations made upon the bills of mortality, London 1662

[1950] Halmos, P.R., and H.E. Vaughan, The marriage problem, Amer. J. Math. 72 (1950), 214-215

[1935] Hall, Ph., On representations of subsets, J. London Math. Soc 10 (1935), 26-30

[1950] Hamming, R.W. Error detecting and error correcting codes, Bell Syst. Tech. J. 22 (1950), 147-160

[1985] Hampel, F.R., P.I. Rousseauw and W.A. Stahel, Robust Statistics, New York (Wiley) 1985

[1971] Hansen, W. und D. Walz, Bemerkungen zur kühnen Strategie, ZfW 20 (1971), 325-331

[1984] Hartung, H. (mit B. Elpelt und K.-H. Klösener), Statistik Lehr- und Handbuch der angewandten Statistik, 2. Aufl. München (Oldenburg) 1984

[1977] Heyde, C.C. and E. Seneta, I.J. Bienaymé, Statistical theory anticipated, Berlin etc. (Springer) 1977

[1981] Hlawka, E., F. Firneis und P. Zinterhof, Zahlentheoretische Methoden in der Numerischen Mathematik, Wien-München (Oldenbourg) 1981

[1957] Jacobs, K., Zur Theorie der Markoffschen Prozesse, Math. Ann. 133 (1957), 375-390

[1957a] Jacobs, K., Fastperiodische diskrete Markoffsche Prozesse von endlicher Dimension, Abh. Math. Sem. Hamburg 21 (1957), 194-246

[1958] Jacobs, K., Konjunkturschwankungen Markoffscher n-Personen-Prozesse mit monomialer Regelung, Math.Z. 69 (1958), 247-270

[1969] Jacobs, K., Das kombinatorische Äquivalenzprinzip und das arcsin-Gesetz von E. Sparre Andersen, in: Selecta Mathematica I, Berlin etc. (Springer) 1969, 53-81

[1969a] Jacobs, K., Rot und Schwarz, in: Selecta Mathematica I, Berlin etc. (Springer) 1969, 28-52

[1970] Jacobs, K., Turing-Maschinen und zufällige 0-1-Folgen, in: Selecta Mathematica II, Berlin etc. (Springer) 1970, 141-167

[1972] Jacobs, K., Markov-Prozesse mit endlichvielen Zuständen, in: Selecta Mathematica IV, Berlin etc. (Springer) 1972, 24-142

[1944] Kakutani, S., Two-dimensional Brownian motion and harmonic functions, Proc. Imp. Acad. Tokyo 20 (1944), 706-714

[1972] Kannappan, P., On Shannon's Entropy, directed divergence and inaccuracy, ZfW 22 (1972), 95-100

[1972a] Kannappan, P., On directed divergence and inaccuracy, ZfW 25 (1972), 49-55

[1969] Knuth, D., The Art of Computer Programming, vol. 2, Reading/Mass. (Addison-Wesley) 1969

[1916] König, D., Über Graphen und ihre Anwendungen, Math. Ann. 77 (1916), 453-465

[1929] Kolmogorov, A.N., Über das Gesetz des iterierten Logarithmus, Math. Ann. 101 (1929), 126-135

[1933] Kolmogorov, A.N., Grundbegriffe der Wahrscheinlichkeitsrechnung, Berlin (Springer) 1933

[1965] Kolmogorov, A.N., Three approaches to the quantitative definition of information, Probl. Pered. Inf. 1 (1965), 1-7

[1959] Kraft, C.H., J.W. Pratt and A. Seidenberg, Intuitive probability on finite sets, AMS 30 (1959), 408-419

[1949] Kraft, L.G., A device for quantizing, grouping and coding amplitude modulated pulses, M.S. Thesis, Dept. of E.E., M.I.T., Cambridge/Mass. 1949

[1985] Krengel, U., Ergodic Theorems, Berlin (de Gruyter) 1985

[1988] Krengel, U., Einführung in die Wahrscheinlichkeitstheorie und Statistik, Braunschweig-Wiesbaden (Vieweg) 1988

[1990] Krengel, U., Wahrscheinlichkeitstheorie, in: Ein Jahrhundert Mathematik, ed. W. Scharlau, Wiesbaden (Vieweg) 1990, 457-489

[1987] Krengel, U. + M. Lin: Order Preserving Nonexpansive Operators in L_1, Isr. J. Math., Vol. 58, 170-192

[1979] Krickeberg, K., + H. Ziezold, Stochastische Methoden, Berlin etc. (Springer) 1979

[1814] Laplace, Pierre Simon, Théorie analytique des probabilités, 2. éd. Paris (Courcier) 1814

[1959] Lehmann, E.L. Testing Statistical Hypotheses, New York (Wiley) 1959

[1973] Lienert, G.A., Verteilungsfreie Methoden in der Biostatistik, 3 Bde., 2. Aufl., Meisenheim (Hain) 1973

[1922] Lindeberg, J.W., Eine neue Herleitung des Exponentialgesetzes in der Wahrscheinlichkeitsrechnung Math. Z. 15 (1922), 211-225

[1978] Lindley, D.V., The Bayesian approach, Scand. J. Statist. 5 (1978), 1-26

[1990] Lindley, D.V., The present position in Bayesian Statistics, Stat. Sci. 5 (1990), 44-89

[1982] Lint, J. van, Introduction to coding theory, New York etc. (Springer) 1982

[1935] Maak, W., Eine neue Definition der fastperiodischen Funktionen, Abh. Math. Sem. Univ. Hamburg 11 (1935), 240-244

[1978] Major, P., On the invariance principle for sums of IID random variables, J. Mult. Anal. 8 (1978)

[1906] Markov, A.A., Extension of the law of large numbers to dependent events (russian), Bull. Soc. Phys. Math. Kazan (2) 15 (1906), 135-156

[1924] Markov, A.A., Calculus of probability (russian), 4nd ed., Moscow 1924

[1966] Martin-Löf, P., The definition of random sequences. Inf. and Control 19 (1966), 602-619

[1989] Martus, P., Asymptotische Eigenschaften nichtstationärer Operatorfolgen im nicht-linearen Fall, 62 pp., Diss. Erlangen 1989

[1971] Meyer, P.L., Introductory probability and statistical applications, Reading/Mass. (Addison-Wesley) 1971

[1911] Minkowski, H., Ges. Abhandlungen, Bd. II, Leipzig (Teubner) 1911

[1919] Mises, R. von, Grundlagen der Wahrscheinlichkeitstheorie, Math. Z. 5 (1919), 52-99

[1718] Moivre, Abraham de, The doctrine of chances, London (Müller) 1718

[1965] Mosteller, F., Fifty challenging problems in probability, with solutions, Reading/Mass. (Addison-Wesley) 1965

[1933] Neyman, J., and E.S. Pearson, On the problem of the most efficient tests of statistical hypotheses, Phil. Trans. Roy. Soc. 231 (1933), 289-337

[1953] Ore, O., Cardan, the gambling scholar, Princeton (UP) 1953

[1954] Pascal, B., Oeuvres, Paris (Pléjade) 1954

[1920] Pólya, G., Über den zentralen Grenzwertsatz der Wahrscheinlichkeitsrechnung und das Momentenproblem, Math. Z. 8 (1920), 171-181

[1954] Pólya, G., Mathematics and plausible reasoning 2 vols.: I. Induction and analogy in mathematics, II. Patterns of plausible inference, Princeton (UP) 1954

[1978] Reichel, G., Carl Friedrich Gauß(30. April 1777 bis 30. April 1977) und die Professoren-, Witwen- und Waisenkasse zu Göttingen, Blätter d.dt.Ges.f. Vers.-Math. 13 (1978), 101-127

[1947] Reichenbach, H., The Theory of Probability, an inquiry into the logical and mathematical foundations of the calculus of probability, 2nd ed., Berkeley (Calif. UP) 1947

[1982] Reid, C., Neyman — from Life, Berlin etc. (Springer) 1982

[1984] Rohatgi, V., Statistical Infercne, New York (Wiley) 1984

[1984] Sachs, L., Applied Statistics: a handbook of techniques, 2nd. ed., Berlin etc. (Springer) 1984

[1970] Schnorr, C.P., Über die Definition von effektiven Zufallstests I, II, ZfW 15 (1970), 297-312, 313-328

[1970a] Schnorr, C.P., Klassifikation der Zufallsgesetze nach Komplexität und Ordnung, ZfW 16 (1970), 1-21

[1977] Schnorr, C.P., A survey of the theory of random sequences, Basic Problems in Methodology and Linguistics, Butts and Hintikka eds., Dordrecht (Reidel) 1977, 193-211

[1981] Seneta, E., Non-negative matrices and Markov chains, Berlin etc. (Springer) 1981

[1948] Shannon, C., A mathematical theory of communication, Bell. Syst. Tech. J. 27 (1948), 379-423, 623-656

[1976] Shafer, G., A mathematical theory of evidence, Princeton (UP) 1976

[1956] Siegel, S., Nonparametric statistics for the behavioral sciences, New York (Mc Graw-Hill) 1956

[1975] Snell, J.L., Introduction to Probability Theory with Computing, Englewood Cliffs/NJ (Prentice-Hall) 1975

[1956] Spitzer, F., A combinatorial lemma and its application to probability theory, TAMS 82 (1956), 323-339

[1964] Strassen, V., An invariance principle for the law of the iterated logarithm, ZfW 3 (1964), 211-226

[1964a] Strassen, V., Asymptotische Abschätzungen in Shannon's Informationstheorie, Trans. IIIrd Prague Conf. Liblice 1962, Prague (Acad. Sci.) 1964, 689-723

[1970] Strehl, V., Elementare kombinatorische Methoden der Fluktuationstheorie, Diplomarbeit Erlangen 1970

[1969] Sudderth, W.D., On measurable, nonleavable gambling houses with a goal, Ann. Math. Stat. 40 (1969), 66-70

[1969a] Sudderth, W.D., On the existence of good stationary strategies, TAMS 135 (1969), 399-414

[1971] Sudderth, W.D., On measurable gambling problems, Ann. Math. Stat. 42 (1971), 260-269

[1986] Székely, G., Paradoxes in Probability Theory and Mathematical Statistics, Dordrecht (Reidel) 1986

[1983] Thompson, T.M., From error-correcting codes through sphere packings to simple groups, MAA 1983

[1959] Trotter, H.F., An elementar proof of the central limit theorem, Arch. Math. 10 (1959), 226-234

[1939] Ville, J., Étude critique de la notion du collectif, Paris (Gauthier-Villars) 1939

[1936] Wald, A., Sur la notion du collectif dans le calcul des probabilités, C.R. Acad. Sci. Paris 202 (1936), 180-183

[1939] Wald, A., Die Widerspruchsfreiheit des Kollektivbegriffs der Wahrscheinlichkeitsrechnung, Erg. e. math. Colloq. 8 (1937), 38-72

[1950] Wald, A., Statistical Decision Functions, New York (Wiley) 1950

[1950a] Wald, A., Basic Ideas of a general theory of statistical decision rules, Proc. ICM 1950, Cambridge/Mass. (Harvard UP) 1952, 231-243

[1958] Wendel, I.G., Spitzer's formula, a short proof, PAMS 9 (1958), 905-908

[1960] Wendel, I.G., Order statistics of partial sums, Ann. Math. Stat. 31 (1960), 1034-1044

[1949] Weyl, H., Almost periodic invariant sets in a matric vector space, Amer. J. Math. 71 (1949) 178-205

[1923] Wiener, N., Differential space, J. Math. Phys. 3 (1923), 131-174

[1987] Wittmann, R., Sufficient moment and truncated moment conditions for the LIL, Prob. Th. 75 (1987), 509-530

[1964] Wolfowitz, J., Coding Theorems of Information Theory, 2nd ed. Berlin etc. (Springer) 1964

[1972] Zielinski, R., Erzeugung von Zufallszahlen, Leipzig (Fachbuchverlag) 1972

Index